Philosophical Studies Series

Volume 122

Editor-in-Chief
Luciano Floridi, University of Oxford, Oxford Internet Institute, United Kingdom

Executive Editorial Board
Patrick Allo, Vrije Universiteit Brussel, Belgium
Massimo Durante, Università degli Studi di Torino, Italy
Phyllis Illari, University College London, United Kingdom
Shannon Vallor, Santa Clara University

Board of Consulting Editors
Lynne Rudder Baker, University of Massachusetts at Amherst
Stewart Cohen, Arizona State University, Tempe
Radu Bogdan, Tulane University
Marian David, University of Notre Dame
John M. Fischer, University of California at Riverside
Keith Lehrer, University of Arizona, Tucson
Denise Meyerson, Macquarie University
François Recanati, Institut Jean-Nicod, EHESS, Paris
Mark Sainsbury, University of Texas at Austin
Barry Smith, State University of New York at Buffalo
Nicholas D. Smith, Lewis & Clark College
Linda Zagzebski, University of Oklahoma

More information about this series at http://www.springer.com/series/6459

Catrin Misselhorn
Editor

Collective Agency and Cooperation in Natural and Artificial Systems

Explanation, Implementation and Simulation

Editor
Catrin Misselhorn
Institute of Philosophy
University of Stuttgart
Stuttgart, Germany

Philosophical Studies Series
ISBN 978-3-319-15514-2   ISBN 978-3-319-15515-9 (eBook)
DOI 10.1007/978-3-319-15515-9

Library of Congress Control Number: 2015948369

Springer Cham Heidelberg New York Dordrecht London
© Springer International Publishing Switzerland 2015
This work is subject to copyright. All rights are reserved by the Publisher, whether the whole or part of the material is concerned, specifically the rights of translation, reprinting, reuse of illustrations, recitation, broadcasting, reproduction on microfilms or in any other physical way, and transmission or information storage and retrieval, electronic adaptation, computer software, or by similar or dissimilar methodology now known or hereafter developed.
The use of general descriptive names, registered names, trademarks, service marks, etc. in this publication does not imply, even in the absence of a specific statement, that such names are exempt from the relevant protective laws and regulations and therefore free for general use.
The publisher, the authors and the editors are safe to assume that the advice and information in this book are believed to be true and accurate at the date of publication. Neither the publisher nor the authors or the editors give a warranty, express or implied, with respect to the material contained herein or for any errors or omissions that may have been made.

Printed on acid-free paper

Springer International Publishing AG Switzerland is part of Springer Science+Business Media (www.springer.com)

Preface

Human-machine interaction (HMI) and human-computer interaction (HCI) became popular in the 1980s parallel to the rise of personal computers. At that point, the approaches that were used for simple technical tools seemed no longer appropriate for complex technical devices like computers. In contrast to simple tools, these complex technical devices confronted their users with a huge amount of information and a lot more affordances for interacting with them. Moreover, it became evident that poorly designed human-machine interfaces can lead to serious trouble. A well-known example in the field is the Three Mile Island accident where, among other factors, failures in human-machine interaction led to an accident in a nuclear power plant. This suggested that a more interactive paradigm of human-machine interaction was required. Properties like usability, ergonomy, and transparency were approached from a multidisciplinary point of view involving computer sciences, psychology, and other fields of study.

Since then, even more complex systems involving humans and machines have arisen. These not only comprise a one-on-one interaction between humans and artificial agents as is the case with computers or robots but also—as the term "artificial systems" indicates—complex dynamical structures such as industrial processes, financial transactions, or transportation networks. These systems create new challenges, because they are more autonomous, intelligent, and dynamical, i.e., human beings are not in full control of them, and they are the subject of uncertainty to a much higher degree. To address these issues, a new research paradigm proved necessary: human-machine cooperation (HMC). The premise that was unique to this new paradigm was that humans have to *cooperate* with machines in order to cope with these complex and dynamical situations.

Although the new paradigm has led to progress in this research area, it has a blind spot. It did not sufficiently take into account the fact that collective agency involving human and artificial agents differs in some important respects from individual agency. The behavior of these systems cannot be reduced to the behavior and intentions of the individuals that constitute them. This is a central insight of the philosophical debate about the collective agency of humans which has been the subject of intense study over the last few decades. Philosophical analyses range from simple forms of "uniform behavior" or "emergent coordination" to complex forms of collective agency which involve shared intentional attitudes like beliefs, intentions, or emotions. However, a shortcoming of these analyses is that they usually do not take into account hybrid systems involving humans and artificial agents.

The aim of this volume is to bring these two separate strands of research together in order to understand the nature of collective agency in natural, artificial, and hybrid systems better and to improve the design and performance of hybrid systems involving human and artificial agents. Problems that arise in the construction of these systems are addressed from a theoretical and practical as well as from an ethical point of view.

The book consists of four sections. The first section is dedicated to the different conceptions of agency that can be found in the discussion. It is concerned with the question under which conditions an object can be considered as an agent and at which point cooperation and collective agency arise. The first chapter serves as an introduction to the topic of the volume. It provides a conceptual framework of different forms of individual as well as collective agency and cooperation in natural and artificial systems in order to help the reader to better understand and situate the goals and arguments of the other contributions. In the second chapter, Fiebich, Schwarzkopf, and Nguyen analyze the notion of cooperation at the intersection of philosophy, psychology, and informatics with the aim of describing different forms of cooperation and collective agency involving humans and robots in a two-dimensional framework. Sabine Thürmel provides a gradual and multidimensional concept of agency that applies to the interaction of human with non-human agents. Her approach combines elements from Bruno Latour's Actor-Network Theory (ANT), the socio-technical approach to distributed agency of Rammert and colleagues, and Luciano Floridi's view of artificial agents.

The second section of the book turns to human-machine cooperation. Ipke Wachsmuth argues for the view that embodied cooperative systems should not just be seen as tools but have to be treated as partners. He then shows how different levels of intentionality (including emotions) – which are necessary for endowing technical systems with collaborative functionality – have been implemented in the humanoid artificial agent Max by his Bielefeld research group.

The example of Max might lead to the assumption that artificial agents should be as human-like as possible in order to make human-machine cooperation natural and intuitive. Yet, as the so-called hypothesis of the Uncanny Valley suggests, this is not the case. Very human-like artificial agents actually produce negative feelings of eeriness in humans. Valentin Schwind tracks in his contribution the cognitive and subconscious mechanisms which are responsible for this phenomenon. He argues that cultural, social, and aesthetic aspects of the phenomenon have so far not received the attention that they deserve.

Hauke Behrendt, Markus Funk, and Oliver Korn provide examples of human-machine cooperation in practice—assistive technologies at workplaces—and discuss their ethical implications. How do they change the individual workplaces and the working environment? May handicapped people profit from these technologies in terms of better inclusion in the world of employment? How do assistive technologies have to be evaluated regarding justice and the meaning of work for the individual good life? Christian Neuhäuser takes up another ethical issue that is becoming more pressing the more autonomous and cognitively powerful artificial systems get. He addresses the question of whether robots can be morally responsible for their doings. His answer to this question is in the negative, but he nevertheless suggests that artificial systems can form part of responsibility networks together with humans.

The focus of the third section is the transition from individual to collective agency. Most of the contributions in this part challenge some of the assumptions that more standard views of collective agency have made so far. One of these assumptions is that collective agency involves collective intentionality or the so-called we-intentions. Olle Blomberg, in contrast, describes forms of joint action that do not require genuine collective intentionality or we-intentions, but only agent-neutral goals.

Another influential view in the collective agency debate is that collective intentionality involves the interconnected planning of several agents. Stephen Butterfill questions this view in his contribution and tries to show that parallel planning can, in certain cases, already be sufficient for genuine collective agency. In contrast to interconnected planning, in parallel planning each agent individually forms a plan of his or her own and the other's actions.

Tom Poljansek argues in his chapter that small-scale models (like going for a walk together) which are mostly used as the paradigm of collective agency are misleading and that one has to include cases of large-scale collective agency in order to understand the phenomenon properly. Among the aspects that have an impact on large-scale collective agency, the "feeling of belonging to a group" figures prominently for him.

Anna Strasser then turns to the question of which conditions are necessary for artificial systems to be part of collective action. She holds that the standard philosophical views of collective agency are cognitively much too demanding and that there are simpler forms of collective agency involving humans and artificial systems. As Strasser argues by drawing parallels to non-human animals and small children, these forms of collective agency amount to more than mere tool use.

Finally, Mog Stapleton and Tom Froese challenge the theoretical framework that underlies many theories of collective agency with the help of the enactivist paradigm in cognitive science. According to them, the fault of the received views is that they abstract away from the biological body as the foundation of cognition and agency. Stapleton and Froese develop a gradual concept of collective agency for biological agents, but remain skeptical when it comes to the possibility of collective agency of computational agents.

The fourth and final section concerns social simulations and their relevance for collective agency. The aim of these simulations is to explain and predict the behavior of collectively acting systems. Moreover, they may also help to improve our understanding of collective agency in general. Meike Tilebein and Maximilian Happach show how simulations are used to model social interaction processes in management science. Eckhart Arnold uses the example of Axelrod's famous simulations of the evolution of cooperation to argue from a philosophical point of view, that one should not trust social simulations naively. Finally, Joanna Bryson discusses social simulations in the broader framework of the cultural evolution of intelligence, particularly the emergence of altruistic behavior. As these examples show, the project of this book is also highly relevant for research and practice in the area of social simulations.

The idea for this book goes back to the conference "Collective Agency and Cooperation in Natural and Artificial Systems" which we organized in July 2013 at the University of Stuttgart. The talks given at the conference provide the basis for the current collection. I would like to thank the University of Stuttgart and the Cluster of Excellence Simulation Technology (SimTech) for the generous support and funding of this conference. I am, moreover, grateful for the extensive help in organizing the conference and preparing the manuscript of this book provided by Hauke Behrendt, Anja Berninger, Jörg Fingerhut, Wulf Loh, Martin Maga, Christoph Michel, Tom Poljansek, Mog Stapleton, and Tobias Störzinger. Last but not least I would like to thank the contributors to this volume who have each written important and innovative contributions to this evolving field of research.

Stuttgart, Germany   Catrin Misselhorn

Contents

Part I Concepts of Agency

1. Collective Agency and Cooperation in Natural and Artificial Systems 3
   Catrin Misselhorn

2. Cooperation with Robots? A Two-Dimensional Approach 25
   Anika Fiebich, Nhung Nguyen, and Sarah Schwarzkopf

3. The Participatory Turn: A Multidimensional Gradual Agency Concept for Human and Non-human Actors 45
   Sabine Thürmel

Part II Human-Machine Cooperation

4. Embodied Cooperative Systems: From Tool to Partnership 63
   Ipke Wachsmuth

5. Historical, Cultural, and Aesthetic Aspects of the Uncanny Valley 81
   Valentin Schwind

6. Ethical Implications Regarding Assistive Technology at Workplaces 109
   Hauke Behrendt, Markus Funk, and Oliver Korn

7. Some Sceptical Remarks Regarding Robot Responsibility and a Way Forward 131
   Christian Neuhäuser

Part III Collective Agency

8 Planning for Collective Agency 149
Stephen A. Butterfill

9 An Account of Boeschian Cooperative Behaviour 169
Olle Blomberg

10 Choosing Appropriate Paradigmatic Examples for Understanding Collective Agency 185
Tom Poljanšek

11 Can Artificial Systems Be Part of a Collective Action? 205
Anna Strasser

12 Is Collective Agency a Coherent Idea? Considerations from the Enactive Theory of Agency 219
Mog Stapleton and Tom Froese

Part IV Simulating Collective Agency and Cooperation

13 Simulation as Research Method: Modeling Social Interactions in Management Science 239
Roland Maximilian Happach and Meike Tilebein

14 How Models Fail 261
Eckhart Arnold

15 Artificial Intelligence and Pro-Social Behaviour 281
Joanna J. Bryson

Index 307

Contributors

Eckhart Arnold Digital Humanities Department, Bavarian Academy of Sciences, Munich, Germany

Hauke Behrendt Institute of Philosophy, University of Stuttgart, Stuttgart, Germany

Olle Blomberg Center for Subjectivity Research, Department of Media, Cognition and Communication, University of Copenhagen, Copenhagen, Denmark

Joanna J. Bryson Department of Computer Science, University of Bath, Bath, UK

Stephen A. Butterfill Department of Philosophy, University of Warwick, Coventry, UK

Anika Fiebich Institute of Philosophie II, Ruhr-University Bochum, Bochum, Germany

Tom Froese Instituto de Investigaciones en Matemáticas Aplicadas y en Sistemas/Centro de Ciencias de la Complejidad, Universidad Nacional Autónoma de México, Mexico City, Mexico

Markus Funk Institute of Visualization and Interactive Systems (VIS), University of Stuttgart, Stuttgart, Germany

Roland Maximilian Happach Institute for Diversity Studies in Engineering, University of Stuttgart, Stuttgart, Germany

Oliver Korn Institute of Visualization and Interactive Systems (VIS), University of Stuttgart, Stuttgart, Germany

Catrin Misselhorn Institute of Philosophy, University of Stuttgart, Stuttgart, Germany

Christian Neuhäuser Institute of Philosophy and Political Science, University of Dortmund, Dortmund, Germany

Nhung Nguyen Artificial Intelligence Group, Faculty of Technology, Bielefeld University, Bielefeld, Germany

Tom Poljanšek Institute of Philosophy, University of Stuttgart, Stuttgart, Germany

Sarah Schwarzkopf Center for Cognitive Science, Institute of Computer Science and Social Research, University of Freiburg, Freiburg, Germany

Valentin Schwind Institute of Visualization and Interactive Systems (VIS), University of Stuttgart/Stuttgart Media University, Stuttgart, Germany

Mog Stapleton Institute of Philosophy, University of Stuttgart, Stuttgart, Germany

Anna Strasser Berlin School of Mind and Brain, Humboldt-Universität zu Berlin, Berlin, Germany

Sabine Thürmel Munich Center for Technology in Society, Technische Universität München, Munich, Germany

Meike Tilebein Institute for Diversity Studies in Engineering, University of Stuttgart, Stuttgart, Germany

Ipke Wachsmuth Bielefeld University, Bielefeld, Germany

Part I
Concepts of Agency

Chapter 1
Collective Agency and Cooperation in Natural and Artificial Systems

Catrin Misselhorn

1.1 Introduction

This volume brings together two so far largely unrelated fields of study: philosophical approaches to cooperation and collective agency and research into human-machine-cooperation and multi-agent systems (MAS) from engineering, robotics, computer science and AI. As will become apparent, both sides may profit from this venture. The operationalization and modeling of human-machine-cooperation and MAS can be improved with the help of philosophical approaches, whereas philosophical hypotheses about cooperation and collective agency might be tested by implementing them in artificial Systems. The aim of this introductory article is to provide a conceptual framework in order to facilitate the dialogue between the different approaches and to situate the contributions to this volume with respect to it.

MAS are systems composed of multiple interacting agents within an environment. There are biological, social as well as artificial multi-agent-systems. One major advantage of MAS is that they can solve problems that are difficult or impossible for an individual agent or a centralized system to solve. An important aspect of MAS is that the agents are acting together in some way to bring about a certain result. At this point research in MAS meets philosophical theories of collective agency. However, the kinds of collective agency involved in different MAS have so far not been sufficiently distinguished and investigated.

The purpose of this article is to provide a systematic distinction of different types of agents and their capacities. Then it will be investigated how cooperation between different kinds of agents arises. Of particular interest is the point at which a group of agents begins to engage in some form of collective agency. As I will argue,

C. Misselhorn (✉)
Institute of Philosophy, University of Stuttgart, Stuttgart, Germany
e-mail: catrin.misselhorn@philo.uni-stuttgart.de

© Springer International Publishing Switzerland 2015
C. Misselhorn (ed.), *Collective Agency and Cooperation in Natural and Artificial Systems*, Philosophical Studies Series 122,
DOI 10.1007/978-3-319-15515-9_1

different kinds of collective agency involve agents with different kinds of capacities. This is particularly important for research in MAS. Therefore, I am proposing a framework of different kinds of collective agency and cooperation in MAS depending on which types of agents are involved. The approaches which are proposed by the contributions to this volume are then going to be situated in this framework.

Two partly alternative and partly overlapping schemes of agency are developed by Fiebich et al. and Thürmel in this volume. Both approaches are primarily concerned with forms of cooperation and collective agency between humans and artificial agents. The framework which I suggest in this chapter is, in contrast, more general. It applies to all kinds of agents including human beings, non-human animals and artificial agents at different levels of complexity. Finally, I will draw some consequences regarding the relevance of this framework for research and practice, specifically in the field of social simulation.

1.2 Types of Agents and Their Capacities

There seems to be an intuitive distinction between the things that merely happen to somebody or something and the things that an agent genuinely does (Wilson and Shpall 2012). The question is which capacities an object must have in order to qualify as an agent. Having the paradigm of human agency in mind, philosophers have often adopted rather high standards for agency. They required, for instance, the capacity for deliberation or even a special kind of knowledge of one's actions (Anscombe 2000; Velleman 1989). In the more technical research on MAS, such sophisticated capacities are not required for agency, as the following selection of quotes shows:

- Most often, when people use the term 'agent' they refer to an entity that functions continuously and autonomously in an environment in which other processes take place and other agents exist. (Shoham 1993, 52)
- This approach involves considering the intelligent entity as an agent, that is to say, a system that senses its environment and acts upon it. (Russell 1997, 59)
- The term agent is used to represent two orthogonal entities. The first is the agent's ability for autonomous execution. The second is the agent's ability to perform domain oriented reasoning. (the MuBot Agent, cited by Franklin and Gasser 1997, 22)
- Intelligent agents are software entities that carry out some set of operations on behalf of a user or another program, with some degree of independence or autonomy, and in so doing, employ some knowledge or representation of the user's goals or desires. (the IBM Agent, cited by Franklin and Gasser 1997, 23)
- An autonomous agent is a system situated within and a part of an environment that senses that environment and acts on it, in pursuit of its own agenda and so as to effect what it senses in the future. (Franklin and Gasser 1997, 25)

In order to do justice to the more technical and the more philosophical views, I suggest distinguishing two dimensions of agency: *autonomy* and *intelligent behavior* which can vary in degree.[1] In the following, these two dimensions will be elaborated by way of highlighting important aspects and stages. As we will see, both dimensions are interdependent in that more demanding forms of intelligent behavior are also coupled to more autonomy and vice versa.

1.2.1 Basic Autonomy and Intelligence

It has been suggested as a minimal criterion of autonomy that an agent "should be able to act without the direct intervention of humans (or other agents)" (Jennings et al. 1998, 8). However, this criterion is not convincing if one includes social systems in the category of MAS. Since the agents involved in social systems are human beings, it is clear that they act on direct intervention by themselves. A reformulation of the underlying idea is that an agent has to be able to act independently of external causes (Shultz 1991). One might be inclined to understand the phrase "independently of external causes" metaphysically as having the capacity to spontaneously start a causal chain by itself. But this would be a very demanding conception of autonomy. A more adequate understanding of basic autonomy is provided by Floridi and Sanders:

> Autonomy means that the agent is able to change state without direct response to interaction: it can perform internal transitions to change its state. So the agent must have at least two states. This property imbues an agent with a certain degree of complexity and independence from its environment. (Floridi and Sanders 2004, 358)

Moreover, to count as an agent, an object has also to show at least minimally intelligent behavior. There are different kinds and degrees of intelligent behavior. First, intelligent behavior is *interactive*. It involves, in its most basic form, interaction with the environment. An interactive agent takes input from its environment and brings about changes in the environment. The environment can be the real world, but it can also be a virtual environment like the web. An interaction is minimally intelligent, if the agent's reaction to the input is appropriate with respect to the agent's survival. More sophisticated forms of intelligent behavior also involve a certain degree of *flexibility*. There must be a variety in the reactions that an agent has at its disposal in a given situation. Another aspect of intelligent behavior is its *adaptivity*, i.e., an agent can modify its reactions to improve its interaction with the environment such that they become more appropriate. Intelligent behavior is, therefore, closely linked to the idea of learning.

---

[1] The term autonomy has a very demanding connotation which stems from Kant's moral philosophy. For Kant, autonomous agents are free rational agents who are the source of the authority behind the moral laws that bind them. The concept of autonomy used when speaking about agents in MAS is, of course, far less demanding.

The criteria for agency discussed so far apply to basic natural organisms and certain artificial agents. A more biologically inspired account is suggested by the enactive theory of agency which is applied to collective agency by Stapleton and Froese in this volume. According to this view, it is essential for agents that they are able to self-organize and self-produce and have certain intrinsic needs. This implies that cells can count as agents, but given the actual state of the art robots or software agents cannot.

1.2.2 Goal-Directedness

Another stage of agency is marked by *goal-directedness*. Interactive, flexible and adaptive behavior has a very general purpose: it serves to ensure the survival of an organism. However, acting often involves a more specific form of *goal-directedness* distinct from merely purposive behavior (Balleine and Dickinson 1998). Goal-directedness requires that agents try to achieve specific goals like getting the ripe fruit lying in the grass in front of them. The term *goal* is, however, systematically ambiguous (Want and Harris 2001).

When it is said, for instance, that the goal of an agent is to open a box, one has to distinguish between the external and the internal goal. The *external goal* is a certain state of the environment, e.g. the open box. The *internal goal* is an internal state guiding the agent's behavior, i.e., a mental representation of the desired state such as the open box (Tomasello et al. 2005). The appropriate control of an agent's behaviour is sufficient for the ascription of an external goal. In order to have internal goals, an agent must also possess mental states. This leads to a new type of agents characterized by *intentionality* which makes for a qualitative difference in intelligent behavior and autonomy.

1.2.3 *Intentionality*

Intentionality is the capacity of the mind to be directed at things, properties or states of affairs in the world or to represent them. It marks an important step in the taxonomy of agents, because it plays a crucial role in rational agency. According to the standard philosophical model which goes back to Hume, rational actions are brought about by agents with intentional states like beliefs and desires: I am going to the library because I want to borrow a certain book and I believe that the library has got it. The model's predictive and explanatory value exploits beliefs and desires as inner states underlying behavioral patterns. Some philosophers (e.g. Davidson 1980) think that beliefs and desires suffice to explain rational action, but others (e.g. Bratman 1984) do not follow this claim. They believe that intentional action additionally involves a specific kind of intentional state called "intention."[2]

---

[2] The so-called "belief-desire-intention" or BDI software model based on Bratman's approach is the standard model for MAS (Rao and Georgeff 1995).

The need to invoke intentions is due to two important facts about practical reasoning (Jennings 1993): The first is that agents have limited resources. Therefore, they cannot continually weigh their competing desires and beliefs when deciding what to do next. Rather, an agent must, at some point, make a commitment to a certain state of affairs for which to aim. The second reason is that intentions are necessary for planning future actions. If an agent has chosen a future action, he or she must form subsequent intentions that will implement the relevant action at a specified time and in a specified way. Take, for instance, the intention to open a box. Doing so will invoke more specific intentions about how to open the box like untying it or cutting it open. These are called plans. Plans are sequences of actions (like recipes) that an agent can perform to achieve one or more of its intentions. Such plans may include other plans: my plan to go for a drive may include a plan to find my car keys.

Two important kinds of rationality constraints can be derived from this description of the role of intentions in action. First, intentions should be both internally coherent and coherent with the agent's other beliefs. Secondly, the agent should be able to monitor whether he or she was able to successfully realize his or her intention and to adapt his or her plans accordingly. Both kinds of rationality constraints imply a new form of autonomy: an intentional agent is not just able to control his or her behavior, but must also have a certain amount of *control over his or her mental states*. One way to control one's mental states is by forming higher-order intentional states.

1.2.4 Higher-Order Intentionality

Whereas first-order intentionality consists in having intentional states like beliefs, desires, and intentions, higher-order intentionality involves intentional states which have other intentional states as their object, i.e., beliefs about beliefs, beliefs about desires, or desires about beliefs. An agent who possesses higher-order intentionality will be able to form beliefs, desires, etc. about his or her own and other agent's mental states.

Higher-order intentionality makes for a qualitative difference in agency and it is often seen as a key element in understanding *free will* which is a very demanding form of autonomy. For Harry Frankfurt human action is distinct from mere animal behavior because we can reflect upon our beliefs and desires by way of forming desires and beliefs about them (Frankfurt 1971). I may have a first-order desire to eat a piece of cake, but I may also have a second-order desire not to want to eat the cake, because I am worried about gaining weight.

The difference between first- and higher-order intentional states can be used to explain freedom of the will and freedom of action (O'Connor 2014): An action is free when the causally relevant desire is the one that I desire to be effective. Freedom of the will is explained by the fact that I am able to choose which first-order desire to act on. I may decide to follow my first-order desire to eat the piece of cake, but I could have also chosen to eat an apple instead in order to stay slim, if that is more important to me. The higher-order desires are the ones with which people identify; they reflect the true self of a person. People like drug addicts lack freedom of the will, because they do not manage to follow the second-order desire with which they identify. Freedom of the will is usually considered to be an important aspect of personhood (see Sect. 1.2.6).

1.2.5 The Affective Aspect of Agency

One might think that with higher-order intentionality one has identified all the relevant capacities for agency. But one should not forget that feelings and emotions might also play a role. Feelings of attraction and aversion might motivate one to act and emotions are often understood as representations of value properties (for instance, fear as a representation of danger) which also have motivational force. Although emotions are intentional states, according to most contemporary views, it has been argued that they are responsible for the fact that human (and maybe animal) agency amounts to something more than being an intentional agent (including higher-order intentionality) narrowly understood (Helm 2000).

This aspect plays a role when it comes to the question of which desires or values one should aim at. For Helm (who is following Hume in this respect), reason can show how to realize a given desire, but it is of no great help in identifying the values or desires worth pursuing. These are provided by "feelings of imports of various kinds" which indicate what a person values or cares about (Helm 2000, 6). This capacity distinguishes, as Helm argues, genuine agents from merely intentional agents:

> Genuine agents are able to engage in activity because they find it worth pursuing—because they *care* about it. In this respect, they differ from what might be called 'mere intentional systems': systems like chess-playing computers that exhibit merely goal-directed behavior mediated by instrumental rationality, without caring. (Helm 2000, 17)

Whereas various kinds of natural and artificial agents can be equipped with basic autonomy and intelligent behavior, goal-directedness, intentionality and maybe even higher-order intentionality, the emotional aspect of agency is for Helm the one which is the mark of genuine agents.

One might consider this a reason to think that genuine agency is in principle unachievable for artificial agents. Yet, there are attempts to endow artificial agents with emotional capacities as Wachsmuth's contribution to this volume shows. He describes the attempt of his research group in Bielefeld to construct an artificial agent (the humanoid Max) that not just possesses various levels of intentionality, but also has emotional capacities. This is supposed to support a natural and intuitively appealing interaction with Max.

One important challenge that this kind of project faces is the topic of Schwind's article in this volume. It is discussed in the literature as the problem of the "uncanny valley." Humans experience the emotional interaction with humanoid artificial agents increasingly positive up to a certain point. When the humanoid agents become very human-like, but not strictly indistinguishable from a human being, this changes all of a sudden, and they elicit negative emotional responses of uncanniness or eeriness. This is an important impediment to smooth and successful human-machine cooperation.

1.2.6 Persons

The most demanding type of agents are persons. Dennett (1976) suggests six necessary conditions of personhood: (1) Persons have to be rational beings, (2) they must possess intentionality, (3) one can take a certain stance or attitude towards persons, (4) persons must be capable of reciprocating in some way, (5) they have to be capable of verbal communication and (6) persons must possess self-consciousness.

The three first conditions are mutually interdependent: Being rational just means possessing intentionality and this is, for Dennett, a matter of being the object of a certain stance, the intentional stance. According to this view, an object qualifies as an intentional system if it makes sense to interpret its behavior by ascribing intentional states like beliefs and desires to it. These three conditions do not just apply to persons, but delimit the much wider class of intentional systems.

Persons have to meet the other three conditions as well. Reciprocity is understood by Dennett as the capacity to take the intentional stance towards other systems, i.e., it amounts to higher-order intentionality. Moreover, persons must also be able to communicate with each other and they have to possess self-consciousness. The last condition, in particular, pertains only to human agents. Self-consciousness is for Dennett the capacity to reflect upon ones beliefs and desires. Although it might seem as if higher-order intentionality is all that is needed for self-consciousness, Dennett thinks that this is not the case. The difference seems to be due to the fact that a person is not confined to given higher-order desires, but is able to deliberate about which ones he or she should adopt. In this process the person takes the stance of a "reason-asker and persuader" (Dennett 1976, 193) towards him- or herself much as towards another person. This is, for Dennett, the "inescapably normative" aspect of personhood (Dennett 1976, 193).

In a similar vein, Christine Korsgaard distinguishes between a natural and a normative conception of agency (Korsgaard 2014).[3] According to the natural

---

[3] This contrast is misleading at first because agency is imbued with normativity almost all the way down. The concept of adaptivity already has a normative dimension. Adaptivity is a value for the agent whose behavior can be adjusted to his or her environment. Another type of normativity comes into play, as we have seen, with the rationality constraints which intentional agents are subject to. Therefore, one has to keep in mind that Korsgaard refers to a specific and particularly demanding kind of normativity.

conception of agency "an action is a movement caused or causally guided by certain mental states, such as belief/desire pairs, or intentions, (…) and the action is attributed to the agent in virtue of the fact that the agent is the one whose mental states brought the action about." (Korsgaard 2014, 190) This view largely conforms to intentional agency as explained above.

Regarding the normative conception of agency, Korsgaard is drawing on two thoughts of Plato and Kant. On the Kantian view, "agency is the exercise of autonomy, and autonomy involves the adoption of a maxim to govern one's action, which Kant understands as making a law for oneself." (Korsgaard 2014, 192) Such a maxim must, of course, for Kant conform to the categorical imperative. This provides the link between autonomy and moral agency in a Kantian framework: genuinely autonomous agency is always morally good. From Plato she takes the idea that normative agency requires unity:

> Unity is essential to agency, whether collective or individual, because an action, unlike other events whose causes in some way run through an agent, is supposed to be a movement, or the effecting of a change, that is backed by the agent as a whole. (Korsgaard 2014, 193)

These two aspects of normative agency have two implications that can, according to Korsgaard, not be captured by the natural conception of agency: The first is that the agent is active in a spontaneous way. This is, for her, incompatible with the idea that actions are simply caused by mental states, as the natural conception has it. The second is that a person's actions reflect something essential about his or her identity. Again, for Korsgaard, the idea of causation by certain mental states cannot capture why actions are an expression of the agent's essential identity in this sense. As we have seen, there exists a close relation between the normative aspect of agency and moral agency. Moreover, normative agency of this type seems to be unachievable for artificial agents, even on Dennett's more moderate account of personhood.

1.3 Collective Agency and Cooperation in Multi-Agent Systems

Based on the types of agents introduced above, one can distinguish between different kinds of MAS. The most basic distinction in the research on MAS is the one between *pure* and *hybrid* systems (Parunak et al. 2007). Pure systems only comprise one type of agent, whereas hybrid systems involve different kinds of agents: The most primitive form of pure MAS are systems which only involve agents with basic autonomy and intelligence, but without intentional states. Pure systems composed from agents capable of goal-directed behavior form a second category. Further there are pure systems with agents possessing first-order intentionality; those which are constituted solely of agents with higher-order intentionality; systems which involve only agents with emotional capacities; and finally the ones that just comprise persons.

Hybrid MAS involve at least two different types of agents. From a technical point of view, hybrid systems are promising when it comes to situations involving different levels of complexity. One might at first glance think that systems composed from agents possessing at least higher-order intentionality are capable of coping with all relevant situations. Yet, although cognitively sophisticated agents are able to solve demanding problems and facilitate communication with users they have their weaknesses, too, since the process of engineering is intensive and time consuming (Parunak et al. 2007). Agents that are cognitively less demanding at the non-intentional level are often computationally more efficient. Therefore, it is also worth investigating the forms of collective agency that can be achieved by cognitively simple agents. Cooperation and collective agency among cognitively simple agents is the topic of Strasser's chapter in this volume. Her primary concern is to show that there are forms of collective action at a sub-intentional level, and that collective actions involving intentional and non-intentional agents do not just amount to tool use.

1.3.1 Swarm Behavior and Emergent Coordination

The most basic type of MAS are non-intentional systems that only involve agents with basic autonomy and intelligence. Can collective agency already occur at this level? One might think that a good starting point for analyzing collective agency in these systems is a group of individuals doing the *same thing*. However, it is clear that merely doing the same thing does not suffice for genuine collective agency as in such cases there does not necessarily have to be any kind of relation between the individuals. Every member of a group can do the same thing, e.g., strive for survival, without acting together with the others, if every member of the group has its own ecological niche independently of the others.

A somewhat more refined notion is *uniform behavior*. Uniform behavior is given when a group of agents shows the same bodily movements, e.g., a group of organisms moving in the same direction. But if there is no interaction between the members of the group this is, again, no case of collective agency. At the next level, *interactive behavior*, the behavior of one agent becomes the input of another agent who then modifies its behavior. The interaction can be *one-sided*, for instance, when one agent takes another agent as an obstacle that he or she tries to avoid. But it can also be *two-sided*, e.g., when the second agent reacts to the first agent's approaching it by getting out of the way of this moving object. This example shows that even two-sided interaction does not automatically amount to genuine cooperation since each agent perceives the other one merely as an object, but not as an agent.

However, there are some cases in which the interaction between several agents leads to a kind of collective behavior. This happens when swarms, herds or flocks of animals (or robots) aggregate and move together. More abstractly, swarm behavior is "the collective motion of a large number of self-propelled entities." (O'Loan and Evans 1998, L99). Swarming is often based on interactive behavior that can be modeled according to quite simple rules. One of the most basic mathematical models of swarm behavior only demands that the agents follow three rules: (1) Collision Avoidance: avoid collisions with your neighbors, (2) Velocity Matching: attempt to match velocity with your neighbors, and (3) Flock Centering: attempt to stay close to your neighbors. These rules were, for instance, used by the well-known boids computer program in order to simulate flocking (Reynolds 1987). What is important in our context is that, although the individual agents follow quite simple rules without being aware of each other as agents and without a centralized control structure, this leads to the emergence of a much more complex and intelligent behavior at the group level, e.g., avoiding an obstacle or escaping a predator.

Yet, the coordination of the behavior of a group of agents with basic autonomy and intelligence does not necessarily have to rely on the representation of rules. Other mechanisms are described by Knoblich et al. (2011) under the label of "emergent coordination". In emergent coordination, several individuals coordinate their behavior based on perception-action couplings. Emergent coordination does not require any joint intentions, plans or knowledge by the agents. It is sufficient that they process perceptual and motor cues similarly. It has, for instance, been observed that when walking people often fall into the same patterns (van Ulzen et al. 2008), and people who are engaged in conversation synchronize their bodily behavior unconsciously (Shockley et al. 2003). This kind of coordinated behavior can occur spontaneously without the agents entering into an agreement, but it can also be part of planned joint actions, or enable these.

Emergent coordination can be based on different processes (Knoblich et al. 2011). One of them is *entrainment*, a form of synchronization which does not require direct mechanical coupling. Another possibility are object affordances, i.e., opportunities that an object provides for a specific kind of agent. An affordance is called *common* by Knoblich et al. (2011) if several agents perceive the same affordances in an object and act accordingly. This leads to emergent coordination, e.g., when a bus is arriving at a bus-stop and people start to enter, taking into account each other's behavior. A *joint affordance* is present if an object does not provide an affordance for one agent, but for two (or more), e.g., a seesaw.

Emergent coordination can also be engendered by a *perception-action matching* (Knoblich et al. 2011). The perception of another agent's action causes in this case a corresponding action tendency in the observer. For instance, seeing somebody dancing may activate dancing representations in people who know how to dance and prompt them to join in. A last mechanism described by Knoblich et al. (2011) is *action simulation*. It relies on the process of perception-action matching which enables the observer to use their own motor representations as an internal model to predict the timing and outcomes of observed actions, e.g., a basketball player who anticipates whether another player's shot will hit or miss the basket.

1.3.2 Collective Goal-Directed Agency

As discussed above, the concept of goal-directed behavior lies in between non-intentional and intentional agency. It can either refer to behavior that is directed at an external state in the environment or that involves an internal state (a mental representation) that guides an agent's behavior as a goal. Let me first turn to collective behavior that is directed at an external goal. It will not suffice that the behavior of a group of agents is directed at the same external goal to count as collective. A group of agents can try to get the same fruit without doing this collectively. For the behavior to be collective, the actions of the individual agents must be directed at the same goal, and their behavior must be coordinated in a specific way. This is what I call *collective goal-directed behavior*.

Let me explain this with the help of an example taken from Knoblich et al. (2011) A tropical ant species called *allomerus decemarticulatus* builds traps from plant hair and fungus in order to capture larger insects. Once an insect gets trapped, the ants come and sting it, carry it away and cut it into pieces (Dejean et al. 2005). In this case, each ant's behavior is individually directed at the same goal—to kill the insect—and their behavior is coordinated. None of the individual ants' behavior is sufficient to produce this result. For this reason, one can speak of collective goal-directed behavior, even though there was (presumably) no inner representation of this goal in any of the ants. There must, of course, be some connection between the individual agent's behavior and the outcome. This connection can be provided, for instance, by a biological function, as in the ants' case: Each ant's behavior has the evolutionarily developed function of contributing to killing the prey. This does not require an internal representation of the goal which is essential for collective goal-directed intentional action.

Only intentional systems can enter into *rational interaction*. However, not any kind of rational interaction amounts to collective agency. Take for example a case where one agent wants to get a banana and believes that the best way to get one is to snatch it away from a conspecific, whereas the other agent wants to keep the banana, and thinks the best way to do this is to hide it. There is rational interaction going on here (in the sense that two individually rational agents are acting onto each other) but no collective agency.

Analogous to the case of collective goal-directed non-intentional behavior, collective goal-directed intentional action involves several agents contributing to the realization of a common goal, for instance, a group of chimpanzees hunting together. This looks similar to the ants' case, but in contrast to the ants, I assume that the chimpanzees each have an internal representation of the goal, i.e., a desire to catch the prey, and they coordinate their behavior. But how does the coordination of their behavior arise? One possibility is that some of the mechanisms already present in non-intentional systems are at work (see Sect. 1.2.1), for instance, action simulation. I am calling this kind of cooperation *simple cooperation*. Alternatively, the chimpanzees might be mutually responsive to each other's beliefs, desires and intentions. In this case, their cooperation would already involve higher-order intentionality and can be called *mutual cooperation*. However, even the partners in mutual cooperation may still treat each other merely as "social tools" (Pacherie 2013) to attain their individual goals.

For this reason, mutual cooperation must be distinguished from *shared or joint cooperation*. Most philosophers nowadays assume that shared cooperation involves some kind of shared or collective intentionality, i.e., "the power of minds to be jointly directed at objects, matters of fact, states of affairs, goals, or values" (Schweikard and Schmid 2013). In shared cooperation, the members of the group have a so-called we-intention and develop joint strategies to realize it. Whether the group of chimpanzees can be credited with collective intentionality or not depends, among other things, on the approach taken towards collective intentionality.[4]

As we will see in the next section, many standard accounts imply that only agents with higher-order intentionality can be credited with shared or joint intentions. Yet, as Blomberg argues in this volume, there are forms of joint agency that are less demanding. According to him, those forms only require that an agent be able to recognize the goals of another agent. But this does not mean that he or she must attribute beliefs to the other. It suffices to see the other's goals as agent-neutral external goals, as "states of affairs towards which the actions are pulled." This kind of joint agency does not require genuine we-intentions and can be achieved by agents who are lacking higher-order intentionality.

1.3.3 Collective Intentionality or We-Intentions

Most standard accounts assume that collective intentionality involves we-intentions that cannot be reduced to individual intentions. But in which sense are we-intentions special and how can they explain shared cooperation? There are three ways in which we-intentions could be distinguished from individual intentions (Pacherie 2013): in terms of their *subject*, in terms of their *content* or in terms of their *mode* (the options are not mutually exclusive).

These three aspects are derived from the fact that we-intentions are a kind of intentional attitude and intentional attitudes generally have a subject (or bearer), a content, and a mode. Take some examples of intentional attitudes: Julia believes that Berlin is the capital of Germany, Paul hopes that it will not rain, Anna desires that the term is over. They all involve a *subject* or bearer (Julia, Paul, Anna), some kind of *content* (that Berlin is the capital of Germany, that it will not rain, that the term is over), and a *mode* in which the subject is related to the content (i.e. believing, hoping, desiring).

Accordingly, we-intentions could be distinguished from individual intentions with respect to their subject, their content or their mode. On the *first* proposal, we-intentions differ from I-intentions because their bearers are literally groups or *collective agents* as separate entities that are dependent, but not reducible to their individual members. Think, for instance, about a soccer team that intends to improve pass-playing. One can

---

[4] For more on the controversy concerning whether hunting chimpanzees are acting jointly or not see Boesch (2005) who thinks that they are and Tomasello and Hamann (2012) who deny this.

argue that the team as a whole can have the intention to improve pass-playing, although not every single member has this intention individually.[5]

The concept of group agents was dismissed for a long time, though, because it seemed to many philosophers ontologically spooky. More recently, there have been attempts to spell out the notion of a group agent in an ontologically innocuous way. One possibility is to apply Dennett's conception of the intentional stance to groups (Tollefsen 2002). For Dennett, something qualifies as an intentional system if it makes sense to interpret its behavior by ascribing intentional states to it. This also holds for groups. If we can interpret the behavior of a soccer team by ascribing the intention to improve its pass playing to it, then we are facing a group agent. A group that can be treated as an intentional system in Dennett's sense need not necessarily be composed from intentional agents. As we have seen above, it might also be possible to interpret a school of fish or an ant colony as an intentional agent, although the individual agents do not possess intentionality.

The disadvantage of this approach is that it does not tell us much about the inner workings of intentional systems. This is unfortunate, particularly if one is concerned with constructing artificial MAS. This makes the view of List and Pettit (2011) attractive; they developed a conception which derives the intentional attitudes of group agents via an aggregation function from the attitudes of the individual group members. If, for instance, a majority rule holds, the group believes or desires what the majority of its members believes or desires. This requires that a group has to be constituted of agents who possess first-order intentionality. Higher-order intentionality is only necessary in so far as the group is supposed to be able to decide upon its own structure and aggregation function.

A *second* way to explain the special character of we-intentions is with reference to their content. This view is defended, for instance, by Michal Bratman (1999, 2006). Collective intentionality arises for him if two agents each have the intention to do something together, are mutually responsive to this intention and adjust their plans to realize the shared intention. The collective aspect resides in the content, because each intention has the form "I intend that *we* J." The subject of the attitude is not a group agent in the literal sense of the term, but it is in some sense collective, because it is a complex of agents, their relations and interlocking intentions. Here is one version of Bratman's analysis illustrated with the help of the example of preparing a sauce hollandaise together following Bratman (1999):

We have a shared intention to prepare a sauce hollandaise if:
(a) I intend that we prepare a sauce hollandaise and (b) you intend that we prepare a sauce hollandaise.
I intend that we prepare a sauce hollandaise in accordance with and because of (1) (a) and (b) and meshing subplans, e.g., my subplan that I am pouring and your subplan that you are stirring. And you intend the same.
(1) and (2) are common knowledge between us.

---

[5]The term "collective intentionality" is sometimes used specifically to describe this view in contrast to shared or joint intentionality that does not presuppose collective agents; but I am not differentiating between shared, joint and we-intentions here.

Bratman's account has the advantage of being metaphysically parsimonious, since it does not rely on anything beyond individual intentions and their relations. It fits, moreover, well together with his BDI framework which is often used in the technical modeling of MAS. Yet, the approach is cognitively costly, since it requires agents with rather sophisticated cognitive capacities who possess higher-order intentionality (cf. Butterfill 2012; Pacherie and Dokic 2006; Tollefsen 2005). Each agent must be able to represent his or her own intentions and the other participants' intentions. Therefore, it might not be suitable for representing forms of collective agency of a more primitive kind.

In his contribution to this volume Butterfill raises the objection that Bratman's interconnected planning is in fact not sufficient for collective agency even if the agents have the required cognitive capacities. However, he does not propose adding even more sophisticated conditions to Bratman's account but rather suggests that parallel planning could do the job. In parallel planning, each agent individually forms one single plan that describes his or her own and the partner's actions. This solution is surprising, because a large part of the debate about collective intentionality concerns the distinction between collective and parallel agency.

More fundamentally, it has been questioned, notably by John Searle (Searle 1990), whether an analysis of collective intentions in terms of individual intentions and their relations can capture genuine collective agency. According to Searle, we-intentions are fundamentally different in type from individual intentions. They involve a different mode, i.e., we-intending instead of I-intending. Interestingly, Searle believes that intending in the we-mode is entirely indifferent to the real existence of other individuals. Even if there was only one individual or just a "brain-in-a-vat", it could have we-intentions, although the basis from which collective intentionality emerges is for him a primitive biological phenomenon which involves:

> a background sense of the other as a candidate for cooperative agency; that is, it presupposes a sense of others as more than mere conscious agents, indeed as actual or potential members of a cooperative activity. (Searle 1990, 414)

Another possibility is to demarcate the difference in mode between individual intentions and we-intentions with the help of their specific normative dimension. As was pointed out by a number of philosophers, one may distinguish the modes of I-intending and we-intending with the help of the specific social commitments and expectations involved in we-intending (Gilbert 2006). Take the case of our two cooks again. If Peter stops to stir then Jenna has a right to rebuke him. This distinguishes we-intentions from I-intentions, where no other individual has a right to rebuke somebody, just because he or she does not comply with his or her intentions.

The normativity of collective intentionality arises for Gilbert from the fact that a group of people forms a joint commitment. This happens if the members of the group commit to act in the same way as a single individual:

> People may jointly commit to accepting, as a body, a certain goal. They may jointly commit to intending, as a body, to do such-and-such. They may jointly commit to believing, or accepting, as a body, that such-and-such. (Gilbert 2006, 136)

In contrast to Searle's and Bratman's views, Gilbert's account of collective intentionality involves the constitution of a group agent. Yet, her view remains ontologically innocuous, since a joint commitment to act as a body only requires that each of the group members expresses his or her willingness to take part in the joint activity. A joint commitment only gets in effect if all the relevant people have agreed to participate. Yet, once it is in place, every member is obligated to performing the joint activity and no one can unilaterally dissolve the joint commitment by simply changing his or her mind. This is why, for Gilbert, sanctions are normatively adequate in the case of non-compliance with a joint intention, but not with an individual intention.

Although one might think that normativity pertains exclusively to human beings, this is not the case. There have been various attempts to build normative multi-agent systems (NorMAS) with artificial agents (Castelfranchi et al. 2000; Hollander and Wu 2010; Mahmoud et al. 2014). Yet, these attempts are not related to Gilbert's view of collective agency. It would have to be investigated whether the kind of normativity that Gilbert's account requires can be implemented in MAS.

There are more approaches to collective intentionality than the ones outlined above, but these are among the most influential ones and they are important for understanding other contributions to this volume. The accounts presented here illustrate that different approaches to collective intentionality depend on different agential capacities as discussed in the last section. Whereas some explanations, like Searle's basic sense of cooperation, may apply to first-order intentional systems (or even non-intentional systems), others (like Bratman's theory) presuppose higher-order intentional systems whose agents are capable to form beliefs and intentions about other agents' beliefs and intentions or agents who are able to incur liabilities (Gilbert's view). However, on some accounts even higher-order intentionality is not enough to explain the specific ways in which humans (and maybe certain kinds of non-human animals) are acting together. One dimension that was found missing in the standard accounts of collective intentionality by Bennett Helm was that genuine agents do not just have inner representations of goals, but they have cares, i.e., they find certain things worth pursuing and they engage in social activity because of these cares.

1.3.4 The Affective Dimension of Collective Agency

Helm's account of individual agency suggested that full-blown human desire involves a "sense of its object as worthwhile, in some way, as having import" (Helm 2008, 4), and this requires having emotions. The same holds for groups. Helm assumes that groups—considered as collective (or as he calls them: plural) agents—can care as well. This requires, according to his view, that the collective agent

> itself has a particular evaluative perspective from within which such import can be disclosed, and this is possible only when this evaluative perspective is in some sense shared by the individuals who constitute that agent. (Helm 2008, 5)

Analogous to the evaluative perspective of the individual which is constituted by the individual's emotions, the collective's evaluative perspective is constituted by the emotions of the group.

But how can a group itself have the relevant emotions? For Helm, the subjects who are experiencing the emotion are the individuals that constitute the group. What makes it the emotion of a plural agent is that each subject is experiencing it as *"one of us* and not merely all on her own." (Helm 2008, 23) In this way, the agent can feel the import that the matter has to the group and not just to him or her individually. As a member of a group of friends I could enjoy playing a soccer game, although I by myself am not particularly fond of playing soccer, and it would not occur to me to look for an occasion to play independently of the group. This seems to be a special case of the idea that collective intentionality is a matter of intending something (in this case: feeling an emotion) in the we-mode.

Apart from caring, feelings or emotions may also play a kind of cohesive role in collective agency. De Jaegher et al. (2010) speak of "engagement" by which they mean "the qualitative aspect of social interaction once it starts to 'take over' and acquires a momentum of its own" (De Jaegher et al. 2010, 442). Tom Poljansek in this volume argues in favor of a "feeling of belonging to a group" which he takes to be particularly important for large scale cases of collective agency. Emotions seem to be, again, difficult to reproduce in artificial MAS. Yet, there have been attempts to create MAS with artificial agents using a model inspired by human emotions. This allows for flexible coordination and cooperation between artificial agents and facilitates the interaction between humans and artificial agents (Steunebrink et al. 2006).

1.3.5 Collective Agency and Responsibility

One of the biggest challenges of collective agency concerns the attribution of responsibility. If persons are acting collectively a dilemma arises. On the one hand, it is not clear that the individuals who make up the group should be considered responsible for the outcome of a collective action. This is due to the fact that their actions might each make a small contribution to the effect without being sufficient for bringing it about. On the other hand, if those who are involved do not seem to be individually responsible then one might be inclined to think that the group as a whole is responsible. Yet, it is not at all obvious that responsibility can be attributed to groups.

A lot depends on how responsibility is defined. Moral responsibility involves more than just causal responsibility. A rock's breaking away from a cliff may be causally responsible for killing a person, but it is not morally responsible. Moral responsibility is often related to being a candidate for blame or approval. If we praise or blame somebody for having done something (e.g. saving or destroying somebodies life) we hold the individual morally responsible for it. List and Pettit suggested three criteria for moral responsibility and discuss them with respect to group agents:

Normative significance The agent faces a normatively significant choice, involving the possibility of doing something good or bad, right or wrong.
Judgmental capacity The agent has the understanding and access to evidence required for making normative judgments about the options.
Relevant control The agent has the control required for choosing between the options.
(List and Pettit 2011, 155)

The first condition does not pose much of a problem. According to List and Pettit group agents can act using the intentional states of the group's members as input. These can, of course, sometimes concern normatively significant choices.

The second condition is somewhat more difficult to meet. It requires the ability to reflect upon beliefs and desires which pertains to persons. Groups do not have this capacity per se, but only via their members. Whether the group members can exercise this capacity depends on the organizational structure of the group. If it involves measures that make sure that the group's beliefs and desires are reflected by the members then a group may satisfy the second condition.

The biggest challenge is, however, the third condition. On the one hand, the actions of a group agent can only be executed by its members. On the other hand, not every group member is in full control of the actions of the group, i.e., the group agent would have to be the controlling instance. Both claims taken together lead to a situation of overdetermination, since one and the same action "cannot be subject both to the control of the group agent and to the control of one or more individuals." (List and Pettit 2011, 160)

List and Pettit try to solve this dilemma by showing that an action can be in the control of the individual members of the group and of the group agent at the same time. The key to their solution lies in the distinction between a "programming cause" and an "implementing cause" (Pettit 1993). They illustrate the difference between the two kinds of causes with the help of the example of a closed flask of water that is breaking when boiled. The temperature of the water would be the programming cause whereas the single molecules which are causing the glass to break are implementing that program. Analogously, a group agent is supposed to be the programming cause of a group's actions whereas the executing members are the implementing causes.

Yet, even if one finds this solution convincing for pure MAS which are constituted solely of persons, new problems arise if one considers hybrid MAS involving persons and artificial agents which are not persons. Christian Neuhäuser takes up this challenge in his contribution to this volume. He applies the concept of a "forward-looking collective responsibility" as recently discussed in political philosophy to hybrid MAS involving humans and robots. Such groups can, according to him, be responsible for bringing about certain results in the future which their individual members cannot achieve on their own. As he argues, this concept is independent of "backward-looking responsibility" which is about finding out who is causally responsible and liable for some harm that has been done in the past.

As these examples show, the capacities of MAS depend on the capacities of their constituent agents. Yet, it is not automatically the case that a collective agent possesses the same capacities as its members. Even a pure system which only involves persons does not necessarily have the same capacities as a collective agent as it would if it were a person, e.g., the capacity to bear moral responsibility.

1.4 Perspectives for Research and Practice

The relevance of the proposed framework of types of agents and collective agency is threefold; it can be brought under the headings: *constructing*, *simulating* and *explaining* collective agency in MAS. (1) In order to construct MAS that are to collectively perform certain tasks one has to know what kind of agents are required and what kinds of collective agency are available in a type of MAS to fulfill the task most efficiently. With respect to hybrid systems, the question is how one can optimize interaction, cooperation and collective agency among different types of agents. This concerns specifically systems involving humans and artificial agents. A good example is provided by the contribution of Behrendt, Funk and Korn to this volume. They present assistive technologies at workplaces and also discuss their ethical implications, e.g., the way in which these technologies might change individual workplaces and the working environment, and the impact that they might have regarding justice and the meaning of work for the individual good life.

(2) Multi-agent systems are also used to simulate various social phenomena. Since MAS are typically very complex, it is useful to be able to simulate them to learn more about their behavior and to investigate the effects of different architectures. Simulations of MAS are—among other things—used to understand biological processes like flocking, to model economic and social processes like financial transactions, social unrest, or scientific paradigm change, and to find technical solutions in areas like driverless transport systems, optimization of production processes or emergency planning. Yet, these simulations can only be adequate, if they involve the right kind of agents. This is where the philosophical discussion of agency at an individual and collective level comes in.

In their contribution to this volume Happach and Tilebein show the value of simulations for management science with respect to dynamically evolving processes involving several agents like decision processes, strategy making, negotiations and operations. They also assess the advantages and drawbacks of the two predominant approaches to social simulations: Agent Based Modeling and System Dynamics.

Despite the ubiquity and explanatory strength of social simulations one also has to take into account various sources of error. These are discussed by Arnold in this volume using the example of Axelrod's simulation-based approach to the evolution of cooperation. To avoid these pitfalls when setting up social simulations it is important to have a clear grasp of the involved agents' relevant capacities, the properties of the social systems they are forming, and their ways to interact with each other collectively. The proposed framework of individual and collective agency is useful for the operationalization of social phenomena when designing social simulations. As Bryson shows in her chapter, carefully set up social simulations can, for instance, contribute to the understanding of the cultural evolution of intelligence, particularly, the emergence of pro-social behavior.

(3) Artificial MAS (soft- or hardware based) can also be used to clarify and test various explanatory hypotheses as to how collective agency and collective intentionality evolve. This would certainly be the philosophically most interesting and challenging area of application. It is based on the idea that computer simulations can be understood as an extension of thought experiments, a method that has been traditionally used in philosophy. Philosophers have developed imaginative scenarios in order to determine what is possible, necessary, and impossible under various assumptions with the aim to test certain philosophical hypotheses. It has, for instance, been traditionally assumed that knowledge consists in justified, true belief, i.e., that these conditions are necessary and jointly sufficient for knowing. In order to test this claim with the help of thought experiments, one has to imagine whether there can be cases of knowledge where one of these conditions is not fulfilled or whether it is possible that these conditions are fulfilled without there being knowledge.

I am inclined to think that simulations may have a similar function for philosophical thinking, only on a much more complex scale. In this respect, I am following Daniel Dennett who endorses a similar view for Artificial Life:

> Artificial Life (…) can be conceived as a *sort* of philosophy–the creation *and testing* of elaborate thought experiments, kept honest by requirements that could never be imposed on the naked mind of a human thinker acting alone. In short, Artificial Life research is the creation of prosthetically controlled thought experiments of indefinite complexity. This is a great way of confirming or disconfirming many of the intuitions or hunches that otherwise have to pass as data for the sorts of conceptual investigations that define the subject matter of philosophy. (Dennett 1994, 291)

Transferred to the topic of social simulation, the idea would be to test various explanatory hypotheses about collective agency and collective intentionality with the help of different simulational settings. The proposed framework of individual and collective agency might help to clarify and operationalize these hypotheses to make them testable.

The reason why I think that simulation is more promising in this area than with respect to other philosophical issues, for instance, in understanding knowledge, is that the latter may be an entirely conceptual issue which can be decided a priori from the philosophical armchair. Understanding collective agency and collective intentionality, in contrast, involves a complex entanglement of conceptual and empirical issues which cannot be scrutinized by purely imaginative a priori methods alone. Computer simulation might, therefore, make a contribution to philosophical inquiry where more traditional methods of philosophy fail. The project initiated in this book—to bring together philosophical approaches to cooperation and collective agency and research into human-machine-cooperation and MAS—promises, therefore, also important insights from a philosophical point of view.

Acknowledgments I am thanking Anja Berninger and Mog Stapleton for helpful comments on former versions of the paper.

References

Anscombe, Elizabeth. 2000. *Intention*. Cambridge, MA: Harvard University Press.
Balleine, Bernard W., and Anthony Dickinson. 1998. Goal-directed instrumental action: Contingency and incentive learning and their cortical substrates. *Neuropharmacology* 37(4–5): 407–419.
Boesch, Christophe. 2005. Joint cooperative hunting among wild chimpanzees: Taking natural observations seriously. *Behavioral and Brain Sciences* 28(5): 692–693.
Bratman, Michael E. 1984. Two Faces of Intention. *The Philosophical Review* 93(3): 375–405.
Bratman, Michael. 1999. I intend that we J. In *Faces of intention: Selected essays on intention and agency*, 142–161. Cambridge: Cambridge University Press.
Bratman, Michael E. 2006. Planning agency, autonomous agency. In *Structures of agency: Essays*, 195–222. Oxford/New York: Oxford University Press.
Butterfill, Stephen. 2012. Joint action and development. *The Philosophical Quarterly* 62(246): 23–47.
Castelfranchi, Cristiano, Dignum Frank, Catholijn M. Jonker, and Treur Jan. 2000. Deliberative normative agents: Principles and architecture. In *Intelligent agents VI. Agent theories, architectures, and languages*, ed. R. Jennings and Lespérance Yves, 364–378. Berlin/Heidelberg: Springer.
Davidson, Donald. 1980. *Essays on actions and events*. Oxford: Oxford University Press.
De Jaegher, Hanne, Ezequiel Di Paolo, and Shaun Gallagher. 2010. Can social interaction constitute social cognition? *Trends in Cognitive Sciences* 14(10): 441–447.
Dejean, Alain, Pascal Jean Solano, Julien Ayroles, Bruno Corbara, and Jérôme Orivel. 2005. Insect behaviour: Arboreal ants build traps to capture prey. *Nature* 434(7036): 973.
Dennett, Daniel C. 1976. Conditions of personhood. In *The identities of persons*, ed. Amelie Oksenberg Rorty, 175–196. Berkeley: University of California Press.
Dennett, Daniel C. 1994. Artificial life as philosophy. *Artificial Life* 1.
Floridi, Luciano, and Jeff W. Sanders. 2004. On the morality of artificial agents. *Minds and Machines* 14(3): 349–379.
Frankfurt, Harry G. 1971. Freedom of the will and the concept of a person. *Journal of Philosophy* 68: 5–20.
Franklin, Stan, and Art Gasser. 1997. Is it an agent, or just a program?: A taxonomy for autonomous agents. In *Intelligent agents III. Agent theories, architectures, and languages – ECAI'96 workshop (ATAL)*, ed. Jörg Müller, Michael J. Wooldridge, and Nicholas R. Jennings, 21–35. Berlin: Springer Verlag.
Gilbert, Margaret. 2006. *A theory of political obligation: Membership, commitment, and the bonds of society*. Oxford: Oxford University Press.
Helm, Bennett W. 2000. Emotional reason how to deliberate about value. *American Philosophical Quarterly* 37: 1–22.
Helm, Bennett W. 2008. Plural agents. *Noûs* 42(1): 17–49.
Hollander, Christopher D., and Annie S. Wu. 2010. The current state of normative agent-based systems. *Journal of Artificial Societies and Social Simulation* 14(2): 6.
Jennings, Nick R. 1993. Commitments and conventions: The foundation of coordination in multi-agent systems. *The Knowledge Engineering Review* 8(3): 223–250.
Jennings, Nicholas R., Katia Sycara, and Michael Wooldridge. 1998. A roadmap of agent research and development. *Autonomous Agents and Multi-Agent Systems* 1(1): 7–38.

Knoblich, Günther, Butterfill Stephen, and Sebanz Natalie. 2011. Psychological research on joint action. In *The psychology of learning and motivation*, vol. 54, ed. Ross Brain, 59–101. Burlington: Academic.

Korsgaard, Christine M. 2014. The normative constitution of agency. In *Rational and social agency: The philosophy of Michael Bratman*, ed. Manuel Vargas and Gideon Yaffe, 190–215. New York: Oxford University Press.

List, Christian, and Philip Pettit. 2011. *Group agency: The possibility, design, and status of corporate agents*. Oxford: Oxford University Press.

Mahmoud, Moamin A., Mohd Sharifuddin Ahmad, Mohd Zaliman Mohd Yusoff, and Aida Mustapha. 2014. A review of norms and normative multiagent systems. *The Scientific World Journal* 2014: 23. doi:10.1155/2014/684587. Article ID 684587.

O'Connor, Timothy. 2014. Free will. In *The Stanford encyclopedia of philosophy*, ed. Edward N. Zalta. http://plato.stanford.edu/entries/freewill. Accessed 13 Aug 2014.

O'Loan, O.J., and M.R. Evans. 1998. Alternating steady state in one-dimensional flocking. *Journal of Physics A: Mathematical and General* 32(8): L99–L105.

Pacherie, Elisabeth. 2013. Intentional joint agency: Shared intention lite. *Synthese* 190(10): 1817–1839.

Pacherie, Elisabeth, and Jérôme Dokic. 2006. From mirror neurons to joint actions. *Cognitive Systems Research* 7(2–3): 101–112.

Parunak, H. Van Dyke, Paul Nielsen, Sven Brueckner, and Rafael Alonso. 2007. Hybrid multiagent systems: Integrating swarming and BDI agents. In *Engineering self-organising systems*, ed. Sven A. Brueckner, Salima Hassas, Márk Jelasity, and Daniel Yamins, 1–14. Berlin/Heidelberg: Springer Verlag.

Pettit, Philip. 1993. *The common mind: An essay on psychology, society and politics*. New York: Oxford University Press.

Rao, Anand S., and Georgeff, Michael P. 1995. BDI-agents: From theory to practice. In *Proceedings of the first international conference on multiagent systems* (ICMAS'95), San Francisco.

Reynolds, Craig. 1987. Flocks, herds and schools. A distributed behavioral model. *Computer Graphics* 1987(21): 25–34.

Russell, Stuart J. 1997. Rationality and intelligence. *Artificial Intelligence: Special Issue on Economic Principles of Multi-agent Systems* 94(1–2): 57–77.

Schweikard, David P., and Hans Bernhard Schmid. 2013. Collective intentionality. In *The Stanford encyclopedia of philosophy*, ed. Edward N. Zalta. http://plato.stanford.edu/entries/collective-intentionality. Accessed 13 Aug 2014.

Searle, John. 1990. Collective intentions and actions. In *Intentions in communication*, ed. Philip R. Cohen, Jerry Morgan, and Martha Pollack, 401–415. Cambridge, MA: MIT Press.

Shockley, Kevin, Marie-Vee Santana, and Carol A. Fowler. 2003. Mutual interpersonal postural constraints are involved in cooperative conversation. *Journal of Experimental Psychology. Human Perception and Performance* 29(2): 326–332.

Shoham, Yoav. 1993. Agent-oriented programming. *Artificial Intelligence* 60(1): 51–92.

Shultz, T.R. 1991. From agency to intention: A rule-based, computational approach. In *Natural theories of mind: Evolution, development and simulation of everyday mindreading*, ed. Andrew Whiten, 79–95. Oxford/Cambridge, MA: Blackwell.

Steunebrink, Bas R., Mehdi Dastani, and John-Jules Ch Meyer. 2006. Emotions as heuristics in multi-agent systems. In *Proceedings of the 1st workshop on emotion and computing-current research and future impact*, ed. Reinhardt, D., P. Levi, and J.Ch. Meyer, 15–18.

Tollefsen, Deborah Perron. 2002. Collective intentionality and the social sciences. *Philosophy of the Social Sciences* 32(1): 25–50.

Tollefsen, Deborah. 2005. Let's pretend!: Children and joint action. *Philosophy of the Social Sciences* 35(1): 75–97.

Tomasello, Michael, and Katharina Hamann. 2012. Collaboration in young children. *The Quarterly Journal of Experimental Psychology* 65(1): 1–12.

Tomasello, Michael, Malinda Carpenter, Josep Call, Tanya Behne, and Henrike Moll. 2005. Understanding and sharing intentions: The origins of cultural cognition. *The Behavioral and Brain Sciences* 28(5): 675–691.
Ulzen, Van, R. Niek, Claudine J.C. Lamoth, Andreas Daffertshofer, Gün R. Semin, and Peter J. Beek. 2008. Characteristics of instructed and uninstructed interpersonal coordination while walking side-by-side. *Neuroscience Letters* 432(2): 88–93.
Velleman, David. 1989. *Practical reflection*, vol. 94, 1. Princeton: Princeton University Press.
Want, S.C., and P.L. Harris. 2001. Learning from other people's mistakes: Causal understanding in learning to use a tool. *Child Development* 72(2): 431–443.
Wilson, George, and Samuel Shpall. 2012. Action. *The Stanford encyclopedia of philosophy*, ed. Edward N. Zalta. http://plato.stanford.edu/entries/action. Accessed 13 Aug 2014.

Chapter 2
Cooperation with Robots? A Two-Dimensional Approach

Anika Fiebich, Nhung Nguyen, and Sarah Schwarzkopf

2.1 The Starting Point

A couple tackles a box. A group of employees forges out a plan how to avoid bankruptcy of their company. An orchestra plays Beethoven's Ninth Symphony. Phenomena that are called 'cooperation' are frequent in the scientific literature. The list is endless. The main aim of this treatise is to characterize cooperation in robots. Therefore we provide a two-dimensional approach to human-human and human-robot cooperation that allows determining where precisely a specific phenomenon that is called 'cooperation' lies on the axis of a 'behavioral dimension' and the axis of a 'cognitive dimension.'[1] The dimensions are orthogonal to each other. Moreover, we discuss the cognitive preconditions that come along with being engaged in cooperation on these two dimensions. This methodological distinction serves as a fruitful means to analyze whether and to what extent

---

All authors contributed equally and author names are listed in alphabetical order.

[1] We refer to embodied artificial agents as robots.

A. Fiebich (✉)
Institute of Philosophie II, Ruhr-University Bochum, Universitätsstraße 150,
Bochum 44780, Germany
e-mail: anifiebich@gmail.com

N. Nguyen
Artificial Intelligence Group, Faculty of Technology, Bielefeld University,
Bielefeld 33594, Germany
e-mail: nhung.nguyen@mail.de

S. Schwarzkopf
Center for Cognitive Science, Institute of Computer Science and Social Research,
University of Freiburg, Friedrichstr. 50, Freiburg 79098, Germany
e-mail: s.schwarzkopf@gmail.com

© Springer International Publishing Switzerland 2015
C. Misselhorn (ed.), *Collective Agency and Cooperation in Natural
and Artificial Systems*, Philosophical Studies Series 122,
DOI 10.1007/978-3-319-15515-9_2

cooperation on a 'behavioral dimension' and on a 'cognitive dimension' is implementable in robots.

The term 'cooperation' is used inflationarily. That is, one and the same phenomenon may be called 'cooperation' under one definition but not under another. Consider two examples: Imagine a group of graduate students who are taught about Adam Smith's 'Hidden Hand' in a business school (Searle 1990). The students form a pact to help humanity by way of each pursuing his or her selfish interests. To an outside observer, their behavior may not look like cooperation at a first glance. However, if the observer knows about the pact that the graduate school students have formed and their shared intention to pursue that pact, he or she may well be in a position to call their behavior cooperation (see Searle 1990 for a discussion of this example).

Now imagine Natalie is the organizer of a conference and arrives at the venue long before the conference starts. She aims to prepare the conference hall and hence starts to carry the many boxes that she finds there out of the hall. Steve is a participant of the conference. He also arrives far too early and wants to relax at the venue before the conference starts. Steve feels disturbed by the many boxes. Hence he starts to carry the boxes out of the hall. Imagine that Natalie observes Steve, and Steve observes Natalie carrying the boxes out of the room. They start to coordinate their behavior by way of Natalie carrying the big boxes and Steve taking the small ones. But neither of them realizes that the other one has recognized him- or herself as partaking in the action as well. Hence no mutual common knowledge about having the same goal is involved. It is not even necessary that they have knowledge about having the same goal. All that is required is that each of them is aware of his or her own goal (i.e., getting rid of the boxes in the room) and to coordinate the own behavior to that of the other in a way that is fruitful to achieve that goal (i.e. carrying the big boxes when the other is carrying the small ones, and vice versa). Now imagine Ipke is sitting in the conference hall. Ipke may see that Natalie and Steve are cooperating. But are they doing so in the same manner as the graduate students?

2.2 A Two-Dimensional Approach to Cooperation

We suggest that the answer is: No. These examples point to two main dimensions of how to describe cooperation that are predominantly used in the contemporary scientific literature: (i) A *behavioral dimension* that focuses on the behavior patterns of the cooperating agents; and (ii) a *cognitive dimension* that focuses on the cognitive states of the agents who are engaged in cooperation.[2] In the following, we exemplify these two dimensions by dipping into the philosophical and empirical literature on cooperation and discussing various examples that lie on different stages of the two dimensions. Moreover, we spell out the cognitive preconditions that being engaged on either

---

[2] In general, a third dimension could be added that discusses the phenomenological experiences of agents whilst cooperating. For the present purposes, however, the two dimensions mentioned here are sufficient.

dimension presuppose in humans and discuss which of these preconditions are implementable in robots. We end with a summary and appeal to future research to address some questions that are still open.

2.2.1 Cognitive Dimension of Cooperation

The cognitive dimension of cooperation is endorsed not only by philosophers (e.g., Searle 1990; Bratman 1993; Tuomela 2010) but also by psychologists (e.g., Carpenter 2009; Tomasello et al. 2005) who emphasize the role that a particular kind of intentionality plays for cooperation. In individual actions, i.e. actions performed by one agent, the conception of 'intentionality' refers to the individual's intention to act (see e.g., Searle 1983) whereas in collective actions, i.e. actions performed by a group of agents, the conception of 'collective intentionality' refers to the intentions of the group. Various accounts have been proposed to determine what the intentions of a group of agents are. Pragmatically, we discuss only those that we consider the dominant theoretical accounts that have been proposed, including Searle's (1990) conception of 'we-intention,' Tuomela's (2010) conception of 'intention in the we-mode,' and Bratman's (1993) conception of 'shared intention.'

Searle (1990) highlights that "the notion of we-intention, of collective intentionality, implies the notion of *cooperation*" (p. 406). In Searle's account, a 'we-intention' is not reducible to a set of I-intentions that is, the sum of the personal intentions of single individuals to aim for the same goal even if aiming for the same goal is mutual common knowledge among the individuals. A 'we-intention' of a single individual represents the contribution of the individual to the joint action; e.g., "we are making the sauce by means of me stirring." Searle defines a 'we-intention' as a special attitude of a single individual involving a 'we-activity.' Notably, in Searle's account, an agent may have a we-intention even in cases where there is no other agent; hence a "brain in a vat" (Searle 1990, p. 407) may have nothing but the illusion of having a body that is engaged in cooperation with another agent.

Searle's conception of 'we-intention' is similar to Tuomela's (2010) conception of 'intention in the we-mode' insofar as both philosophers emphasize a special kind of attitude as being distinctive for having such an intention. However, Tuomela not only discusses the intentional content of this intention but also the reasons for and the implications of having such an intention with respect to commitments and the specific phenomenal experiences in group activities. Tuomela (2010) distinguishes between group activities that agents perform in the 'we-mode' opposed to the 'pro-group I-mode.' For example, two drivers may happen to arrive from different directions at a tree trunk lying on the road and blocking their way. To continue their drive, they jointly remove the trunk. This is an example of a pro-group I-mode joint action. Agents who have an intention in the pro-group I-mode to be engaged in a group activity accept a shared goal of the group (e.g., to remove a trunk) due to private reasons (e.g., you want to continue your drive to be in time for your grandma's coffee party and I want to escape the police officers who are chasing me after I robbed the city bank). In joint actions performed in the we-mode, in contrast, agents pursue a shared goal on the basis of group reasons (maybe determined by a group ethos) and they are collectively committed to each other to pursue that goal. That is, none of the agents is in a position to stop pursuing the goal without the agreement of the other agents. Tuomela's explorations of joint commitments are compatible with and partly draw on Gilbert's (2009) account (see Sect. 2.3 for a discussion of Gilbert's account).

Bratman (1993, 1992) accounts for what he calls a 'shared intention' not as an attitude or intention in the mind of a single agent (such as proposed by Searle, see above) but as a "state of affairs that consists primarily in attitudes (none of which are themselves the shared intention) of the participants and interrelations between those attitudes" (pp. 107–8). That is, at least two individuals are required for there to be a shared intention. Mutual common knowledge of aiming for the same goal,[3] intending to participate in a cooperative action to pursue that goal in accordance with and because of the other's intention, being willed to compromise and to mesh individual sub-plans are distinctive features of how the individual attitudes of the participants are interrelated in the case of a shared intention. Imagine two people having the shared intention of painting a house together: One has the sub-plan of painting it red whereas the other has the sub-plan of painting it blue. If both of them do not compromise and mesh the individual sub-plans, they won't succeed in pursuing their shared intention. Notably, Bratman (2009) also emphasizes that the intentions of the single individuals involve a we-activity in the Searlean sense; "a Searlean we-intention is, then, a candidate for the intentions of individual participants that together help to constitute a shared intention, though Searle himself does not say how the we-intentions of different participants need to be inter-related for there to be a shared intention" (p. 41). Having a shared intention is, by Bratman's (1992) account, essential for 'shared cooperative activities' that are characterized by mutual responsiveness, commitment to the joint activity and commitment to mutual support of the agents.

For the present purposes, we make use of a hybrid account of the shared intention of a group. That is, we postulate that such an intention needs to involve a we-activity (as proposed by Searle); is characterized by the joint commitment of the group members to pursue the intention until its end (as emphasized by Gilbert and Tuomela); and presupposes mutual common knowledge of aiming for the same goal, intending to participate in a cooperative action to pursue that goal in accordance with and because of the other's intention as well as being willed to compromise and

---

[3] One remark needs to be made with respect to this definition. We deliberatively do not adopt Bratman's conception of 'mutual common knowledge' but rather prefer to use a cognitively less demanding definition to stress that mutual common knowledge of aiming for the same goal neither requires an infinite number of overlapping embedded mental states nor do these mental states need to be linguistic in content (Wilby 2010) and hence can be involved in basic cooperative phenomena such as joint attention (Fiebich and Gallagher 2013) that are already present in preverbal infants (Tollefsen 2005).

to mesh individual sub-plans (as highlighted by Bratman). Finally, we follow Bratman in considering at least two agents necessary for there to be a shared intention since we strive for an analogy to the behavioral dimension of cooperation that presupposes the participation of at least two agents (see below).

In our account, being engaged in cooperation on the cognitive dimension presupposes minimally that

(a_i) two (or more) agents perform actions to pursue the same goal, and
(a_{ii}) the agents know that they have the same goal.

Cooperation on the cognitive dimension occurs in various degrees. For example, when the agents have mutual common knowledge of having the same goal (i.e., each agent knows about having the same goal like the other agent and also knows that the other agent knows about that), the phenomenon lies on a higher point on the axis than if the agents only have knowledge themselves of having the same goal like the other agent. Pursuing the same goal on the basis of having a shared intention marks the endpoint (i.e. the highest stage) on the cognitive dimension of cooperation; we call this point 'cognitive cooperation supreme.' These are the necessary criteria that are together sufficient for a phenomenon to be 'cognitive cooperation supreme':

(a_i) Two (or more) agents perform actions to pursue the same goal, and
($a_{ii'}$) the agents have a shared intention, i.e., a common goal involving we-intentions.

In general, a phenomenon may be described as cooperation on the cognitive dimension to a more moderate degree, ranging from the agents having knowledge of having the same goal over having mutual common knowledge of having the same goal (i.e., having a 'common goal') to pursuing the same goal on the basis of having a shared intention.

2.2.2 The Behavioral Dimension of Cooperation

Cooperation described on the behavioral dimension is typically not explicitly spelled out but often underlies the experimental paradigms that investigate various aspects of cooperative behavior in social interactions. In our account, 'cooperation' on the behavioral dimension is characterized by the coordinated behavior of the interacting agents. The more complex the coordinated behavior is to an outside observer, the higher is the degree of the described phenomenon on the continuum of the behavioral dimension.

Sebanz et al. (2006) define joint action as "any form of social interaction whereby two or more individuals coordinate their actions in space and time to bring about a change in the environment" (p. 70). By this definition, the goal that both agents have is to change the environment. A variety of cognitive mechanisms (such as joint attention, action observation, task-sharing and action coordination) may play a role in the success of joint actions, however, no shared intention or mutual common knowledge of having the same goal need be involved. This view is implicitly shared among a number of psychologists (e.g., Isenhower et al. 2010; Chaminade et al. 2012) as well as computer scientists (e.g., Sofge et al. 2005; Lenz et al. 2008).

In psychology, for example, studies with physically coupled dyads (Reed et al. 2006a, b) interacting independently of each other with only haptic feedback suggest that the participants develop cooperative strategies without knowing that they are cooperating. Here the term cooperation is used without presupposing that the agents have a shared intention to cooperate. Likewise, in other studies, participants are asked to play sequential turn-taking games (McCabe et al. 2001; Decety et al. 2004) in which they have only few possibilities to interact. In Decety et al.'s (2004) study, for example, participants play a game with the experimenter. Both strive towards the goal of the game, i.e. the creation of a specific pattern on a computer screen, in alternating turns. In most of these board-game-like operationalizations of cooperation, the participant interacts with a confederate or with a computer script.

Of course, cooperation that is described on the behavioral dimension may also involve specific cognitive factors. For example, in other studies, mutual common knowledge of having the same goal is present among the participants and cooperation is defined as the coordination of actions to achieve a common goal. Chaminade et al. (2012) manipulate the 'amount of cooperation' by a changing number of individual action possibilities in a joint motor task. Shockley et al. (2003) use a similar principle: Each participant needs to coordinate his or her behavior to that of the other participant by helping each other to achieve a particular goal. Here, as in those studies in which two real participants interact with each other and do so simultaneously (Isenhower et al. 2010; Newman-Norlund et al. 2008), mutual common knowledge of having the same goal is involved. However, it is unclear whether a shared intention is involved or the participant's interaction with the other person is nothing but using the other person as a 'social tool' (Moll and Tomasello 2007), i.e., a means to an end. Also, the motivation to cooperate by promising extra money for accurate and quick joint performance (e.g., Böckler et al. 2011) or by simulating an in-group and letting team members work together against an out-group (e.g., Vonk 1998) does not control for whether participants are engaged in a 'joint action in a we-mode' on the basis of a group ethos or whether they just participate for their own benefit (i.e., a 'joint action in a pro-group I-mode,' see Sect. 2.2.1). In Isenhower et al.'s (2010) study, participants are not explicitly asked to cooperate. In the beginning they move wooden planks independently of each other. Then, the participants start to carry the planks together at the latest when they encounter difficulties carrying a plank alone. This situation may, but does not need to, evoke the formation of a shared intention.

In our view, Sebanz et al.'s (2006) notion of joint action serves as a fruitful starting point to capture the three necessary criteria that together are sufficient for there to be 'cooperation' on the behavioral dimension, which are:

(A_1) Two (or more) agents coordinate their behavior in space and time,
(A_2) which is observable from the outside,
(A_3) to bring about a change in the environment.

Some remarks need to be added with respect to these criteria: in (A_3), 'to bring about a change in the environment' amounts to bring about a change in the physical or in a virtual environment by either achieving a particular end state (e.g., to locate the table together in front of the TV), or an end product (e.g., to build a tower together). Notably, (A_1) does not necessarily require that the agents coordinate their behavior in a real-interactive setting. As the conference example illustrated (see Sect. 2.1), one agent A may coordinate his or her behavior to the behavior of the other agent B with a time delay. To coordinate his or her behavior to that of B it is not even required that A perceives B's behavior. It is sufficient that A coordinates his or her behavior to what A detects to be the result of B's behavior. For example, Steve does not need to observe Natalie carrying all the big boxes out of the conference hall. It is sufficient that Steve realizes that Natalie has carried the big boxes so that he takes the small ones. For there to be cooperation, however, Steve needs to perceive the disappeared big boxes in the conference hall as the result of Natalie's action.

Moreover, rather than referring to particular cognitive (e.g., motivational) states that guide the behavior of the agents, the term 'to' in 'to bring about a change in the environment' in (A_3) refers to the efficiency of the more or less complex coordinated behaviors of two agents as a means to an end that is detected by an outside observer (independent of whether or not the goal recognized by the observer is the same goal as pursued by the agents). Such complexity can be analyzed solely on the behavioral dimension without the need to refer to cognitive notions such as shared goal (i.e., having mutual common knowledge of aiming for the same goal) or shared intention. Complexity is defined by the amount of observable coordinative action sequences within the cooperation. For example, randomly taking any box out of the room is less complex than taking the boxes out of the room in a specific order. Finally, we emphasized that the coordinated behavior of the agents needs to be observable from the outside (A_2) since we draw here on a conception of cooperation as it is used in experimental paradigms.

We propose that there is no end point of the continuum of the behavioral dimension of cooperation analog to the end point of the cognitive dimension of cooperation. There is no 'behavioral cooperation supreme.' Rather, the axis of the behavioral dimension has an open end which is characterized by highly complex coordinated behavior of the agents to achieve a particular goal as it is recognized from the outside. This requires the agents to have the same goal with neither knowledge about having the same goal nor a shared intention being necessarily involved.

2.2.3 Two Dimensions

As illustrated in Fig. 2.1, 'cognitive cooperation supreme' is the end of the axis of the cognitive dimension. At this stage, two agents perform particular actions to pursue a shared intention. In the business school example, the graduate students

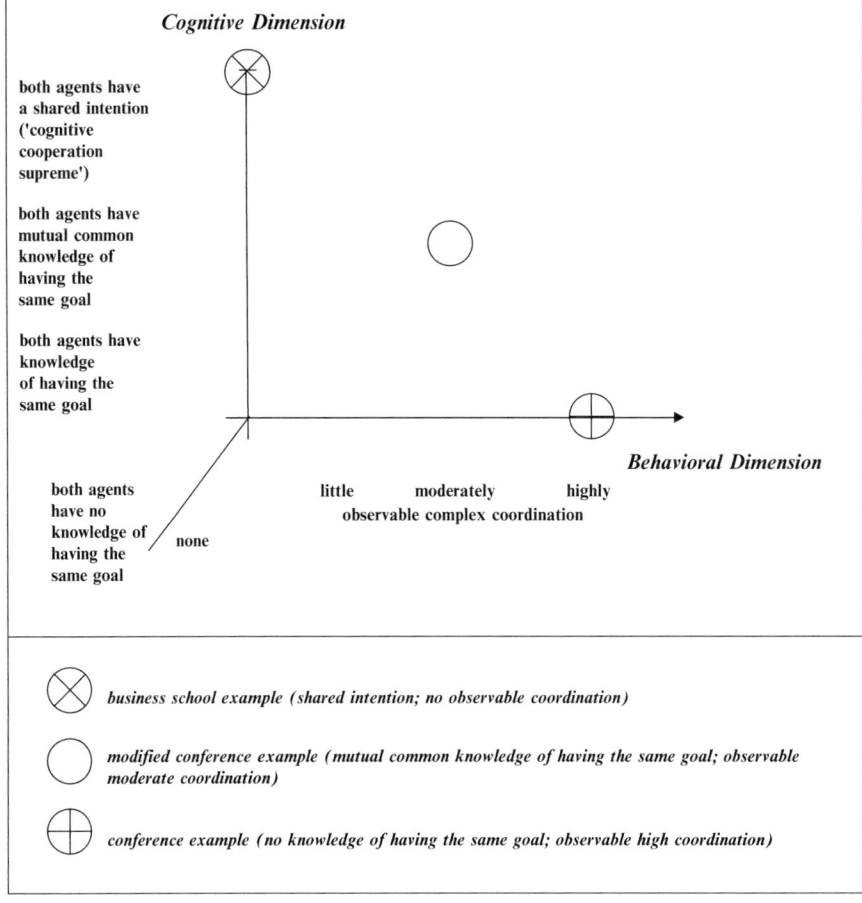

Fig. 2.1 Two dimensions of cooperation

behave in a selfish manner to pursue their shared intention of helping humanity. This is a case of 'cognitive cooperation supreme.' Moreover, the coordinated behavior patterns of the graduate students are not at all observable from the outside. Hence the phenomenon lies on a zero point of the behavioral axis. We propose calling such a phenomenon that lies on the zero point of the behavioral axis and not on the zero point of the cognitive axis 'pure cognitive cooperation.'

In contrast, given that Steve and Natalie pursue their goal of emptying the room (see example in Sect. 2.1) by carrying the boxes out of the room, their behavior is coordinated in order to bring about a change in the environment but no knowledge of having the same goal has to be involved. Hence the phenomenon lies on the zero point of the cognitive axis. We propose calling a phenomenon that lies on the zero point of the cognitive axis and not on the zero point of the behavioral axis 'pure behavioral cooperation.'

Of course, a variety of options exist in between the extreme cases of 'pure cognitive cooperation' and 'pure behavioral cooperation.' Two agents may pursue a shared intention in a highly coordinated fashion. Furthermore, the behavior of the agents may be more or less coordinately structured. That is, coordination can be regarded as a matter of degree. The cognitive states of the agents can also be regarded as a matter of degree, ranging from having a shared intention to having mutual common knowledge of having the same goal to having knowledge of having the same goal (see Fig. 2.1).

A phenomenon that lies in the middle of both axes could be a modified version of the conference example in which Natalie and Steve have mutual common knowledge of having the same goal of emptying the conference hall by carrying the boxes out of the room but without having a shared intention and hence without joint commitments to pursue that goal until its end. Moreover they may coordinate their behavior in a moderately complex way. Moderate complex coordination applies to behavior in which agents are not consistently mutually responsive over time whereas little complex coordination means that the agents coordinate their behavior only occasionally.

These two dimensions shall serve as a fruitful means for scientists not only to locate themselves in the debate but also to facilitate the dialogue on cooperation among different disciplines and traditions. Moreover, the two-dimensional approach to cooperation proposed here has implications concerning the robustness of a particular cooperative phenomenon (see Sect. 2.3), as well as for determining whether and to what extent cooperation on either dimension is implementable in artificial systems such as robots (see Sect. 2.4).

2.3 Implications of the Two-Dimensional Approach for the Robustness of Cooperation

Cooperation is robust when it is resistant against various confounding factors that may prevent the agents from pursuing their goal until its end. The role of the robustness of cooperation has been discussed by various philosophers with respect to commitments; all are in agreement that commitments play a central role for cooperation. Bratman (1992), for example, emphasizes that commitment to the joint activity (i.e., pursuing the joint activity until its end) as well as commitment to mutual support (i.e., helping each other in situations that demand for that) are central features of what he calls a 'shared cooperative activity.' He also adds mutual responsiveness to the list. That is, he proposes that each agent is obliged to be responsive to the other's actions by coordinating his or her own behavior to the actions of the other; "each seeks to guide his behavior with an eye to the behavior of the other, knowing that the other seeks to do likewise" (p. 328). As indicated above, observable coordinated behavior patterns are only necessarily involved when describing cooperation on the axis of the behavioral dimension but not of the cognitive dimension. Moreover, 'pure behavioral cooperation' requires coordinated behaviors of the agents (to various degrees) but does not presuppose knowledge about aiming for the same goal.

Following a number of philosophers (Bratman 1992; Gilbert 2009; Tuomela 2010), we propose that joint commitments are only involved when the cooperating agents have a shared intention.[4] That is, joint commitments are only involved in what we have called 'cognitive cooperation supreme.' In Gilbert's (2009) account, agents who have a shared intention are obliged to pursue that intention due to the following criteria: (i) the *concurrence criterion* according to which "absent special background understandings, the concurrence of all parties is required in order that a given shared intention be changed or rescinded, or that a given party be released from participating in it"; and (ii) the *obligation criterion* according to which "each party to a shared intention is obligated to each to act as appropriate to the shared intention in conjunction with the rest" (p. 175).

The concurrence criterion ensures that the agents are not in the position to rescind from the joint commitment of pursuing a shared goal unless the concurrence of all parties is given. To illustrate this recall the conference example. Natalie and Steve have the same goal to empty the room without mutual common knowledge being involved. They structure their carrying of the boxes out of the room in a coordinated fashion. But if Steve were to change his mind spontaneously and stop carrying the boxes because he prefers to have a coffee with Ipke, Natalie would not be in a position to complain. She would also not be entitled to complain if having the same goal of carrying the boxes out of the room is mutual common knowledge between them. However, if Natalie and Steve have the shared intention of carrying the boxes out of the room in order to prepare the venue for the conference and Steve were to change his mind under these conditions, Natalie would be in a position to complain. As the obligation criterion states, Steve would 'owe' Natalie the future action of carrying the boxes. Of course, Natalie may kindly allow Steve to go for a coffee with Ipke if he asks. But according to the concurrence criterion, concurrence of Natalie and Steve is required to release him from the joint commitment to carry the boxes. These two criteria spelled out by Gilbert (2009) refer to what Bratman (1992) has called 'commitment to joint activity.' Bratman emphasizes that 'commitment to mutual support' plays a crucial role in shared cooperative activities. For example, if somebody shuts the door in Natalie's face when she wants to go through, Steve is obliged to open the door for her since Natalie is carrying a box and hence unable to do so.

All these examples illustrate that 'cognitive cooperation supreme' is more robust than any cooperation that is described on a lower stage of the cognitive dimension

---

[4] Contrary to Bratman (1992), Gilbert proposes that (2009) "people share an intention when and only when they are jointly committed to intend as a body to do such-and-such in the future" (p. 167). We keep neutral to this. Hence we do not discuss Gilbert's 'disjunction criterion' according to which the shared intention of a plural subject to pursue a goal continues to exist even when the agents who constitute the plural subject no longer have personal contributory intentions. Whether or not cooperating agents act as a 'plural subject' or not is not relevant for the present purposes.

since the joint commitment to the joint activity and the joint commitment to mutual support that naturally comes along with having a shared intention enhances the chances that the agents' goal will be pursued till its end. In all other cooperative phenomena, the agents are only privately committed to achieve their individual goal, which is the same goal as that of the other agent. Hence, no joint commitment to pursue the goal until its end is involved.

Joint commitments are involved in long-term joint actions as well as in short-term joint actions. However, they are more likely to be called into action to reduce the risk of the shared intention failing in long-term joint actions; the longer the joint action lasts, the more likely it is that a disturbance will emerge (e.g., an agent is not able to contribute his or her part to the joint action in a given situation and requires help from the cooperator).

Joint commitments may also render meshing sub-plans redundant. As pointed out by Bratman (1993), when agents pursue a shared intention, willingness to mesh individual sub-plans may be required. For example, when Natalie and Steve have the shared intention to carry the boxes out of the conference hall, Natalie may have the sub-plan of carrying the bigger boxes while Steve carries the smaller ones. A problem arises if Steve has the same sub-plan like Natalie and is only willed to carry the big boxes. If Natalie and Steve find no agreement, their shared intention runs the risk of failing. However, if Natalie and Steve have the joint commitment from the beginning that one of them has to carry the big boxes, no meshing of sub-plans is required anymore.

'Cognitive cooperation supreme' may be even more robust when the shared intention the agents have involves a joint commitment to pursue particular rules and regularities to achieve the goal. Of course, the very same effect may be yielded when the agents feel only privately committed to the very same rules and regularities. Notably, however, rules and regularities do not need to make a cooperation more robust. Agreed upon rules and regularities may reduce the cognitive workload insofar as (some) decision making processes on how to pursue the shared intention lapse. In some cases, however, where the rules are very complex and novel to the agents, rules and regularities may make the achievement of the shared goal more difficult.

Notably, we do not propose that joint commitments are the only factor that determines the robustness of a cooperative phenomenon. Joint commitments ensure that agents pursue the shared goal until its end. Highly complex coordinated behavior, in turn, may (but does not need to) enhance the success of the cooperation to lead to the aimed goal. Moreover, rules and regularities that may make cooperation more robust could also be involved in a cooperative phenomenon that is described as 'pure behavioral cooperation.' Indeed, we can imagine a situation in which two agents do not know they have the same goal but in which each agent coordinates his or her behavior to that of the other agent by pursuing the very same rules as the other agent, and vice versa, which makes the cooperation more robust due to the efficiency of those rules for the given goal – but again without either agent knowing the other is pursuing these very rules.

2.4 Applications to Human-Robot Cooperation

There exists a wide range of intelligent agents[5] (Goodrich and Schultz 2007; Fong et al. 2003), such as robots, that are capable of interacting with humans. In order to build *cooperative robots*, technical models are typically designed on the basis of empirical findings and theoretical considerations from philosophy or psychology. Our aim is to discuss whether and if so, to what extent cooperation described on the behavioral and cognitive dimension that we postulated in Sect. 2.2 is implementable in robots. In order to address this question, we now spell out the cognitive preconditions that come along with cooperation on either dimension to determine which preconditions robots satisfy to be engaged in cooperation on various stages of the axes of the cognitive or behavioral dimension. Therefore we translate the cognitive preconditions into technical preconditions, which describe the abilities that robots need to possess in order to satisfy the criteria of our two dimensions of cooperation. We conclude with a discussion of the implications of human-robot cooperation compared to human-human cooperation with respect to joint commitments and rules and regularities.

2.4.1 Preconditions of Human-Robot Cooperation

We start by discussing the cognitive preconditions of the agents in cooperation on the behavioral dimension in which minimally (A_1) two (or more) agents coordinate their behavior in space and time, (A_2) which is observable from the outside, (A_3) to bring about a change in the environment.

There are robotic systems that satisfy the minimal criteria of cooperation on the behavioral dimension. In order to coordinate their behavior, each agent needs to perceive the (result of the) other agent's behavior, and to react to the other's behavior in a way that is fruitful to achieve the own goal. Therefore the agent has to represent a goal and an action plan of how to achieve it efficiently.

Lenz et al. (2008) and Grigore et al. (2012) presented work on human-robot coordination of planning and acting within a joint physical task. In these studies, Sebanz et al.'s (2006) notion of joint action was explicitly adopted. The cognitive system, in this case the robot, is able to represent its own goal and an action plan. Moreover, the robot can perceive the movements of the other agent in order to coordinate its own actions to those of the other. The work of Lenz et al. (2008) involved a joint construction task with toy blocks between a robot and a human. With the camera sensor system, which acts as artificial 'eyes,' the robot is able to

---

[5] We refer to the definition of intelligent agents as introduced by Wooldridge and Jennings (1995). An intelligent agent in their definition has one of the following abilities to satisfy predefined objectives (p. 116): reactivity (i.e. perceiving the environment, responding to changes), proactiveness (i.e. taking initiative in goal-directed behavior) and social ability (i.e. interacting with other agents, e.g. humans).

detect the bodily movements of the human partner. By pattern recognition methods, the robot is able to match the movement patterns with a database of predefined body actions, e.g., hand-grasping an object. The work of Grigore et al. (2012) included a physical task of preparing a drink and handing it over to a human. In both cases, the robot is able to perceive and to recognize the actions of the human partner as well as to react on them.

In more complex behavioral cooperation, the agents coordinate their behavior to bring about a change in the environment in a highly complex way. This requires a large amount of attention and reactivity towards the other's movements. For example, the system of Gray et al. (2005) recognizes the human's action of pushing a button by detecting the human's body motion through hardware sensors attached to the human body. The robot uses this information to track the human's success and to evaluate how to best help the human. These studies show that robots are capable of cooperation on a behavioral dimension.

Now we go further to specify the preconditions that are needed to describe cooperation on the cognitive dimension. Let us consider the minimal criteria: (a_i) two (or more) agents perform actions to pursue the same goal, and (a_{ii}) the agents know that they have the same goal.

This requires not only representing the own goal but also the goal of the other agent as well as recognizing that these goals are one and the same. Knowledge of having the same goal is modeled by establishing a goal for the robot (e.g., grasping an object, handing it over to a human partner), detecting the human's actions (e.g., that the human opens his or her hand and approaches object), recognizing the partner's goal (e.g., the partner is ready to receive the object), and matching the robot's goal with that of the human (i.e., both goals are congruent). Once the same goal is established, the robot starts to pursue its goal (i.e. grasping the object, handing it over to human partner). In the work of Foster and Matheson (2008) the robot JAST is capable of representing a joint assembly goal[6] that is to be achieved by the robotic system and the human. The human interaction with JAST is based around jointly assembling wooden construction toys. JAST fulfills the minimal criteria of a cooperation on the cognitive dimension.

Recall that in a 'cognitive cooperation supreme,' the second criterion is that (a_{ii}) the agents have a shared intention, i.e., a common goal involving we-intentions. The presupposition of having a common goal is having mutual common knowledge of having the same goal, which makes it a cognitive precondition of 'cognitive cooperation supreme.' Another cognitive precondition is representing the goal involving we-intentions. These two preconditions are required for having a shared intention.

Mutual common knowledge of having the same goal can be very basic, e.g., when achieving joint attention together (Fiebich and Gallagher 2013). Joint attention is implementable in robots. One example of signaling joint attention to the human partner was presented by Pfeiffer-Lessmann and Wachsmuth (2009). In this study, the robot is equipped with the ability to establish joint attention using gaze

---

[6] A goal within an assembling task (e.g. a building task with toy blocks) that is to be reached by the involved partners.

cueing, gaze following and eye contact in a context where both the human and the robot are looking at the same object.

The crucial question is whether robots are capable of having a shared intention or whether cooperation under a cognitive description is system-specific and dedicated only to humans. Having a shared intention goes beyond having mutual common knowledge of having the same goal: It also involves we-intentions of both agents. The challenge is to translate a we-intention into technical terms. Following Searle's (1990) conception of 'we-intention,' Hoffman and Breazeal (2004) made an attempt to implement a we-intention in the robot Leonardo. In their technical setup, Leonardo performs the task of pressing three buttons. The task is represented by the agent and divided into subtasks with specific actions that are to be performed. These actions are only performed when the predefined conditions of the action are satisfied. All actions are performed until the end of the task, i.e. until the conditions of the goal (in this case of switching all buttons on) are met. Leonardo is able to engage in performing this task jointly with a human and is able to have a common goal involving what the authors have modeled as we-intention. The idea is that in cooperation, a task is not only performed by Leonardo himself but together with a human partner. Recall that according to Searle (1990), the content of a 'we-intention' represents the agent's contribution to the joint action. The task should be divided between the robot and the human. That is, Leonardo needs to take the human's actions into account, when he decides what to do next. For each action that is to be performed in order to reach the common goal, Leonardo negotiates who is taking the turn for the next action. If he is able to do it, he will suggest taking the turn by using his hand and showing a pointing gesture to himself. He then waits for the human's reaction. The human can acknowledge or refuse the turn-suggestion. The task is then assigned as being performed by Leonardo himself or the human partner. Leonardo represents the common goal not only as his individual goal, but also as the goal of the partner. If an action is assigned to the human, Leonardo monitors and evaluates the execution of the action with respect to the common goal as if he were performing the action himself. This is understood as Leonardo's commitment to the partner's action.

According to Bratman (1993), a 'shared intention' involves intending to participate in a cooperative action to pursue a common goal in accordance with and because of the other's intention. In the technical system underlying Leonardo's behavior, this is realized during the whole interaction. Leonardo continually tracks a joint plan (i.e. the currently executed task) which is to be performed in order to reach that common goal. He derives his own intentions based on the common plan. He also derives the human partner's actions and changes made in the world that apply to the sub-plans.

In a more recent study, Dominey and Warneken (2011) created a similar implementation of a 'we-intention' in a robotic system. In this study, the robot stored a particular action sequence that the experimenter demonstrated. This action sequence included different agents (i.e., the robot and the experimenter) for different actions in a cooperative game (e.g., 'chasing the dog with the horse'). The robot was not only capable of storing this 'we-intention' but also of switching flexibly between the roles of the agents in the action sequence, dependent on whether the experimenter made the first move or not.

In sum, Hoffman and Braezael's (2004) as well as Dominey and Warneken's (2011) studies can be regarded as successful attempts of implementing a shared intention and therefore a 'cognitive cooperation supreme' in an artificial system. Pfeiffer-Lessmann (2011) presented similar work that will be discussed below in more detail. Whether the implications of having a shared intention are also addressed is an open question.

2.4.2 Implications of Human-Robot Cooperation

In Sect. 2.3, we have discussed two factors that may make cooperation more robust: rules and joint commitments. We will now discuss to what degree these two factors have already been successfully implemented in robots.

For a robot it is crucial to determine appropriate coordinated behavior patterns in order to achieve a common goal. Thus the main question is how to achieve these patterns. One method of acquiring knowledge about how to structure and select coordinated behavior patterns is by first collecting empirical data from human-human interaction where people cooperate to solve tasks in a specific domain. Characteristic features of human behavior are identified and adopted for implementation in the robotic systems. In such systems, these empirical data are used to model rules, in order to coordinate and plan interaction structures that resemble human cooperative behavior in task-solving (Allen et al. 2001). That means empirical data help engineers to program rules in advance which enable robots to exhibit what seem to be appropriate behavioral patterns in cooperation. The goal of this research area is to build robots that act in a way that, from a human's point of view, is judged to be like human behavior. The more rules are predefined, the more human-like the behavior might appear. But situations may occur that were not considered in the programming. The technical consequences of having cooperation with rules and regularities that are *not* defined in advance are that unpredicted and new situations might occur when the human partner does not react in the presupposed way. In dynamically changing environments, rule-based systems are prone to errors, as there are no predefined rules for dealing with the new situation. Then the robot chooses the wrong action to respond or is not even able to continue the interaction at all. That means the less rules and regularities are defined in advance, the more ambiguities are there to choose the appropriate actions – and the more difficult it is for artificial systems to engage in cooperation. Therefore, the more rules are defined in advance, the more robust is the human-robot cooperation. In order for robots to deal with situations in which rules are not explicitly defined, they need sophisticated learning mechanisms. But these mechanisms are still deficient and a huge amount of engineering work (Goodfellow et al. 2013) needs still to be done to reach human abilities.

The other implication concerns joint commitments. It remains an open question whether Hoffman and Breazeal (2004) implemented having a joint commitment in the robot. Leonardo is privately committed to the joint goal of pressing all the buttons because he continues to pursue the goal once the other agent signals him to join the task. This communicative signal could be regarded as an indicator of a joint commitment between the agents. However, Hoffman and Breazeal (2004) do not discuss Leonardo's reaction when the human partner suddenly stops contributing to the joint action task. Does he complain, or try to reengage the partner? This case does not seem to be implemented in the robotic system Leonardo. In Dominey and Warneken's (2011) study, in contrast, if the experimenter does not take over his or her action part himself or herself and denies the robot's question "Is it OK?" after a 15 s delay, the robot starts to help the experimenter. Pfeiffer-Lessmann (2011, pp. 235–236), in turn, presented cMax, a robot that is able to deal with a situation in which a human partner stops reacting during an ongoing cooperation. In the presented example cMax tries to give a conversational turn to the human partner, but the human does not react and turns away. cMax is able to recognize that the human is not taking the current turn according to discourse rules. In this case cMax expresses another giving-turn gesture with his hand in order to provoke a reaction from the human. If the human still does not react, cMax makes another attempt with a more articulated action. When the human does not react, cMax then assumes that the human wants to stop the interaction. cMax takes over and closes the conversation by saying: "Let's leave it at that." cMax builds expectations during the cooperation according to contact management rules. As these rules were broken by the human, cMax's simulated emotional state is affected.[7] cMax becomes more and more annoyed. Since joint commitments play a crucial role for the robustness of cooperation, future research in computer science should put more effort in implementing full-fledged joint commitments in robots.

2.5 General Discussion

In this treatise, we proposed a two-dimensional approach to cooperation according to which a phenomenon can be described as lying on the axes of two dimensions; an axis of a behavioral dimension that is defined by the complexity of coordinated behaviors of the agents as observable from the outside, and an axis of a cognitive dimension that is defined by the cognitive states of the agents. This approach not only enables scientists from different disciplines and traditions to locate themselves in the debate when investigating what they call 'cooperation,' it also provides a framework to spell out the cognitive preconditions that being engaged in cooperation on either dimension involves. Identifying such preconditions served as a fruitful means to address the leading question of the present treatise, which is to determine whether and to what extent behavioral and cognitive cooperation are implementable in robots.

We would like to draw scientists' attention to the methodological distinction of a behavioral and a cognitive dimension of cooperation. The object of investigation differs depending on the dimension used to characterize cooperation. Whereas the

---

[7] The underlying simulated emotional system was presented by Becker et al. (2004).

latter dimension is defined by the cognitive states of the agents, the former is defined by the observability of the agents' behavior patterns as being coordinated or not. This differentiation enables cognitive scientists and behavioral researchers to discuss research phenomena in a more specific way. Moreover, the distinction between a behavioral and a cognitive dimension builds a fruitful framework for accurate operationalizations in empirical studies and for defining exact phenomena in order to implement them in robots. While cognitive psychological accounts as well as philosophy of mind accounts are more interested in the cognitive dimension, e.g., there are two paradigms of research in computer science: The first one aims at simulating human-like cognitive processes in robots interacting with humans. This again helps further understanding of the functioning of the human mind. Computer scientists in this field are more interested in the cognitive dimension of cooperation. The second paradigm of research in computer science tries to model robots that are perceived as acting human-like. This method intends to increase the human's willingness to cooperate with the robot in the same way as with a human cooperation partner. Computer scientists working in this field are more interested in the behavioral dimension of cooperation.

The analysis has shown that robots are capable of being engaged in human-robot cooperation on either dimension. However, the implications of a shared intention with respect to joint commitments being involved are only partly implemented in the robotic systems that we have discussed above. Recall that when one cooperator does not fulfill his or her part in the planned action sequence, the other cooperator is in the position to rebuke (Gilbert 2009, see Sect. 2.3). This has been successfully implemented in the robot cMax by Pfeiffer-Lessmann (2011). Recall also that joint commitments involve a commitment to mutual support (Bratman 1993). That is, if one cooperator is not capable of performing his or her action part, the other is obliged to help out. Dominey and Warneken (2011) have accounted for this case. They have modeled a robotic system that helps the human partner by performing his or her part in a situation in which he or she does not do anything when it is his or her turn and denies the question 'Is it OK?.' Note, however, that the robot is taking over the whole action of the partner. That is, Dominey and Warneken (2011) do not account for the case that part of the action has already been performed by the experimenter and the robot has to *supplement* the action to its end. Accounting for such a case is a challenge for future research. Moreover, future research may account for the robot's sensitivity to whether the human partner is not able or not willed to perform his or her action part in situations where this is not explicitly stated. In addition, future research may address the question whether and to what extent knowing that one is engaged in cooperation with a robot rather than with another human has an impact on the experienced feeling or awareness of being a fellow of this group of agents (so-called 'fellow-feeling') and how this, in turn, may influence other factors such as e.g., standing to the joint commitment or willingness to help. By combining a robot's ability to cooperate with an artificial emotion system, Pfeiffer-Lessmann's work on the robot cMax may be a starting point for future research on the implementation of a fellow-feeling in robots.

Acknowledgements We would like to thank Olle Blomberg, Cédric Paternotte, and Nadine Pfeiffer-Lessmann for helpful comments on a previous version of this treatise. Moreover, we thank Stephanie Nicole Schwenke for proofreading this piece in grammar and spelling. This research has been supported by the VW-project "The Social Mind".

References

Allen, James F., Donna K. Byron, Myroslava Dzikovska, George Ferguson, Lucian Galescu, and Amanda Stent. 2001. Towards conversational human-computer interaction. *AI Magazine* 22(4): 27–37.

Becker, Christian, Stefan Kopp, and Ipke Wachsmuth. 2004. Simulating the emotion dynamics of a multimodal conversational agent. In *Proceedings of affective dialogue systems: Tutorial and research workshop (ads 2004)*, 154–165. Berlin/Heidelberg: Springer.

Böckler, Anne, Günther Knoblich, and Natalie Sebanz. 2011. Giving a helping hand: Effects of joint attention on mental rotation of body parts. *Experimental Brain Research* 211: 531–545.

Bratman, Michael E. 1992. Shared cooperative activity. *The Philosophical Review* 101(2): 327–341.

Bratman, Michael E. 1993. Shared intention. *Ethics* 104(1): 97–113.

Bratman, Michael E. 2009. Shared agency. In *Philosophy of the social sciences: Philosophical theory and scientific practice*, ed. C. Mantzavinos, 41–59. Cambridge: Cambridge University Press.

Carpenter, Malinda. 2009. Just how joint is joint action in infancy? *Topics in Cognitive Science* 1: 380–392.

Chaminade, Thierry, Jennifer L. Marchant, James Kilner, and Christopher D. Frith. 2012. An fMRI study of joint action – Varying levels of cooperation correlates with activity in control networks. *Frontiers in Human Neuroscience* 6: 1–11 (Art. 179).

Decety, Jean, Philip L. Jackson, Jessica A. Sommerville, Thierry Chaminade, and Andrew N. Meltzoff. 2004. The neural bases of cooperation and competition: An fMRI investigation. *NeuroImage* 23: 744–751.

Dominey, Peter, and Felix Warneken. 2011. The basis of shared intentions in human and robot cognition. *New Ideas in Psychology* 29: 260–274.

Fiebich, Anika, and Shaun Gallagher. 2013. Joint attention in joint action. *Philosophical Psychology* 26(4): 571–587.

Fong, Terrence, Illah Nourbakhsh, and Kerstin Dautenhahn. 2003. A survey of socially interactive robots. *Robotics and Autonomous Systems* 42(3–4): 143–166.

Foster, Mary E., and Colin Matheson. 2008. Following assembly plans in cooperative, task-based human-robot dialogue. In *Proceedings of the 12th workshop on the semantics and pragmatics of dialogue (Londial 2008)*. London.

Gilbert, Margaret. 2009. Shared intention and personal intentions. *Philosophical Studies* 144: 167–187.

Goodfellow, Ian J., Dumitru Erhan, Pierre L. Carrier, Aaron Courville, Mehdi Mirza, Ben Hamner, and Yoshua Bengio. 2013. Challenges in representation learning: A report on three machine learning contests. In *Neural information processing*, ed. M. Lee, A. Hirose, Z.-G. Hou, and R.M. Kil, 117–124. Berlin/Heidelberg: Springer.

Goodrich, Michael A., and Alan C. Schultz. 2007. Human-robot interaction: A survey. *Foundations and Trends in Human-Computer Interaction* 1(3): 203–275.

Gray, Jesse, Cynthia Breazeal, Matt Berlin, Andrew Brooks, and Jeff Lieberman. 2005. Action parsing and goal inference using self as simulator. *Robot and human interactive communication, ROMAN 2005*, 202–209.

Grigore, Elena C., Kerstin Eder, Anthony G. Pipe, Chris Melhuish, and Ute Leonards. 2012. Joint action understanding improves robot-to-human object handover. *IROS 2013*, 4622–4629.

Hoffman, Guy, and Cynthia Breazeal. 2004. Collaboration in human-robot teams. In *Proceedings of AIAA first intelligent systems technical conference*, Chicago (AIAA), 1–18.
Isenhower, Robert W., Michael J. Richardson, Claudia Carello, Reuben M. Baron, and Kerry L. Marsh. 2010. Affording cooperation: Embodied constraints, dynamics, and action-scaled invariance in joint lifting. *Psychonomic Bulletin & Review* 17(3): 342–347.
Lenz, Claus, Suraj Nair, Markus Rickert, Alois Knoll, Wolfgang Rösel, Jürgen Gast, Alexander Bannat, and Frank Wallhoff. 2008. Joint action for humans and industrial robots for assembly tasks. In *Proceedings of the IEEE international symposium on robot and human interactive communication*, 130–135. Munich.
McCabe, Kevin, Daniel Houser, Lee Ryan, Vernon Smith, and Theodore Trouard. 2001. A functional imaging study of cooperation in two-person reciprocal exchange. *PNAS* 98(20): 11832–11835.
Moll, Henrike, and Michael Tomasello. 2007. Cooperation and human cognition: The Vygotskian intelligence hypothesis. *Philosophical Transactions of the Royal Society B* 362(1480): 1–10.
Newman-Norlund, Roger D., Jurjen Bosga, Ruud G.J. Meulenbroek, and Harold Beckering. 2008. Anatomical substrates of cooperative joint-action in a continuous motor task: Virtual lifting and balancing. *NeuroImage* 41: 169–177.
Pfeiffer-Lessmann, Nadine. 2011. *Kognitive Modellierung von Kooperationsfähigkeiten für einen künstlichen Agenten*. Bielefeld: Universitätsbibliothek.
Pfeiffer-Lessmann, Nadine, and Ipke Wachsmuth. 2009. Formalizing joint attention in cooperative interaction with a virtual human. In *KI 2009: Advances in artificial intelligence*, ed. B. Mertsching, M. Hund, and Z. Aziz, 540–547. Berlin: Springer.
Reed, Kyle, Michael Peshkin, Mitra J. Hartmann, Marcia Grabowecky, James Patton, and Peter M. Vishton. 2006a. Haptically linked dyads. Are two motor-control systems better than one? *Psychological Science* 17(5): 365–366.
Reed, Kyle B., Michael Peshkin, Mitra J. Hartmann, James Patton, Peter M. Vishton, and Marcia Grabowecky. 2006b. Haptic cooperation between people, and between people and machines. In *Proceedings of the 2006 IEEE/RSJ (international conference on intelligent robots and systems)*, Oct 2006, Beijing, 2109–2114.
Searle, John. 1983. *Intentionality. An essay in the philosophy of mind*. Cambridge: Cambridge University Press.
Searle, John. 1990. Collective intentions and actions. In *Intentions in communication*, ed. P. Cohen, J. Morgan, and M. Pollack, 401–415. Cambridge, MA: MIT Press.
Sebanz, Natalie, Harold Bekkering, and Günther Knoblich. 2006. Joint action: Bodies and minds moving together. *Trends in Cognitive Sciences* 10(2): 70–76.
Shockley, Kevin, Marie Santana, and Carol A. Fowler. 2003. Mutual interpersonal postural constraints are involved in cooperative conversation. *Journal of Experimental Psychology. Human Perception and Performance* 29(2): 326–332.
Sofge, Donald, Magdalena D. Bugajska, J. Gregory Trafton, Dennis Perzanowski, Scott Thomas, Marjorie Skubic, Samuel Blisard, Nicholas L. Cassimatis, Derek P. Brock, William Adams, and Alan C. Schultz. 2005. Collaborating with humanoid robots in space. *International Journal of Humanoid Robotics 01/2005* 2: 181–201.
Tollefsen, Deborah. 2005. Let's pretend! Joint action and young children. *Philosophy of the Social Sciences* 35: 75–97.
Tomasello, Michael, Malinda Carpenter, Josep Call, Tanja Behne, and Henrike Moll. 2005. Understanding and sharing intentions: The origins of cultural cognition. *Behavioral and Brain Sciences* 28: 675–735.
Tuomela, Raimo. 2010. *The philosophy of sociality*. Oxford: Oxford University Press.
Vonk, Roos. 1998. Effects of cooperative and competitive outcome dependency on attention and impression preferences. *Journal of Experimental Social Psychology* 34: 265–288.
Wilby, Michael. 2010. The simplicity of mutual knowledge. *Philosophical Explorations* 13(2): 83–100.
Wooldridge, Michael, and Nicholas R. Jennings. 1995. Intelligent agents: Theory and practice. *Knowledge Engineering Review* 10(2): 115–152.

Chapter 3
The Participatory Turn: A Multidimensional Gradual Agency Concept for Human and Non-human Actors

Sabine Thürmel

3.1 Introduction to the Participatory Turn in Socio-technical Systems

New varieties of interplay between humans, robots and software agents are on the rise: virtual companions, self-driving cars and the collaboration between humans and virtual agents in emergency response systems exemplify this development. Humans have evolved from "naturally born cyborgs" (Clark 2003) to adaptive, co-dependent, socio-technical agents. Computer-based artefacts are no longer mere tools but may be capable of individual and joint action, too. Turkle characterises this development as follows: "Computational objects do not simply do things *for* us, but they do things *to* us as people, to our ways of seeing ourselves and others. Increasingly, technology puts itself into a position to do things *with* us" (Turkle 2006, 1). This insight was gained when Turkle studied the nascent robotics culture. It is equally valid for software agents. The starting point of the evolution of software agents is constituted by interface agents providing assistance for the user or acting on his or her behalf. As envisioned by (Laurel 1991) and (Maes 1994), they have evolved into increasingly autonomous agents in virtual environments. Moreover software agents may be found in cyber-physical systems (Mainzer 2010, 181). While classical computer systems separate physical and virtual worlds, cyber-physical systems (CPS) observe their physical environment by sensors, process the information, and influence their environment with so-called actuators while being connected by a communication layer. Collaborative software agents can be embedded in cyber-physical systems if the different nods in the cyber-physical system need to coordinate. Examples include distributed rescue systems (Jennings 2010), smart

---

S. Thürmel (✉)
Munich Center for Technology in Society, Technische Universität München, Munich, Germany
e-mail: sabine@thuermel.de

© Springer International Publishing Switzerland 2015
C. Misselhorn (ed.), *Collective Agency and Cooperation in Natural and Artificial Systems*, Philosophical Studies Series 122,
DOI 10.1007/978-3-319-15515-9_3

energy grids (Wedde et al. 2008), and distributed health monitoring systems (Nealon and Moreno 2003). These systems are first simulated and then deployed to control processes in the material world. Humans may be integrated for clarifying and/or deciding non-formalized conflicts in an ad hoc manner. Most agent-based cyber-physical systems aim at enhancing process automation. However, there exist also systems focusing on optimizing the collaboration between humans, robots and software agents. In such environments each participant plays a specific role and contributes in a specific way to the overall problem solution. Examples include diverse areas as self-organizing production systems offering customized products in highly flexible manufacturing environments or managed health care systems, where humans and virtual carers collaborate (Hossain and Ahmed 2012). Thus software agents have been promoted from assistants to interaction partners. The socio-technical fabric of our world has been augmented by these collaborative systems.

Current collaborative constellations between humans and technical agents are asymmetric: their acts are based on different cognitive systems, different degrees of freedom and only partially overlapping spheres of experience. However, new capabilities may emerge over time on the techno level. Self-organization and coalition forming on the group level can occur. New cultural practices come into being. The enactment of joint agency in these heterogeneous constellations is well past its embryonic stage. Therefore it becomes vital to understand agency and inter-agent coordination in purely virtual and cyber-physical systems.

The potential of agent-based virtual or cyber-physical systems becomes actual in testbed environments and real-time deployments. This perspective is elaborated in Sect. 3.2 which is dedicated to the potentiality and actuality of social computing systems. This section is included because many sociologists focus solely on the actuality of socio-technical systems that is on "agency *in medias res*" neglecting the fact that the potential of technical systems is determined by their design. Technical agents are not black boxes just to be observed but may be analysed in detail by computer scientists and engineers.

Agency in socio-technical systems may be attributed in different ways: two of the most relevant ones for attributing agency to both humans and non-humans are presented in Sect. 3.3. These approaches take a technograph's approach aiming at describing agency as it unfolds. This paper does not intend to evaluate agency in socio-technical systems from an observer's standpoint. In Sect. 3.4 the agential perspective is characterized as a certain level of abstraction when analysing a system.

In Sect. 3.5 a novel approach to attribute agency in socio-technical systems is presented. A multidimensional gradual agency concept for human and non-human actors is introduced. In this framework individual and joint agency may be taken into view. The framework is applicable to constellations of distributed and collective agency. Scenarios where solely humans act can be compared to testbed simulations.

Finally, an outlook is given on how the framework presented here may support further both the software engineer and the philosopher when modelling and analysing role-based interaction in socio-technical systems.

3.2 Potentiality and Actuality of Computing Systems

Computer simulations let us explore the dynamic behaviour of complex systems. Today they are not only used in natural sciences and computational engineering but also in computational sociology. Social computing systems focus on the simulation of complex interactions and relationships between individual human and/or non-human agents. If the simulations are based on scientific abstractions of real-world problem spaces they enable us to gain new insights. For example "crowd simulation" systems are useful if evacuation plans have to be developed. Collaborative efforts may be simulated, too. A case in point is the coordination of emergency response services in a disaster management system based on so-called electronic market mechanisms (Jennings 2010). The humanities, social and political science, behavioral economics, and studies in law have discovered agent-based modeling (ABM) too. Academics have applied ABM to study the evolution of norms (Muldoon et al. 2014), and to explore the impact of different social organization models on settlement locations in ancient civilizations (Chliaoutakis and Chalkiadakis 2014). Since agent-based models may provide a better fit than conventional economic models to model the "herding" among investors, early-warning systems for the next financial crisis could be built based on ABM (Economist 2010). Even criminal behavior, deliberate misinterpretations of norms or negligence can be studied. Therefore it is hardly surprising that the Leibniz Center for Law at the University of Amsterdam had been looking – although in vain – for a specific Ph.D. candidate in legal engineering: He or she should be capable of developing new policies in tax evasion scenarios. These scenarios were planned to be based on ABM (Leibnizcenter for Law 2011). The novel technical options of "social computing" not only offer to explain social behaviour but they may also suggest ways of changing it.

Social simulation systems are similar to numerical simulations but use different conceptual and software models. Both approaches may complement each other. Numerical methods based on non-linear equation systems support the simulation of quantitative aspects of complex discrete systems (Mainzer 2007). In contrast, multi-agent systems (MAS) enable collective behaviour to be modelled based on the local perspectives of individuals, their high-level cognitive processes and their interaction with the environment (Woolridge 2009). Current agent-based software systems range from swarm intelligence systems, based on a bionic metaphor for distributed problem solving, to sophisticated e-negotiation systems (Woolridge 2009).

Simulations owe their attractiveness to the elaborate rhetoric of the virtual (Berthier 2004): "It is a question of representing a future and hypothetical situation as if it were given, neglecting the temporal and factual dimensions separating us from it – i.e. to represent it as actual" (Berthier 2007, 4). Social computing systems are virtual systems modelled, e.g. by MAS, and realized by the corresponding dynamic computer-mediated environments. Computational science and engineering as well as computational sociology systems benefit from these computer-based interaction spaces. The computer is used both as a multipurpose machine and a unique tool used to store and deliver information.

Virtuality in technologically induced contexts is even better explained if Hubig's two-tiered presentation of technology in general as a medium (Hubig 2006) is adopted. He distinguishes between the "potential sphere of the realization of potential ends" and the "actual sphere of realizing possible ends" (Hubig 2010, 4). Applied to social computing systems – or IT systems in general – it can be stated that their specification corresponds to the "potential sphere of the realization of potential ends" and any run-time instantiation to a corresponding actual sphere. In other words: due to their nature as computational artefacts, the potential of social computing systems becomes actual in a concrete instantiation. Their inherent potentiality is actualized during run-time. "A technical system constitutes a potentiality that only becomes a reality if and when the system is identified as relevant for agency and is embedded into concrete contexts of action" (Hubig 2010, 3).

Since purely computational artefacts are intangible, i.e. existing in time but not in space, the situation becomes even more challenging: one and the same social computing program can be executed in experimental environments and in real-world interaction spaces. The demonstrator for the coordination of emergency response services may go live and coordinate human and non-human actors in genuine disaster recovery scenarios. With regard to its impact on the physical environment, it possesses a virtual actuality in the testbed environment and a real actuality when it is employed in real time in order to control processes in the natural world.

In the case of social computing systems, the "actual sphere of realizing possible ends" can either be an experimental environment composed exclusively of software agents or a system deployed to control processes in the material world. Humans may be integrated for clarifying and/or deciding non-formalized conflicts in an ad hoc manner. Automatic collaborative routines or new practices for ad hoc collaboration are established. Novel, purely virtual or hybrid contexts realizing collective and distributed agency materialize.

3.3 Attributing Agency in Socio-technical Systems

In order to exemplify the state of the art in attributing collective and distributed agency in socio-technical systems, two schools are briefly summarized: the Actor Network Theory (ANT) and the socio-technical approach of attributing distributed agency of Rammert and colleagues. Both are aimed at analysing constellations of collective inter-agency by attributing agency both to human and non-human actors but they differ in essential aspects.

The ANT approach introduces a flat concept of agency and a symmetrical ontology applicable both to human and non-human actors (e.g. (Latour 2005)) whereas the distributed agency approach of Rammert et al. promotes a levelled and gradual concept of agency based on the "practical fiction of technologies in action" (Rammert and Schulz-Schaeffer 2002; Rammert 2011). Latour focuses on "interobjectivity" (Latour 1996) that is links, alliances, and annexes between all kinds of objects whereas Rammert takes a more nuanced view on inter-agency based on Anthony Giddens' stratification model of action (1984).

3.3.1 The Actor Network Theory (ANT)

As a practitioner of science and technology studies and a true technograph, Bruno Latour was the first to attribute agency and action both to humans and non-humans (Latour 1988). Together with colleagues such as Michel Callon, a symmetric vocabulary was developed that they deemed applicable both to humans and non-humans (Callon and Latour 1992, 353). This ontological symmetry led to a flat concept of agency where humans and non-human entities were declared equal. Observations gained in laboratories and field tests were described as so-called actor networks, heterogeneous collectives of human and non-human entities, mediators and intermediaries. The Actor Network Theory regards innovation in technology and sciences as largely depending on whether the involved entities – whether they be material or semiotic – succeed in forming (stable) associations. Such stabilizations can be inscribed in certain devices and thus demonstrate their power to influence further scientific evolution (Latour 1990). All activity emanates from so-called actants (Latour 2005, 54). The activity of forming networks is called "translation" (Latour 2005, 108). Statements made about actants as agents of translation are snapshots in the process of realizing networks (Schulz-Schaeffer 2000, 199). The central empirical goal of the actor network theory consists in reconstructively opening up convergent and (temporarily) irreversible networks (Schulz-Schaeffer 2000, 205). Thus the ANT approach could more aptly be called a "sociology of translation", an "actant-rhyzome ontology" or a "sociology of innovation" (Latour 2005, 9).

It should be noted that Latour has quite a conventional, tool-oriented notion of technology. This may be due to the fact that smart technology and agent systems are nowhere to be found in his studies.

Latour only focuses on actual systems and their modes of existence. However, one may (and should) clearly distinguish between agency (potentiality) and action (actuality) – especially if the investigations are led by techno-ethicists. Moreover, virtual actuality does not equate with real actuality in most circumstances. A plane crash in reality is very different from one in a simulator.

3.3.2 Distributed Agency and Technology in Action

The conditions under which we can attribute agency and inter-agency to material entities and how to identify such entities as potential agents are important to Werner Rammert and Ingo Schulz-Schaeffer (Rammert and Schulz-Schaeffer 2002, 11). Therefore they developed a gradual concept of agency in order to categorize potential agents regardless of their ontological status as machines, animals or human beings. Rammert is convinced that "it is not sufficient to only open up the black box of technology; it is also necessary and more informative to observe the different dimensions and levels of its performance" (Rammert 2011, 11). The model is inspired by Anthony Giddens' stratification model of action (1984). The approach distinguishes between three levels of agency:

- causality ranging from short-time irritation to permanent restructuring,
- contingency, i.e. the additional ability "to do otherwise", ranging from choosing pre-selected options to self-generated actions, and, in addition, on the highest level
- intentionality as a basis for rational and self-reflective behaviour (Rammert and Schulz-Schaeffer 2002, 49; Rammert 2011, 9).

The "reality of distributed and mediated agency" is demonstrated, e.g. based on an intelligent air traffic system (Rammert 2011, 15). Hybrid constellations of interacting humans, machines and programs are identified.

Moreover, a pragmatic classification scheme of technical objects depending on their activity levels is developed. This enables classification of the different levels of "technology in action". It starts with passive artefacts, and continues with reactive ones, i.e. systems with feedback loops. Next come active ones, then proactive ones, i.e. systems with self-activating programs. It ranges further up to co-operative systems, i.e. distributed and self-coordinating systems (Rammert 2008, 6). The degrees of freedom in modern technologies are constantly increasing. Therefore the relationship between humans and technical artefacts evolves "from a fixed instrumental relation to a flexible partnership" (Rammert 2011, 13). Rammert identifies three types of inter-agency: "interaction between human actors, intra-activity between technical agents and interactivity between people and objects" (Rammert 2008, 7). These capabilities do not unfold "*ex nihilo*" but "*in medias res*". "According to [this] concept of mediated and situated agency, agency arises in the context of interaction and can only be observed under conditions of interdependency" (Rammert 2011, 5).

These reflections show how "technology in action" may be classified and how constellations of collective inter-agency can be evaluated using a gradual and multilevel approach. Similar to Latour, these authors are convinced that artefacts are not just effective means but must be constantly activated via practice (enactment) (Rammert 2007, 15).

Since this approach focuses exclusively on "agency *in medias res*", i.e. on snapshots of distributed agency and action, the evolution of any individual capabilities, be they human or non-human, are not accounted for. Even relatively primitive cognitive activities such as learning via trial and error, which many machines, animals and all humans are capable of, are not taken into account by Rammert's perspective on agency. A clear distinction between human agency, i.e. intentional agents, and technical agency, a mere pragmatic fiction, remains. In Rammert's view, technical agency "emerges in real situations and not in written sentences. It is a practical fiction that has real consequences, not only theoretical ones" (Rammert 2011, 6). In his somewhat vague view, the agency of objects built by engineers "is a practical fiction that allows building, describing and understanding them adequately. It is not just an illusion, a metaphorical talk or a semiotic trick" (Rammert 2011, 8).

3.4 Levels of Abstraction

This paper does not intend to analyse agency in socio-technical systems from an observer's standpoint. The agency of technology is not considered a "pragmatic fiction" as Rammert did (2011). In my view the agential perspective on technology should be characterized as a certain level of abstraction when analysing a socio-technical system.

In the following, the agency of technology is perceived as a (functional) abstraction corresponding to a level of abstraction (LoA) as defined by Floridi. An LoA "is a specific set of typed variables, intuitively representable as an interface, which establishes the scope and type of data that will be available as a resource for the generation of information" (Floridi 2008, 320). For a detailed definition see (Floridi 2011, 46).

An LoA presents an interface where the observed behaviour – either in virtual actuality or real actuality – may be interpreted. Under an LoA, different observations may result due to the fact that social computing software can be executed in different run-time environments, e.g. in a testbed in contrast to a real-time environment. Different LoAs correspond to different abstractions of one and the same behaviour of computing systems in a certain run-time environment. Different observations under one and the same LoA are possible if different versions of a program are run. Such differences may result when software agents are replaced by humans.

Conceptual entities may also be interpreted at a chosen LoA. Note that different levels of abstraction may coexist. Since levels of abstractions correspond to different perspectives, the system designer's LoA may be different from the sociologist's LoA or the legal engineer's LoA of one and the same social computing system. These LoAs are related but not necessarily identical.

The basis of technology in action is not a pragmatic fiction of action but a conceptual model of the desired behaviour. From the designer's point of view, metaphors often serve as a starting point to develop, e.g. novel heuristics to solve NP-complete (optimization) problems, that is problems for which no fast solution is known. Such metaphors may be borrowed from biology, sociology or economics. Research areas such as neural nets, swarm intelligence approaches and electronic auction procedures are products of such approaches. In the design phase, ideas guiding the modelling phase are often quite vague at first. In due course, their concretization results in a conceptual model (Ruß et al. 2010, 107) which is then specified as a software system. From the user's or observer's point of view, during run-time the more that is known about the conceptual model, the better its potential for (distributed) agency can be predicted and the better the hybrid constellations of (collective) action, emerging at run-time, may be analysed. Thus the actuality of agential behaviour is complemented by a perspective on the system model determining the potential of technical agents. The philosophical value added by this approach not only lies in a reconstructive approach as intended by Latour and Rammert but also in the conceptual modelling and engineering of the activity space. Under an LoA for agency and action, activities may be observed as they unfold. Moreover, the system may be analysed and educated guesses about its future behaviour can be made. Both the specifics of distinct systems and their commonalities may be compiled.

3.5 Multidimensional Gradual Agency

3.5.1 Introduction

The following proposal for a conceptual framework for agency and action was first introduced in (Thürmel 2012) and expanded in (Thürmel 2013). It is intended to provide a multidimensional gradual classification scheme for the observation and interpretation of scenarios where humans and non-humans interact. It enables appropriate lenses to be defined, i.e. levels of abstraction, under which to observe, interpret, analyse and judge their activities. This does justice to Floridi's dictum that the task of the "philosophy of information" is "conceptual engineering, that is, the art of identifying conceptual problems and of designing, proposing, and evaluating explanatory solutions" (Floridi 2011, 11). In our case, the conceptual problem is how to characterize agency and interagency between humans, robots and software agents such that all current forms of interplay can be analysed. Moreover the framework should be so flexible to allow future technical developments to be included but to be not more complex than necessary. The proposed solution is a multidimensional gradual classification scheme which is presented in the following.

In contrast to observing "agency *in medias res*" (Rammert 2011, 15), the potential of smart, autonomous technology is the focus of this model. The engineering perspective makes it possible to design the potential and realize the actuality of computer-mediated artefacts. While technographs such as Latour strive to observe and analyse the interactions without prejudices by "opening up black boxes", this paper advocates making use of computational science and engineering know-how in order to enhance the understanding of socio-technical environments. Thus Latour's flat and symmetric concept of agency, which he applies to both humans and non-humans, is not used. Rammert's fiction of technical agency (Rammert 2011) is substituted by Floridi's "method of levels of abstraction" (Floridi 2008, 2011). Thus the underdetermined so-called "pragmatic fiction of technical agency" need not be contrasted with the "reality of distributed agency" (Rammert 2011) in socio-technical environments. Both the potential of individual and distributed agency and its actualization may be described by domain-specific levels of abstraction. A multidimensional perspective on the individual and joint capabilities of human and non-human actors replaces the one-dimensional layered model of Rammert and Schulz-Schaeffer (2002).

As Rammert states, "agency really is built into technology" but – as demonstrated above – not "as it is embodied in people" (Rammert 2011, 6) but by intelligent design performed by engineers and computer scientists. In order to demonstrate the potential for agency, not only the activity levels of any entities but also their potential for adaptivity, interaction, personification of others, individual action and joint action has to be taken into account.

Being at least (re)active is the minimal requirement for being an agent. Higher activity levels allow the environment to be influenced. Being able to adapt is a gradual faculty. It starts with primitive adaption to environment changes and ranges up to the adaption of long-term strategies and the corresponding goals based on past experiences and (self-reflective) reasoning of human beings. As shown below, acting may be discerned from just behaving based on activity levels and on being able to adapt in a "smart" way.

The potential for interaction is a precondition of any collaborative performance. The potential of the personification of others enables agents to integrate predicted effects of own and other actions. "Personification of non-humans is best understood as a strategy of dealing with the uncertainty about the identity of the other … Personifying other non-humans is a social reality today and a political necessity for the future" (Teubner 2006, 497). Personification is similar to Dennett's intentional stance (1995) since it is a pragmatic attribution. It starts with the attribution of simple dispositions up to perceiving the other as a human-like actor. This capability may affect any tactically or strategically motivated individual action. Moreover, it is a prerequisite of any form of defining joint goals and joint (intentional) commitment in any ensemble of agents. The capabilities for individual action and joint action may be defined based on activity levels, the potential for adaptivity, interaction and personification of others possessed by the involved actor(s).

Any type of an agential entity may be classified according to its characteristics in these dimensions. For any entity type the maximum potential (in these dimensions) is defined by a distinct value tuple. It may be depicted by a point in the multidimensional space spanned by the dimensions introduced above.

Any instantiation of an agent may be characterized by a distinct value tuple at a moment in time, i.e. by its actual time-stamped value. In agent-based systems, the changes over time correspond to changes of state of each agent.

Note that in the following the granularity on the different axes is only used as an example and can be adjusted according to the systems to be analysed and/or compared.

3.5.2 The Multidimensional Framework for Individual and Distributed Agency

The conceptual framework for agency and action offers a multidimensional gradual classification scheme for the observation and interpretation of scenarios where humans and non-humans interact. The "activity level" axis aims at the dimension "technology in action" (Rammert 2008) providing a scale for the grade of active behaviour a technical object may display. The degree of adaptivity describes the plasticity of the phenotype. Individual agency may be described based on the potential for activity and adaptivity. Interaction is needed for coordination and control via communication. The personification of others may serve as a basis for joint action. Joint agency may be defined based on these dimensions.

The activity level allows individual behaviour to be characterized depending on the degree of self-inducible activity potential. It starts with passive entities such as road bumpers, hammers, and nails. Entities that display a certain predefined behaviour once they are started may be called semi-active (Rammert 2011, 7) or active without alternatives. Examples include hydraulic pumps or software artefacts such as algorithms searching in batch mode, compilers or basic help assistants. Reactive objects demonstrate the next level. These technical elements display identical output to identical input, e.g. realized by simple feedback loops or other situated reactions such as heating systems, swarm intelligence systems or ant colony optimization algorithms. Active entities permit individual selection between alternatives resulting in changes in the behaviour. From an internal perspective, this corresponds to Rammert's level of contingency, ranging from choosing between preselected options to self-generated actions (Rammert 2011, 9). The minimal requirement for active entities is: perceive-plan-act. Examples are to be found in robotics, in sophisticated software agents like automatic bid agents in high-frequency trading systems, and in certain multi-agent systems realizing e-negotiation as well as in cyber-physical systems. Proactive entities try to anticipate the behaviour of other entities and act accordingly. The minimal requirement for their internal organization is: perceive-predict-(re)plan-act. Such technical modules are part of many cyber-physical systems where processes are controlled, e.g. in traffic control systems. Multi-agent systems in the above-mentioned emergency control systems may also display proactive behaviour in a dynamically changing environment. The next level corresponds to the ability to set one's own goals and pursue them. It requires self-regulation based on self-monitoring and self-control. Intrinsic motivation may support such a process management, at least in humans. For the foreseeable future, self-conscious intentional behaviour will be reserved for humans.

These capabilities depend on an entity-internal system for information processing linking input to output. In the case of humans, it equals a cognitive system connecting perception and action. For material artefacts or software agents, an artificial "cognitive" system couples (sensor) input with (actuator) output.

Based on such a system for (agent-internal) information processing, the level of adaptivity may be defined. It characterizes the plasticity of the phenotype, i.e. the ability to change one's observable characteristics including any traits that may be made visible by a technical procedure, in correspondence to changes in the environment. Models of adaptivity and their corresponding realizations range from totally rigid to simple conditioning up to impressive cognitive agency, i.e. the ability to learn from past experiences and to plan and act accordingly. A wide range of models coexist enabling study and experimentation with artificial "cognition in action". This dimension is important to all who define agency as situation-appropriate behaviour and who deem the plasticity of the phenotype as an essential assumption of the conception of man.

The potential for interaction, i.e. coordination by means of communications, is the basis of most if not all social computing systems and approaches to distributed problem solving. It may range from non-existent to informing others via bulletin boards such as those used in social networks and other forms of asynchronous communication to hard-wired cooperation mechanisms. Structured communication based on predefined scripts constitutes the next level. Examples are found in electronic auctioning. Unstructured, ad hoc communication of arbitrary elements may be found in demand-oriented coordination.

Inter-agency between technical agents and humans ranges from fixed instrumental relations, as found in cyborgs, to temporary instrumental relations like those between a human and an exoskeleton, a virtual servant or a robopet. Principal-agent relations are realized when tasks are delegated to others. Simple duties may be delegated to primitive software assistants such as mailbots, sellbots or newsbots. More complex ones may be performed by robots or software agents that are bound by directives. Flexible partnerships may be seen in hybrid multi-agent systems, where humans interact with technical agents. Technical agents are currently not stakeholders to whom legal or moral responsibility is delegated. Nevertheless, these options and their implications for our legal systems are already discussed in the literature (Pagallo 2013; Chopra and White 2011).

Primitive mechanisms for coordination as in swarm intelligence systems do not need the personification of interaction partners. Ad hoc cooperation is a different case. The personification of others lies in the foundation for interactive planning, sharing strategies and adapting actions. This capability is non-existent in most material and software agents.

Assuming that another agent possesses a certain disposition to behave or act may be considered as the most fundamental level of personification. It may be found in theoretic game approaches or in so-called minimal models of the mind (Butterfill and Apperly 2013). Such models are used in robotics, e.g. in Hiatt et al. (2011). Many technical tools display purely passive or only reactive behaviour without the ability to adapt to changes in the environment. However, many technical agents such as automatic bid agents of electronic auctioning systems or cars on autopilot are able to learn from past experiences. The dynamics of social interaction and action-based learning and concept forming may lie at the foundation for "bootstrapping the cognitive system" of robots. Cangelosi and colleagues present a "roadmap for developmental robotics" based on such an approach (Cangelosi et al. 2010). Dominey and Warneken aim to explore "the basis of shared intentions in human and robot cognition". They demonstrate how "computational neuroscience", robotics and developmental psychology may stimulate each other (2011). These research projects may serve as circumstantial evidence for an evolutionary path in robotics.

The work done by Tomasello and colleagues motivates engineers to strive towards lessening the gap between the cognitive and agential capabilities of current technical agents and humans. According to the latest scientific findings, "chimpanzees understand others in terms of a perception-goal psychology, as opposed to a full-fledged, human-like belief-desire psychology" (Call and Tomasello 2008, 187). This provides the basis for topic-focused group decision making based on egoistical behaviour: "they [the chimpanzees] help others instrumentally, but they are not so inclined to share resources altruistically and they do not inform others of things helpfully" (Warneken and Tomasello 2009, 397). Thus great apes may display so-called joint intentionality (Call and Tomasello 2008). In contrast, young children seem to have a "biological disposition" for helping and sharing. It may even be shown that "collaboration encourages equal sharing in children but not in chimpanzees" (Hamann et al. 2011). Understanding the other as an intentional agent allows even infants to participate in so-called shared actions (Tomasello 2008). Understanding others as mental actors lays the basis for interacting intentionally and acting collectively (Tomasello 2008). Engineers do not expect technical agents to evolve as humans did but they may profit from the insights gained in evolutionary anthropology.

Currently there is quite a gap between non-human actors and human ones in terms of their ability to take the "shared point of view" (Tuomela 2007) and to interact intentionally. This strongly limits the scope of social computing systems when they are used to predict human behaviour or if they are aimed at engineering and simulating future environments.

3.5.3 Individual Agency and Inter-agency

The potential for both individual action and for joint action may be defined based on the above-mentioned capabilities for activity, adaptivity, interaction and personification of others. The individual agency ranges from the individual potential for behaving to the individual potential for acting: disjunct levels may be defined based on the dimensions "activity level" and ability to adapt. The activity level of technical artefacts defines whether an artefact may only be used as a tool or may actively interact with its environment. Adaptivity is crucial for the individual regulation of the behaviour and subtle execution control.

In order to stress the communalities between human and non-human agents, an agent is counted as being capable of acting (instead of just behaving) if the following conditions concerning its ontogenesis hold: "the individual actor [evolves] as a complex, adaptive system (CAS), which is capable of rule-based information processing and based on that able to solve problems by way of adaptive behaviour in a dynamic process of constitution and emergence" (Kappelhoff 2011, 320). Thus the ability to learn and adapt is deemed crucial for acting.

Disjunct levels of inter-agency and distributed agency may be characterized based on activity levels and the corresponding range of the capability for interaction. Current technical agents mostly interact based on predefined scripts. Ad hoc communications often remain a technical challenge. However, the robotics learns from evolutionary anthropologists so that service robots can be taught to move in households and participate in basic cooperative actions (CoTeSys 2014) or be taught verbal interaction based on exemplary communication events (Fischer et al. 2011). These projects may provide the first tentative examples of "shared cooperative activity" (Bratman 1992), a kind of shared activity which is based on mutual responsiveness, commitment to joint activity and commitment to mutual support. Rational, self-organising teams consisting of collaborating humans, robots and software agents can be found in current research projects as self-organizing production systems or managed health care systems. Such teams display a "modest sociality" (Bratman 2014) emerging from structures of interconnected planning agency.

Constellations of inter-agency and distributed agency in social computing systems or hybrid constellations, where humans, machines and programs interact, may be described, examined and analysed using the above-introduced classification scheme for agency and action. These constellations start with purely virtual systems like swarm intelligence systems and fixed instrumental relationships between humans and assistive software agents where certain tasks are delegated to artificial agents. They continue with flexible partnerships between humans and software agents. They range up to loosely coupled complex adaptive systems. The latter may model such diverse problem spaces as predator–prey relationships of natural ecologies, legal engineering scenarios or disaster recovery systems. Their common ground and their differences may be discovered when the above-outlined multidimensional, gradual conceptual framework for agency and action is applied.

A subset of these social computing systems, namely those that may form part of the infrastructure of our world, provides a new form of "embedded governance". Their potential and limits may also be analysed using the multidimensional agency concept.

3.6 Conclusions and Future Work

The proposed conceptual framework for agency and action offers a multidimensional gradual classification scheme for the observation and interpretation of scenarios where humans and non-humans interact. It may be applied to the analysis of the potential of social computing systems and their virtual and real actualizations. The above-introduced approach may also be employed to describe situations where decisions to act are delegated to technical agents. It can be used both by the software engineer and the philosopher when role-based interaction in socio-technical systems is to be defined and analysed during execution.

Proto-ethical agency in social computing systems may be explored by adapting (Moor 2006) to the framework. Profiting from work done by Darwall (2006), the framework could be expanded in order to potentially attribute commitments to diverse socio-technical actors. Shared agency, a "planning theory of acting together" as defined by Bratman (2014), could be investigated in socio-technical contexts where technical elements are not mere tools but interaction partners. Last but not least, social relations to technical agents could be evaluated similarly to (Misselhorn et al. 2013) by making use of the framework when characterizing potentiality and actuality of the technical agents.

References

Berthier, Denis. 2004. *Médiations sur le réel et le virtuel*. Paris: Editions L'Harmattan.
Berthier, Denis. 2007. Qu'est-ce que le virtuel. *La Jaune et la Rouge*, June 2007, http://denis.berthier.pagesperso-orange.fr/Articles/JR-Qu-est-ce-que-le-virtuel.pdf. Accessed 27 Apr 2014.
Bratman, Michael. 1992. Shared cooperative activity. *The Philosophical Review* 101(2): 327–341.
Bratman, Michael. 2014. *Shared agency: A planning theory of acting together*. Cambridge, MA: Harvard University Press.
Butterfill, Stephen, and Ian Apperly. 2013. How to construct a minimal theory of mind. *Mind & Language* 28(5): 606–637.
Call, Joseph, and Michael Tomasello. 2008. Does the chimpanzee have a theory of mind? 30 years later. *Trends in Cognitive Sciences* 12(5): 187–192.
Callon, Michel, and Bruno Latour. 1992. Don't throw the baby out with the bath school! A reply to Collins and Yearley. In *Science as practice and culture*, ed. Andrew Pickering, 343–368. Chicago: Chicago University Press.
Cangelosi, Angelo, Giorgio Metta, Gerhard Sagerer, Gerhard Nolfi, Chrystopher Nehaniv, Kerstin Fischer, Jun Tani, Tony Belpaeme, Giulio Sandini, Francesco Nori, Luciano Fadiga, Britta Wrede, Katharina Rohlfing, Elio Tuci, Kerstin Dautenhahn, Joe Saunders, and Arne Zesche. 2010. Integration of action and language knowledge: A roadmap for developmental robotics. *IEEE Transactions on Autonomous Mental Development* 2(3): 167–195.
Chliaoutakis, Angelos, and Georgios Chalkiadakis. 2014. Utilizing agent-based modeling to gain new insights into the ancient Minoan civilization. In *Proceedings of the 2014 international conference on autonomous agents and multi-agent systems,* 1371–1372. Richland: SC.
Chopra, Sami, and Laurence White. 2011. *A legal theory for autonomous artificial agents*. Ann Arbor: The University of Michigan Press.
Clark, Andy. 2003. *Naturally born cyborgs*. Oxford: Oxford University Press.
CoTeSys. 2014. Cluster of excellence cognition in technical systems. http://www.cotesys.de/. Accessed 28 Apr 2014.
Darwall, Stephen. 2006. *The second person standpoint: Morality, respect and accountability*. Cambridge, MA: Harvard University Press.
Dennett, Daniel. 1995. *The intentional stance*. Cambridge, MA: MIT Press.
Dominey, Peter, and Felix Warneken. 2011. The basis of shared intentions in human and robot cognition. *New Ideas in Psychology* 29: 260–274.
Economist. 2010. Agents of change, 22 July 2010. http://www.economist.com/node/16636121/print. Accessed 27 Apr 2014.
Fischer, Silke, Denis Schulze, Pia Borggrebe, Martina Piefke, Sven Wachsmuth, und Katharina Rohlfing. 2011. Multi-modal anchoring in infants and artificial systems. Poster presented at IEEE-ICDL-EPIROB conference 2011, Frankfurt, 24–27 Aug 2011.
Floridi, Luciano. 2008. The method of levels of abstraction. *Minds and Machines* 18(3): 303–329.
Floridi, Luciano. 2011. *The philosophy of information*. Oxford: Oxford University Press.
Giddens, Anthony. 1984. *The constitution of society, outline of the theory of structuration*. Cambridge, UK: Polity Press.
Hamann, Katharina, Felix Warneken, Julia Greenberg, and Michael Tomasello. 2011. Collaboration encourages equal sharing in children but not in chimpanzees. *Nature* 476: 328–331. doi:10.1038/nature10278.
Hiatt, Laura, Anthony Harrison, and Gregory Trafton. 2011. Accommodating human variability in human–robot teams through theory of mind. *Proceedings of the International Joint Conference on Artificial Intelligence 2011* 3: 2066–2071.
Hossain, M. Anwad, and Dewan Ahmed. 2012. Virtual caregiver: An ambient-aware elderly monitoring system. *IEEE Transactions on Information Technology in Biomedicine: A publication of*

the IEEE Engineering in Medicine and Biology Society 16(6): 1024–1031. doi:10.1109/TITB.2012.2203313.
Hubig, Christoph. 2006. *Die Kunst des Möglichen I – Technikphilosophie als Reflexion der Medialität*. Bielefeld: Transcript Verlag.
Hubig, Christoph. 2010. Technik als Medium und "Technik" als Reflexionsbegriff. Manuscript. http://www.philosophie.tu-darmstadt.de/institut/mitarbeiterinnen_1/professoren/a_hubig/downloadbereich/downloadsprofhubig.de.jsp. Accessed 30 Apr 2014.
Jennings, Nicolas. 2010. *ALADDIN end of project report*. http://www.aladdinproject.org/wp-content/uploads/2011/02/finalreport.pdf. Accessed 27 Apr 2014.
Kappelhoff, Peter. 2011. Emergenz und Konstitution in Mehrebenenselektionsmodellen. In *Emergenz – Zur Analyse und Erklärung komplexer Strukturen*, ed. Jens Greve and Annette Schnabel, 319–345. Berlin: Suhrkamp Taschenbuch Wissenschaft.
Latour, Bruno. 1988. Mixing humans and nonhumans together. The sociology of a door-closer. *Social Problems* 35(4): 298–310.
Latour, Bruno. 1990. Drawing things together. In *Representation in scientific practice*, ed. Michael Lynch and Steve Woolgar, 19–68. Cambridge, MA: MIT Press.
Latour, Bruno. 1996. On interobjectivity. *Mind, Culture, and Activity* 3: 228–245.
Latour, Bruno. 2005. *Reassembling the social – An introduction to actor-network-theory*. Oxford: Oxford University Press.
Laurel, Brenda. 1991. *Computers as theatre*. Reading: Addison-Wesley.
Leibnizcenter for Law. 2011. Multi-agent PhD position available. www.leibnitz.org/wp-content/uploads/2011/02/wervering.pdf. Accessed 24 Jan 2012.
Maes, Pattie. 1994. Adaptive autonomous agents. *Artificial Life Journal* 1(1): 135–162.
Mainzer, Klaus. 2007. *Thinking in complexity. The complex dynamics of matter, mind, and mankind*, 5th ed. Heidelberg: Springer.
Mainzer, Klaus. 2010. *Leben als Maschine?: Von der Systembiologie zur Robotik und Künstlichen Intelligenz*. Paderborn: Mentis Verlag.
Misselhorn, Catrin, Ulrike Pompe, and Mog Stapleton. 2013. Ethical considerations regarding the use of social robots in the fourth age. *GeroPsych* 26(2): 121–133.
Moor, James. 2006. The nature, importance, and difficulty of machine ethics. *IEEE Intelligent Systems* 21(4): 18–21. doi:10.1109/MIS.2006.80.
Muldoon, Ryan, Chiara Lisciandra, Christina Bicchieri, Stephan Hartmann, and Jan Sprenger. 2014. On the emergence of descriptive norms. *Politics, Philosophy and Economics* 13(1): 3–22.
Nealon, John, and Antonio Moreno. 2003. Agent-based health care systems. In *Applications of software agent technology in the health care domain*, ed. John Nealon and Antonio Moreno, 3–18. Basel: Birkhäuser Verlag.
Pagallo, Ugo. 2013. *The law of robots: Crimes, contracts, and torts*. New York: Springer.
Rammert, Werner. 2007. *Die Techniken der Gesellschaft: In Aktion, in Interaktivität und in hybriden Konstellationen*. Berlin: The Technical University Technology Studies Working Papers.
Rammert, Werner. 2008. *Where the action is: Distributed agency between humans, machines, and programs*. Berlin: The Technical University Technology Studies Working Papers.
Rammert, Werner. 2011. *Distributed agency and advanced technology or: How to analyse constellations of collective inter-agency*. Berlin: The Technical University Technology Studies Working Papers.
Rammert, Werner, and Ingo Schulz-Schaeffer. 2002. Technik und Handeln: Wenn soziales Handeln sich auf menschliches Verhalten und technische Abläufe verteilt. In *Können Maschinen handeln? Soziologische Beiträge zum Verhältnis von Mensch und Technik*, ed. Werner Rammert and Ingo Schulz-Schaeffer, 11–64. Frankfurt: Campus.
Ruß, Aaron, Dieter Müller, and Wolfgang Hesse. 2010. Metaphern für die Informatik und aus der Informatik. In *Menschenbilder und Metaphern im Informationszeitalter*, ed. Michael Bölker, Mathias Gutmann, and Wolfgang Hesse, 103–128. Berlin: LIT Verlag.

Schulz-Schaeffer, Ingo. 2000. Akteur-Netzwerk-Theorie. Zur Koevolution von Gesellschaft, Natur und Technik. In *Soziale Netzwerke. Konzepte und Methoden der sozialwissenschaftlichen Netzwerkforschung*, ed. Johannes Weyer, 187–209. München: R. Oldenburg Verlag.

Teubner, Gunther. 2006. Rights of non-humans? Electronic agents and animals as new actors in politics and law. *Journal of Law and Society* 33: 497–521.

Thürmel, Sabine. 2012. A multi-dimensional agency concept for social computing systems. In *Proceedings of the AISB/IACAP world congress social computing, social cognition, social networks and multiagent systems*, ed. Gordana Dodig-Crnkovic, Antonino Rotolo, Giovanni Sartor, Judith Simon, and Clara Smith, 87–91.

Thürmel, Sabine. 2013. Die partizipative Wende: Ein multidimensionales, graduelles Konzept der Handlungsfähigkeit menschlicher und nichtmenschlicher Akteure. Dissertation, München: Technische Universität München.

Tomasello, Michael. 2008. *Origins of human communication*. Cambridge, MA: MIT Press.

Tuomela, Raimo. 2007. *The philosophy of sociality: The shared point of view*. Oxford: Oxford University Press.

Turkle, Sherry. 2006. A nascent robotics culture: New complicities for companionship, *American Association for Artificial Intelligence (AAAI) technical report series,* July 2006, 21st national conference on artificial intelligence human implications of human-robot interaction (HRI), Boston July 2006.

Warneken, Felix, and Michael Tomasello. 2009. Varieties of altruism in children and chimpanzees. *Trends in Cognitive Science* 13(9): 397–402.

Wedde, Italy, Horst, Sebastian Lehnhoff, Christian Rehtanz, and Olav Krause. 2008. Bottom-up self-organization of un-predictable demand and supply under decentralized power management. In *Proceedings of the 2nd IEEE international conference on self-adaptation and self-organization* (SASO'08), 10–20. Venice: Italy.

Woolridge, M. 2009. *An introduction to multiagent systems*, 2nd ed. Italy: Venice.

Part II
Human-Machine Cooperation

Chapter 4
Embodied Cooperative Systems: From Tool to Partnership

Ipke Wachsmuth

4.1 Introduction

The idea of embodied cooperative systems pursues a vision of systems that are helpful to humans by making interaction between humans and artificial systems natural and efficient. The long-term objective of our research is a thorough understanding of the processes and functional constituents of cognitive interaction in order to replicate them in artificial systems that can communicate and cooperate with humans in a natural way. While "cooperative system" could be said to mean a pair (or group) of individuals acting together in the attempt to accomplish a common goal, we prefer the notion of a system exhibiting cooperative behavior by taking on (some of) the goals of another individual and acting together with the other to achieve these shared goals. Cooperation thus involves, as we shall explain further below, some kind of joint intention, which means the ability to represent coordinated action plans for shared goals. Crucial for such cooperation is communication, and when we speak of embodied systems here, the idea is that these systems "by nature" can also employ nonverbal behaviors in cooperative dialogue when coordinating actions between agents.

This contribution is written from the perspective of artificial intelligence which, as an academic discipline, is concerned with building machines – artificial agents – that model human intelligent behaviors and exploit them in technical applications. Such behaviors typically include the functions of perceiving, reasoning, and acting. Research has been moving on towards envisioning artificial agents (e.g. autonomous robots) as partners rather than tools with whom working 'shoulder-by-shoulder' with humans can be effective (Breazeal et al. 2004). Then such systems will also need to incorporate capacities which enable them to align with their human

I. Wachsmuth (✉)
Bielefeld University, Bielefeld, Germany
e-mail: ipke.wachsmuth@uni-bielefeld.de

interactant through shared beliefs and intentions. When we thus view agents as intentional systems, a central idea is that their behavior can be understood by attributing them beliefs, desires and intentions (see Sect. 4.2). The questions particularly addressed from this perspective in this chapter are the following:

1. How can joint intentions and cooperation be modeled and simulated?
2. Can we attribute joint intention to a system or team involving both, a human and an artificial agent?

One of the most basic mental skills is inferring intentions – the ability to see others as intentional agents and to understand what someone else is doing. Intentions are not directly observable, thus they need to be inferred from the interactant's overt behaviors. The types of information exploited to infer intentions are comprised by the interactant's verbal behavior, gaze and facial expression, gestures, as well as the perceived situation and prior knowledge. Hence inferring intentions is not a monolithic mental faculty, but a composite of different mechanisms including attentive processes (i.e. processes enabling a system to actively focus on a target) and more general cognitive processes such as memory storage or reasoning. Both, understanding others' intentions and representing them as being able to understand intentions, are relevant factors in cooperation.

Human beings (and certain animals) can develop a mental representation of the other, making assumptions (possibly false ones) about the other's beliefs, desires, intentions and probable actions – a 'Theory of Mind' (Premack and Woodruff 1978). Theory of Mind (ToM) refers to the ability to understand others as rational agents, whose behavior is lead by intentional states like beliefs and desires. There are two aspects of Theory of Mind: 'cognitive' ToM (inferring intentional states of the other), and 'affective' ToM (inferring emotional states of the other), referring to an understanding of what the other is likely to do resp. what are the other's feelings. These ideas may also be a valuable prerequisite in modeling communication with virtual humans (Krämer 2008). While our own work has included artificial agents that can infer another agent's emotional state (e.g. Boukricha et al. 2011), we shall in the following mainly focus on the intentional states involved in cooperation.

To endow artificial systems with cooperative functionality, they need to be enabled to adopt (some of) the goals of another individual and act together with the other to achieve these shared goals.[1] Acting together requires that intentional agents engage with one another to form a joint intention, i.e. represent coordinated action plans for achieving their common goals in joint cooperative activity (Tomasello et al. 2005; Bratman 1992). The activity itself may be more simple (e.g. engaging in a conversation; see Sect. 4.2) or complex (like constructing a model airplane; see Sect. 4.3). For collaborative engagement, Tomasello et al. (2005) further stress the importance of *joint attention* (mutual knowledge of interactants sharing their focus of attention; see Sect. 4.4) as well as interactants' ability to reverse action roles and help the other if needed.

---

[1] The aspect of "mutual benefit" often included in definitions of cooperation is not taken up here because artificial systems do not seem to have genuine interests that could benefit from cooperation; see (Stephan et al. 2008).

If we want to construct artificial systems that are helpful to humans – interacting with us like "partners" –, then such systems should be able to understand and respond to the human's wants in order to be assistive in a given situation. Technically, this challenge involves the implementation of a range of skills such as: processing language, gaze and gestures, representing intentional states for self and other, detecting and manipulating the other's attention, responding to bids for joint attention, accomplishing goals in joint activity.

In the remainder of this chapter, we will outline these ideas taking the virtual humanoid agent "Max" as a model for a communicative and cooperative agent. We will first look at Max as a "conversational machine" and describe its cognitive architecture, then move on to cooperation by examining details of a cooperative construction scenario, focus on the coordination of attention, and conclude with a view of how the perception of artificial systems may change from tool to partnership.

4.2 Conversational Machines

The development of conversational machines, i.e. machines that can conduct human-like dialogue, has been a goal of artificial intelligence research for long (Schank 1971; Cassell et al. 2000). Why would we want to build such machines in general? On the one hand, there is the motive that learning to generate certain intelligent behaviors in artificial systems will help us to understand these behaviors in detail. That is, in our research we devise explanatory models in the form of computer simulations to obtain a better understanding of cognitive and social factors of communication. On the other hand, building conversational machines is expected to help make communication between humans and machines more intuitive.

4.2.1 Machines as Intentional Systems

Building a machine that can exhibit or simulate rational behavior (*as if* it were an agent acting rationally to further its goals in the light of its beliefs) leads us to look at machines as intentional systems, i.e. systems that perceive changes in the world, represent mental attitudes (like beliefs, goals, etc.), and reason about mental attitudes in order to arrive at decisions on how to plan actions and act.

Research approaches towards modeling mental attitudes and practical reasoning are frequently based on functional models of planning and choosing actions by means-ends analysis, mainly in versions of the belief-desire-intention paradigm (BDI) (Rao and Georgeff 1991).[2] The basic idea is the description of the internal working state of an agent by means of intentional states (beliefs, desires, intentions) as well as the layout of a control architecture that allows the agent to choose

---
[2] The BDI approach comes from Michael Bratman (Bratman 1987); one of its fundamentals can be traced back to the work of Daniel Dennett (Dennett 1987) on the behavior of intentional systems.

rationally a sequence of actions on the basis of their representations. By recursively elaborating a hierarchical plan structure, specific intentions are generated until, eventually, executable actions are obtained (Wooldridge 2002).

Modeling intentional states is based on their symbolic representation. One of its assets is the flexibility it provides for planning and reasoning. In beliefs, for instance, facts concerning the world may be stored that an agent is not (or no longer) able to perceive at the moment, which, however, have effect on the agent's further planning. It is a difference, though, whether an agent draws conclusions simply on the basis of his beliefs and desires or whether he makes use of them – with a corresponding cognitive representation – recognizing them as his own. In many cases such differentiation may not have functional advantages. An agent should be expected, however, to represent intentional states explicitly as being his own ones, if he must also record and deal specifically with other agents' intentional states.

4.2.2 Bielefeld Max Project

In our research laboratory at Bielefeld taking a cognitive modeling approach scientific enquiry and engineering are closely intertwined. Creating an artificial system that replicates aspects of a natural system can help us understand the internal mechanisms that lead to particular effects. Special for our approach is that we are not just building and studying intelligent functions in separate. Over many years, we have attempted to build coherent comprehensive systems integrating both symbolic and dynamic system paradigms, one of them "Max".

Max is a "virtual human" – an artificial agent embodied in virtual reality with a person-like appearance (see Fig. 4.1, top). By means of microphones and video cameras or tracker systems, Max can "hear" and "see" his human interlocutors and process verbal instructions and gestures. Max is equipped with verbal conversational abilities and can employ his virtual body to exhibit nonverbal behaviors in face-to-face interaction. With a modulated synthetic voice and an animated face and body, Max is able to speak and gesture, and to mimic emotions (Kopp and Wachsmuth 2004). The face of Max is computer-animated by simulated muscle effects and displays lip-synchronous speech, augmented by eyebrow raises and emotional facial expression. Max's articulated body is driven by a kinematic skeleton (comprising roughly one-third of the degrees of freedom of the human skeleton), with synchronized motion generators giving a realistic impression for his body movements. Emotional expression (which also includes voice modulation) is driven by a dynamic system, which responds to various kinds of stimuli (external: seeing faces, bad words, etc.; internal: goal achievement or failure, resulting in positive resp. negative emotions), and which defines the agent's explicit emotional state over time in pleasure-arousal-dominance (PAD) space (Becker et al. 2004). The agent is controlled by a cognitive architecture (Sect. 4.2.3) which is based on the symbolic belief-desire-intention (BDI) approach to modeling rational agents, while integrating concurrent reactive behaviors and emotions. Thus the mental state of Max is comprised by an intentional as well as an emotional state (see Fig. 4.1, bottom).

Fig. 4.1 Virtual human Max: outer appearance and schematic view of internal state

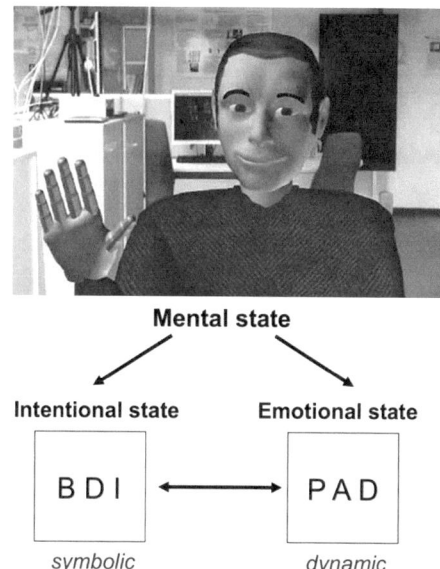

With the Bielefeld Max project we investigate the details of face-to-face interaction and how it is possible to describe them – in parts – so precisely that a machine can be made to simulate them. This means that collecting insights about the functioning of human cognitive interaction is an important focus of our work. A technical goal is also the construction of a system as functional and convincing as possible that may be applied in different ways.

4.2.3 Cognitive Architecture

To organize the complex interplay of sensory, cognitive and actoric abilities, a cognitive architecture has been developed for Max (Leßmann et al. 2006), aiming at making his behavior appear believable, intelligent, and emotional. Here, 'cognitive' refers to the structures and processes underlying mental activities, including attentive processes. Bearing a functional resemblance to the links that exist between perception, action, and cognition in humans, the architecture has been designed for performing multiple activities simultaneously, asynchronously, in multiple modalities, and on different time scales. It provides for reactive and deliberative behaviors running concurrently, with a mediator resolving conflicts in favor of the behavior with the highest utility value.

Figure 4.2 gives an outline of the cognitive architecture of Max. Explicitly represented goals (desires), which may be introduced through internal processing as well as by external influences, are serving as "inner motivation" triggering behavior. Max can have several desires at the same time, the highest-rated of which is selected by a utility function to become the current intention. The BDI interpreter determines

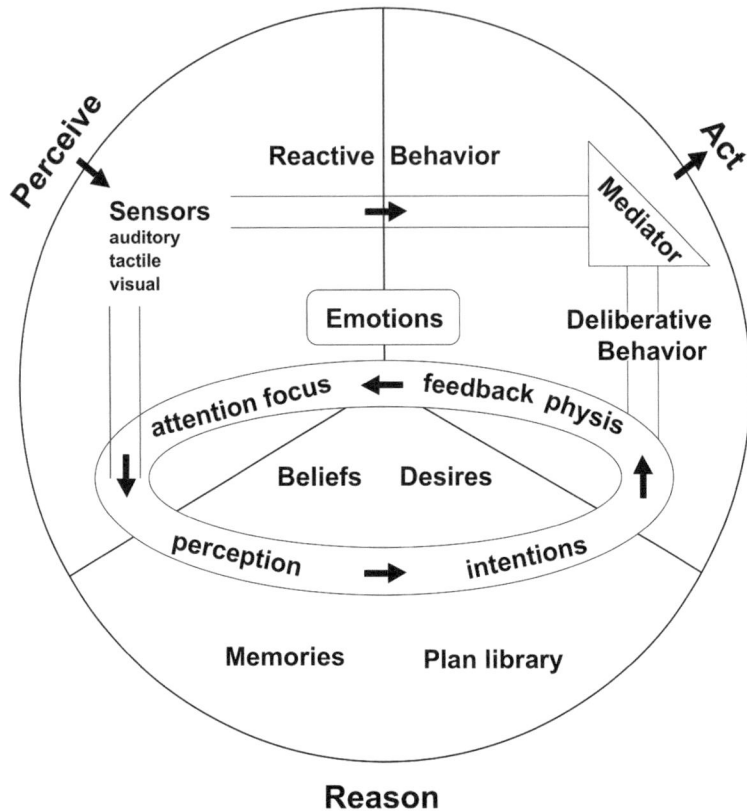

Fig. 4.2 Outline of the cognitive architecture of Max (Reproduced from Leßmann et al. 2006)

the current intention on the basis of existing beliefs, current desires as well as options for actions.

Options for actions are available in a plan library in the form of abstract plans that are described by preconditions, context conditions, consequences that may be accomplished, and a priority function. Plan selection is further influenced by the current emotional state (in PAD space, see above) in that the emotion is used as precondition and context condition of plans to choose among alternative actions. If a concrete plan drawn up on the basis of these facts has been executed successfully, the related goal will become defunct.

The conduct of dialogue is based on an explicit modeling of communicative functions related to, but more specific than, multimodal communicative acts (Poggi and Pelachaud 2000) and generalizing speech act theory (Searle and Vanderveken 1985). A communicative function explicates the functional aspects of a communicative act on three levels (interaction, discourse, content) and includes a performative reflecting the interlocutor's intention, with a dialogue manager controlling reactive behaviors and creating appropriate utterances in response (Kopp et al.

2005). Dialogue is performed in accordance with the mixed initiative-principle, this means, for instance, that in case the human fails to answer, Max himself takes the initiative and acts as the speaker. The plan structure of the BDI system makes it possible to assert new goals during the performance of an intention that may replace the current intention, provided it has a higher priority. If the previous intention is not specifically abandoned in this process and its context conditions are still valid, it will become active again after the interruption.

4.2.4 Max as a Museum Guide

Since 2004 Max has been employed as a museum guide in a public computer museum (the Heinz Nixdorf MuseumsForum in Paderborn), taking the step from a research prototype to a system being confronted with many visitors in a rich real-world setting (Kopp et al. 2005). Displayed on a large projection screen (Fig. 4.3), Max provides the visitors with various information and engages them in a conversation. For instance, greeting a group of visitors, he could say: "Max, that's me. I'm an artificial person that can speak and gesture. I am artificial, but I can express myself just like you…" A visitor might ask Max "How is the weather?" and Max would then access the current weather forecast in the internet and read it to the visitor. Altogether, this research has embarked on the goal of building embodied agents that can engage with humans in face-to-face conversation and demonstrate many of the same communicative behaviors as exhibited by humans.

A screening was done after the first 7 weeks of Max's employment in the computer museum (Kopp et al. 2005). Statistics was based on log files anonymously

Fig. 4.3 Max interacting with visitors in the Heinz Nixdorf MuseumsForum

recorded from dialogues between Max and visitors to the museum. Among other aspects, the data were evaluated with respect to the successful recognition of a communicative function, that is, whether Max could associate a visitor's want with an input. We found that Max recognized a communicative function in nearly two-thirds of all cases. Even when Max had sometimes recognized an incorrect communicative function (as humans may also do), we may conclude that in these cases Max conducted sensible dialogue with the visitors. In the other one-third of cases, Max did not turn speechless but simulated "small talk" by employing commonplace phrases, still tying in visitors with diverse kinds of social interaction.

In some sense, Max could be attributed rational behavior (*as if* he were an agent acting rationally), namely, "minimal rationality" (Dretske 2006): This notion requires not only that behavior be under the causal control of a representational state, but that it be explained by the content of that representational state. Minimal rationality, so to say, requires that what is done is done for a reason (not necessarily a good reason). In light of the above statistics on Max's service as a museum guide, Max's answers were given for *a reason* (in the fulfillment of communicative goals Max associated with visitors' inputs) in many cases. That is, his behavior might be termed "minimally rational" in Dretske's sense (at least in an *as-if* sense – by the way the agent was programmed).

4.3 From Conversation to Cooperation

The above explained ideas about Max as a conversational machine are relevant also for embodied cooperative systems, i.e. systems acting together with others to accomplish shared goals by employing verbal and nonverbal behaviors in coordinating their actions. Such systems may be embodied as robotic agents (e.g. Breazeal et al. 2004) or (such as Max) as humanoid agents projected in virtual reality. If we want to achieve that Max and a human interlocutor mutually engage and coordinate action in solving a joint problem, a central question is how the processes involved interact and how their interplay can be modeled. For example, inter-agent cooperation relies very much on common ground, i.e. the mutually shared knowledge of the interlocutors. Nonverbal behaviors such as gaze and gestures are important means of coordinating attention between interlocutors and therefore related to both inferring intentions and coordinating actions. Note that the conduct of dialogue is a form of cooperation, because participants have to coordinate their mental states.

In one of our research settings Max was employed to study cooperative dialogue in a construction task, where Max and a human interlocutor solve the joint problem of constructing a model airplane from a 'Baufix' wooden toykit (Leßmann et al. 2006). In this setting the human interlocutor and Max stand opposite each other at a table (see Fig. 4.4, top). With the exception of the person shown left, the pictured scene is projected virtual reality. On the table, there are different building bricks: bars with three holes or five holes, several screws with colored caps, a cube with holes on all sides, etc., all items will be assembled in the course of the

4 Embodied Cooperative Systems: From Tool to Partnership 71

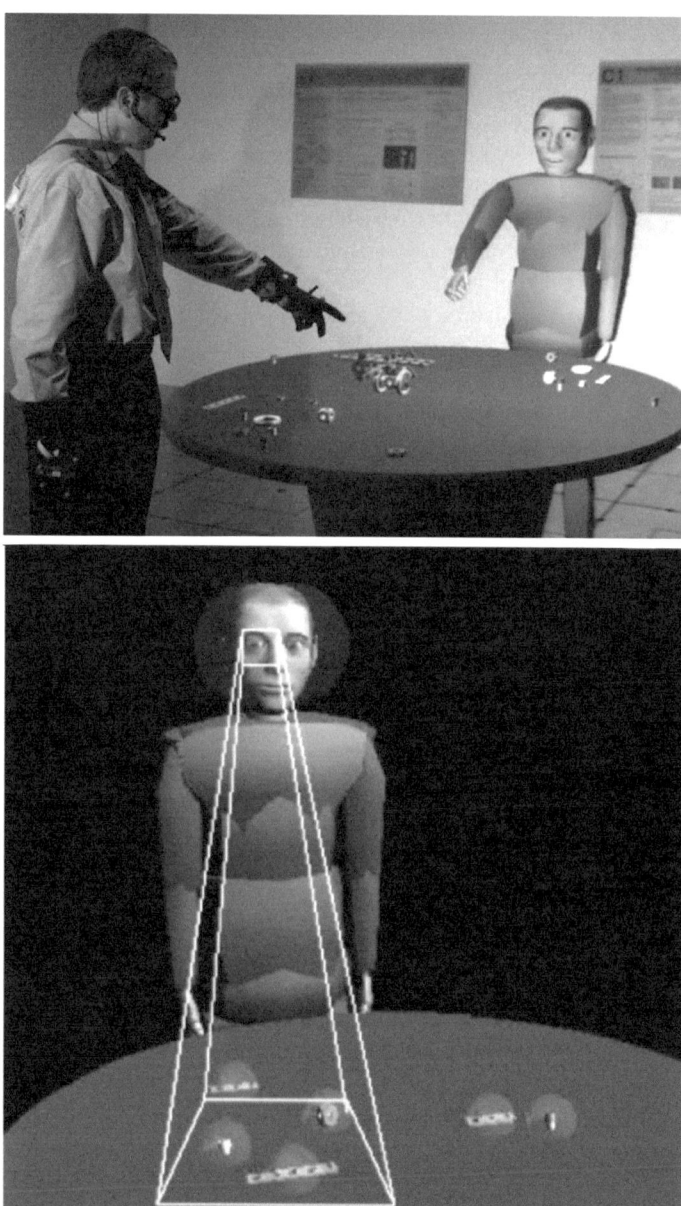

Fig. 4.4 Cooperative dialogue with Max (*top*); virtual world perception of Max (*bottom*)

dialogue. The human sees Max and the entire setting three-dimensionally by means of stereo glasses and hears the synthetic voice of Max due to surround sound from hidden speakers.

To enable the artificial agent to participate in the interaction, Max is equipped with both view sensors for the perception of the virtual environment (see Fig. 4.4, bottom) and diverse sensors to capture the signals of the human in the real world. Max 'sees' the human's gaze direction as well as hand and arm movements by means of an infrared tracking system and data gloves, and – via a radio microphone – he 'hears' the human's speaking which he processes with a speech recognizer. The percepts of both worlds (real and virtual) are integrated to form a coherent perception of the ongoing interaction. Both the human and Max are able to have individual parts assembled by natural language instructions and gestures, carried out in simulation with realistic sounds added. While the human interlocutor in general has the role of proposing the next construction step, Max has knowledge of construction details and helps the other with his role if needed. Thus action roles may switch between the interactants according to their competences. This is done, for example, as follows[3]:

Max	Hello, I'm Max. How can I help you? [*smiling*].
Human	Explain the assembly of a propeller.
Max	Let's do it together. First, a screw has to be inserted into the middle of a three-hole-bar.
Human	Insert the yellow screw into the bar [*points to a three-hole-bar, the first screw is inserted into the first hole*].
Max	No, that was the wrong hole [*screw slips out*].
Max	I'll explain it to you again. A screw has to be inserted into the middle of a three-hole-bar.
Human	Insert the screw into the middle of the three-hole-bar [*screw is inserted into the middle of the three-hole-bar chosen before*].
Max	Okay, that's correct. Now insert this [*points to the bar with the screw*] into the three-hole-bar.
Human	You do it.
Max	Okay, I will perform the action [*the objects are connected as instructed*].
Max	And now the bars need to be turned crosswise to each other [*shows the rotation using a gesture*], etc.

Fitted to this trial scenario, Max is equipped with limited knowledge of the 'Baufix world' and is capable of planning and reasoning such that he may act as an intelligent assistant. Specialized planners (for constructing Baufix objects) and memories (for storing dynamically updated representations for the state of constructed objects) have been integrated into the cognitive architecture. Further, Max has some grammatical rules and a semantic lexicon for processing the meaning of dialogue inputs. Within a limited vocabulary Max is able to talk – including the generation of appropriate gestures –, producing verbal utterances from a repertoire of stereotype statements. These also include the term 'I', without Max having a

---

[3] Dialogue translated from German to English.

4 Embodied Cooperative Systems: From Tool to Partnership

notion of himself at the current time.[4] Independent of that it could be demonstrated how Max can cope with changing situations that require language, perception, and action to be coordinated so that cooperation between the human and the artificial system takes place with natural efficiency.

4.4 Coordinating Attention

As was said in the introduction, one of our questions is how joint intentions and cooperation can be modeled and simulated. Attentive processes and sharing attention are important precursors for cooperative interaction in which interactants pursue shared goals by coordinated action plans (joint intentions). Inter-agent cooperation relies much on common ground, one aspect being whether interactants know together that they share a focus of attention (e.g. know that they are both looking at the same target object as illustrated in Figs. 4.5 and 4.6 below). This kind of intentionally sharing a focus of attention is referred to as "joint attention" below. It would presuppose interactants to mutually perceive one another and perceive the

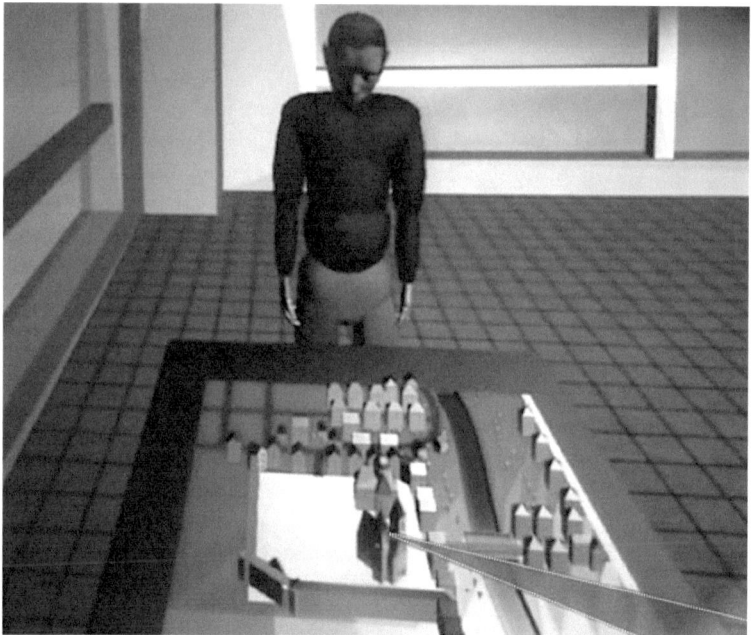

Fig. 4.5 Max can pick up the human's gaze by means of eye-tracking

---

[4] On how to configure an artificial agent so as to enable him to adopt a first-person perspective see Wachsmuth (2008).

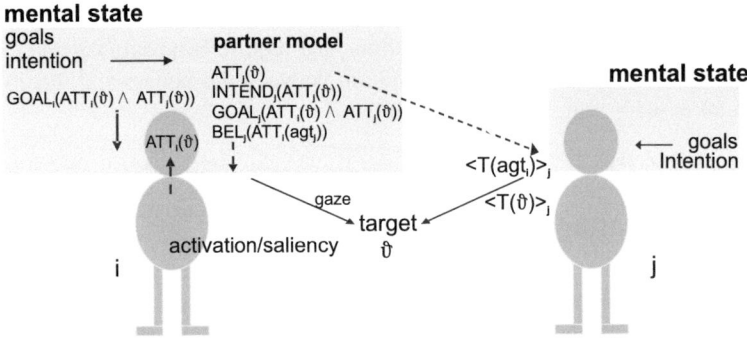

Fig. 4.6 Joint attention focusing on a target object ϑ from artificial agent i's perspective (Reproduced from Pfeiffer-Leßmann and Wachsmuth 2009)

perceptions of the other, that is, attend to each other. An important means of coordinating attention between interlocutors are nonverbal behaviors such as gaze and gestures. For example, following each other's direction of gaze allows interlocutors to share their attention.

Judged to be crucial for goal-directed behavior, attention has been characterized as an increased awareness of something (Brinck 2003), intentionally directed perception (Tomasello et al. 2005), or as "the temporally-extended process whereby an agent concentrates on some features of the environment to the (relative) exclusion of others" (Kaplan and Hafner 2006). A foundational skill in human social interaction, joint (or shared) attention can be defined as simultaneously allocating attention to a target as a consequence of attending to each other's attentional states, or "re-allocating attention to a target *because* it is the object of another person's attention" (Deák et al. 2001). In contrast to joint perception (the state in which interactants are just perceiving the same target object without further constraints concerning their mental states), the intentional aspect of joint attention has been stressed, in that interlocutors have to deliberately focus on the same target while being mutually "aware" of sharing their focus of attention (Tomasello et al. 2005). If virtual humans are to engage in joint attention with a human interactant, they need to be enabled to meet conditions as described above. For instance, they would need to infer the human interactant's focus of attention from the interactant's overt behaviors. Prerequisites for this are attention detection (e.g. by gaze following) as well as attention manipulation (e.g. by issuing gaze or pointing gestures).

We have investigated joint attention in a cooperative interaction scenario (different from the one described in Sect. 4.3) with the virtual human Max, where again the human interlocutor meets the agent face-to-face in virtual reality. The human's body and gaze are picked up by infrared cameras and an eye-tracker (mounted on the stereo glasses for three-dimensional viewing), informing Max where the interlocutor is looking at; this way Max can follow the human's gaze (see Fig. 4.5). For instance, when the human focuses on an object, Max can observe the human's gaze alternating between an object and Max's face and attempt to establish joint attention, by focusing on the same object. Or, initiating a bid for joint attention, Max can choose an object and attempt to draw the attention of his interlocutor to the object by gaze and pointing gestures until joint attention is established.

So, which inferences exactly need to be drawn to establish joint attention by aligning the mental states of cooperating agents? Pfeiffer-Leßmann and Wachsmuth (2009) describe a formal model which specifies the conditions and cognitive processes that underlie the capacity for joint attention. In accordance with Tomasello et al. (2005) joint attention is conceived of as an intentional process. Our model provides a theoretical framework for cooperative interaction with a virtual human (in our case: Max). It is specified in an extended belief-desire-intention modal logic, which accounts for the temporally-extended process of attention (Kaplan and Hafner 2006) in interaction between two intentional agents, during which agents' beliefs may change. To account for such dynamics of an agent's beliefs in our model, the logic was extended to include activation values. The idea is that activation values influence the beliefs' accessibility for mental operations, resulting in an overall saliency of a belief (and likewise other intentional states). For instance "increased awareness" (Brinck 2003) can be modeled by use of activation values.

In order to account for these ideas, the above described cognitive architecture (Sect. 4.2.3) adopting the BDI paradigm of rational agents (Rao and Georgeff 1991) has been augmented by incorporating a partner model to account for the agent's perspective on its interlocutor, as well as a dynamic working memory. The working memory stores the changing beliefs (and other intentional states) of the agent, and also the target objects that may be in the agent's focus of attention. Activation values are used as a measure for saliency, i.e. an object with a higher activation value is more salient than one with a lower activation value. Whenever an object gets in the agent's gaze focus (see Fig. 4.4, right) or is subject to internal processing, activation values are increased.

To establish joint attention, an agent must employ coordination mechanisms of understanding and directing the intentions underlying the interlocutor's attentional behavior, such as: tracking the attentional behavior of the other by gaze monitoring; deriving candidate objects the interlocutor may be focusing on; inferring whether attentional direction cues of the interlocutor are uttered intentionally; reacting instantly, as simultaneity is crucial in joint attention; and in response employing an adequate overt behavior which can be observed by the interlocutor. Meeting these conditions, Pfeiffer-Leßmann and Wachsmuth (2009) describe the mental state required for an agent i to believe in joint attention while focusing conjointly with its interlocutor j on a certain target ϑ (theta). While we won't go into detail here, see Fig. 4.6 and brief explanations following for an illustration of these ideas.

Figure 4.6 illustrates the following (BEL, GOAL, and INTEND are modal connectives in the logic used for modeling beliefs, goals, intentions and attentional states):

If agent i (here artificial agent Max) attends (focuses attention: ATT) to a target ϑ and has the goal that both, agents i (Max) and j (human interactant) jointly attend to the same target ϑ, then agent i needs to infer (and assert in the partner model)

- that agent j (also) attends to the target ϑ
- that agent j intends (INTEND) to attend (intentionally attends) to the target ϑ

- that agent j adopts the goal that both agents jointly attend to the target ϑ, and
- that agent j (human interactant) believes (BEL) that (artificial agent) i attends to (human interactant) j

T (test-if) pertains to test-actions that are to infer if human interactant j focuses attention on agent i while simultaneously (observed by gaze-alternation) allocating attention to target ϑ. If agent i (Max) perceives the interlocutor's behavior as a test-action and is able to resolve a candidate target object, the agent infers that the interlocutor's focus of attention resides on that target.[5] The formalization provides a precise means as to which conditions need to be met and which inferences need to be drawn to establish joint attention by aligning the mental states of cooperating agents.

Going from this formal model to the implemented system, some heuristics had to be used, for instance when the human interlocutor focuses several times (or for an extended duration) on an object, the agent interprets this as the attention focus being intentionally drawn upon the target, or that the addressee's response to an agent initiating joint attention needs to take place within a certain time frame. To follow this up, an eye-tracker study was conducted (Pfeiffer-Leßmann et al. 2012) examining dwell times (fixation durations) of referential gaze during the initiation of joint attention, the results of which further contribute to making our formal model of joint attention operational.

4.5 Conclusion: From Tool to Partnership

From the perspective of artificial intelligence, this contribution has addressed embodied cooperative systems, i.e. embodied systems exhibiting cooperative behavior by taking on (some of) the goals of another individual and acting together with the other to accomplish these goals employing verbal and nonverbal communicative behaviors. The questions we set out to address from this perspective were the following: (1) How can joint intentions and cooperation be modeled and simulated? (2) Can we attribute joint intention to a system or team involving both, a human and an artificial agent?

Cooperation, as was said in Sect. 4.1, involves adopting (some of) the goals of another individual and acting together with the other to achieve these shared goals. Joint intention refers to the ability of interactants to represent coordinated action plans for achieving their common goal in joint activity.

Taking the virtual humanoid agent "Max" as an example of an embodied cooperative system, we introduced first steps towards how joint intentions and cooperation can be modeled and simulated. We described a cognitive architecture, based on the BDI paradigm of rational agents and augmented by a partner model accounting

---

[5] For further detail, the formal definition of joint attention, and the specification of epistemic actions that lead to the respective beliefs and goals see Pfeiffer-Leßmann and Wachsmuth (2009).

for the agent's perspective on its interlocutor (representing the inferred intentional states in the sense of 'cognitive' ToM), as well as a dynamic working memory storing the changing intentional and attentive states of the agent. This cognitive architecture enables Max to engage in joint activities (conducting conversation, solving a joint problem) with human interlocutors. Further we have outlined which inferences need to be drawn to establish joint attention (the common ground of interactants knowing together that they share a focus of attention) by aligning the mental states of cooperating agents. While a complete model of joint intention remains to be done, we used the case of joint attention to lay out how an artificial agent can represent goals and intentions of his human interactant in a partner model that could be employed for representing coordinated action plans in the plan structure of the BDI system. Thus we could provide some preliminary insights on question (1) as to how joint intentions and cooperation can be modeled and simulated.

Our second question (can we attribute joint intention to a system or team involving both, a human and an artificial agent?) is more complicated since it involves the idea of partnership, i.e. a relationship between two (or more) individuals working or acting together. There is accumulating evidence that in the context of cooperative settings, the view that humans are users of a certain "tool" has shifted to that of a "partnership" with artificial agents, insofar they can in some sense be considered as being able to take initiative as autonomous entities (Breazeal et al. 2004; Negrotti 2005). According to Negrotti (2005), a true partner relationship is to be conceived as a peer-based interaction, wherein each partner can start some action unpredicted by the other. Looking at Max, we note that the cooperative interaction in the above described scenarios is characterized by a high degree of interactivity, with frequent switches of initiative and roles of the interactants. In consequence, though being goal-directed, the interaction with Max appears fairly unpredictable. Thus Max appears to be more than a tool (thing) entirely at our disposal and under our control.

So, in light of what has been said above, can we attribute joint intention to a system or team involving both, a human and an artificial agent?

Perhaps one day. There is a long way to go, though. We have to acknowledge that state-of-art agent technology is still far from being sufficiently sophisticated to implement all the behaviors necessary for a cooperative functionality and in particular joint intention in a coherent technical system.[6] But it has to be realized that artificial systems may increasingly take on functions which were reserved to human beings so far and thus seem to become more human-like. For instance, in recent work (Mattar and Wachsmuth 2014), a person memory was developed for Max which allows the agent to use personal information remembered about his interlocutors from previous encounters which, as evaluations have shown, makes him a better conversational partner in the eyes of his human interlocutors.

---

[6] Note that, even when we have attempted to build a coherent comprehensive system, not all aspects described in this article have been integrated in one system, that is, different versions of "Max" were used to explore the above ideas in implemented systems.

Also to be noted, the desires of Max do not originate in "real needs" that Max might have; they were programmed functionally equivalent to intentional states we would attribute to a real person, resulting in behaviors that appear somewhat rational. Another programmed "need" of Max is that he demands his conversational partners to be polite. The emotional state of Max (see Sect. 4.2) is negatively influenced by inputs containing ungracious or politically incorrect wordings ("no-words") which, when repeated, can eventually trigger a plan causing the agent to leave the display and stay away until the emotion has returned to a balanced state (an effect introduced to de-escalate rude visitor behavior in the museum). The period of absence can either be shortened by complimenting Max or extended by insulting him again, see (Becker et al. 2004). Altogether, this kind of behavior of Max may be taken as beginnings of moral judgement.

In conclusion: If we want to construct artificial systems that are helpful to humans and interact with us like "partners", then such systems should be able to understand and respond to the human's wants – infer and share our intentions – in order to be assistive in a given situation. It may be asked if it makes a big difference for embodied cooperative systems to be helpful whether their understandings and intentions (and the intentions they share with us) are real or "*as-if*" (Stephan et al. 2008).

Acknowledgments The research reported here draws on the contributions by the members of Bielefeld University's artificial intelligence group which is hereby gratefully acknowledged. Thanks also to Catrin Misselhorn and an anonymous referee for helpful comments to improve the text. Over many years this research has been supported by the Deutsche Forschungsgemeinschaft (DFG) in the Collaborative Research Centers 360 (Situated Artificial Communicators) and 673 (Alignment in Communication), the Excellence Cluster 277 CITEC (Cognitive Interaction Technology), and the Heinz Nixdorf MuseumsForum (HNF).

References

Becker, Christian, Stefan Kopp, and Ipke Wachsmuth. 2004. Simulating the emotion dynamics of a multimodal conversational agent. In *Affective dialogue systems*, ed. E. André, L. Dybkjaer, W. Minker, and P. Heisterkamp, 154–165. Berlin: Springer.
Boukricha, Hana, Nhung Nguyen, and Ipke Wachsmuth. 2011. Sharing emotions and space – Empathy as a basis for cooperative spatial interaction. In *Proceedings of the 11th international conference on intelligent virtual agents* (IVA 2011), ed. Kopp, S., S. Marsella, K. Thorisson, and H.H. Vilhjalmsson, 350–362. Berlin: Springer.
Bratman, Michael E. 1987. *Intention, plans, and practical reason*. Harvard: Harvard University Press.
Bratman, Micheal E. 1992. Shared cooperative activity. *Philosophical Review* 101(2): 327–341.
Breazeal, Cynthia, Andrew Brooks, David Chilongo, Jesse Gray, Guy Hoffman, Cory Kidd, Hans Lee, Jeff Lieberman, and Andrea Lockerd. 2004. Working collaboratively with humanoid robots. In *Proceedings of humanoids 2004*, Los Angeles.
Brinck, Ingar. 2003. The objects of attention. In *Proceedings of ESPP 2003 (European Society of Philosophy and Psychology)*, Torino, 9–12 July 2003, 1–4.
Cassell, Justine, J. Sullivan, S. Prevost, and E. Churchill (eds.). 2000. *Embodied conversational agents*. Cambridge, MA: MIT Press.
Deák, Gedeon O., Ian Fasel, and Javier Movellan. 2001. The emergence of shared attention: Using robots to test developmental theories. In *Proceedings of the first international workshop on epigenetic robotics*, Lund University Cognitive Studies, vol. 85, 95–104.

Dennett, Daniel C. 1987. *The intentional stance*. Cambridge, MA: MIT Press.
Dretske, Fred I. 2006. Minimal rationality. In *Rational animals?* ed. Susan L. Hurley and Matthew Nudds, 107–116. Oxford: Oxford University Press.
Kaplan, Frédéric, and Verena V. Hafner. 2006. The challenges of joint attention. *Interaction Studies* 7(2): 135–169.
Kopp, Stefan, and Ipke Wachsmuth. 2004. Synthesizing multimodal utterances for conversational agents. *Computer Animation and Virtual Worlds* 15: 39–52.
Kopp, Stefan, Lars Gesellensetter, Nicole C. Krämer, and Ipke Wachsmuth. 2005. A conversational agent as museum guide – Design and evaluation of a real-world application. In *Intelligent virtual agents*, ed. Themis Panayiotopoulos, Jonathan Gratch, Ruth Aylett, Daniel Ballin, Patrick Olivier, and Thomas Rist, 329–343. Berlin: Springer.
Krämer, Nicole C. 2008. Theory of mind as a theoretical prerequisite to model communication with virtual humans. In *Modeling communication with robots and virtual humans*, ed. Ipke Wachsmuth and Günther Knoblich, 222–240. Berlin: Springer.
Leßmann, Nadine, Stefan Kopp, and Ipke Wachsmuth. 2006. Situated interaction with a virtual human – Perception, action, and cognition. In *Situated communication*, ed. Gert Rickheit and Ipke Wachsmuth, 287–323. Berlin: Mouton de Gruyter.
Mattar, Nikita, and Ipke Wachsmuth. 2014. Let's get personal: Assessing the impact of personal information in human-agent conversations. In *Human-computer interaction*, ed. M. Kurosu, 450–461. Berlin: Springer.
Negrotti, Massimo. 2005. Humans and naturoids: From use to partnerships. In *Yearbook of the artificial*, Cultural dimensions of the user, vol. 3, ed. Massimo Negrotti, 9–15. Bern: Peter Lang European Academic Publishers.
Pfeiffer-Leßmann, Nadine, and Ipke Wachsmuth. 2009. Formalizing joint attention in cooperative interaction with a virtual human. In *KI 2009: Advances in artificial intelligence*, ed. B. Mertsching, M. Hund, and Z. Aziz, 540–547. Berlin: Springer.
Pfeiffer-Leßmann, Nadine, Thies Pfeiffer, and Ipke Wachsmuth. 2012. An operational model of joint attention – Timing of gaze patterns in interactions between humans and a virtual human. In *Proceedings of the 34th annual conference of the Cognitive Science Society*, ed. N. Miyake, D. Peebles, and R.P. Cooper, 851–856. Austin: Cognitive Science Society.
Poggi, Isabella, and Catherine Pelachaud. 2000. Performative facial expression in animated faces. In *Embodied conversational agents*, ed. J. Cassell, J. Sullivan, S. Prevost, and E. Churchill, 155–188. Cambridge, MA: MIT Press.
Premack, David, and Guy Woodruff. 1978. Does the chimpanzee have a theory of mind? *Behavioral and Brain Sciences* 4: 512–526.
Rao, A.S., and M.P. Georgeff. 1991. Modeling rational agents within a BDI-architecture. In *Principles of knowledge representation and reasoning*, ed. J. Allen, R. Fikes, and E. Sandewall, 473–484. San Mateo: Morgan Kaufmann.
Schank, Roger C. 1971. Finding the conceptual content and intention in an utterance in natural language conversation. In *Proceedings of IJCAI 1971 (International joint conference on artificial intelligence)*, London 1–3 Sept 1971, 444–454.
Searle, John R., and Daniel Vanderveken. 1985. *Foundations of illocutionary logic*. Cambridge: Cambridge University Press.
Stephan, Achim, Manuela Lenzen, Josep Call, and Matthias Uhl. 2008. Communication and cooperation in living beings and artificial agents. In *Embodied communication in humans and machines*, ed. Ipke Wachsmuth, Manuela Lenzen, and Günther Knoblich, 179–200. Oxford: Oxford University Press.
Tomasello, Michael, Malinda Carpenter, Josep Call, Tanya Behne, and Henrike Moll. 2005. Understanding and sharing intentions: The origins of cultural cognition. *Behavioral and Brain Sciences* 28: 675–691.
Wachsmuth, Ipke. 2008. 'I, Max' – Communicating with an artificial agent. In *Modeling communication with robots and virtual humans*, ed. Ipke Wachsmuth and Günther Knoblich, 279–295. Berlin: Springer.
Wooldridge, Michael. 2002. *An introduction to multiagent systems*. Chichester: Wiley.

Chapter 5
Historical, Cultural, and Aesthetic Aspects of the Uncanny Valley

Valentin Schwind

5.1 Introduction

The basis of collaborative agency and mutual understanding between all kinds of cooperating systems and entities is communication. Our most common and natural way of communication is communicating with other humans. Artificial systems are often designed according to our expectations and simulate human appearance or human behavior in order to improve communication and cooperation with artificial systems. For example, Alan Turing's (1950) famous test is only passed by an artificial intelligence that can convince a certain number of people that they are communicating with a human.

This humanization – or anthropomorphization – is described as "the tendency to attribute human characteristics to inanimate objects, animals, and others with a view to helping us rationalize their actions" (Duffy 2003). Artificial entities with an anthropomorphic appearance or behavior produce significantly more positive reactions than a purely functional approach. Thus, it became important or even necessary to enable an unreserved communication with a socially interacting machine (Riek et al. 2009; Breazeal 2004; Scassellati 2001).

Dautenhahn (1999) emphasizes: "Artificial social agents (robotic or software), which are supposed to interact with humans are most successfully designed by imitating life, i.e. making the agents mimic as closely as possible animals, in particular humans." Meanwhile, many differently motivated disciplines are engaged with the creation of more realistic anthropomorphic agents and figures. Technical enhancements in computer animations, construction of androids, forensics, facial

V. Schwind (✉)
Institute of Visualization and Interactive Systems (VIS), University of Stuttgart/Stuttgart Media University, Stuttgart, Germany
e-mail: valentin.schwind@vis.uni-stuttgart.de; schwindv@hdm-stuttgart.de

reconstructions, etc. enable human images, movements, and facial expressions that can hardly be distinguished from real human ones.

One might assume now that a more realistic, human-like image leads to more familiarity and thus to higher acceptance and improved emotional access. But a certain phenomenon ensures that observers of such realistic figures no longer accept an artificial representation. On the contrary, the representation will be rejected and thus a smooth communication or an emotional bonding is made impossible. This seemingly paradoxical phenomenon is known today as the "Uncanny Valley."

In 1970 Masahiro Mori, Professor of Engineering at Tokyo Institute of Technology, presented the hypothesis according to which extremely realistic, human-like robots or prostheses provoke negative emotional sensation (Mori 1970). Mori predicted that the more human-like a robot is, the more accepted it will be. This is true until a certain point of realism is exceeded, and then the acceptance suddenly drops. In Japanese Mori called this phenomenon "bukimi no tani." Today it is known in its English translation "Uncanny Valley." The translation goes back to the book *Robots: Facts, Fiction and Prediction* by Jasia Reichardt (1978) where the Japanese word "bukimi" is translated as "uncanny" (Pollick 2010).

This term probably refers to two German articles by Ernst Anton Jentsch and Sigmund Freud at the beginning of the twentieth century. In *About the Psychology of the Uncanny* (orig. *Zur Psychologie des Unheimlichen*) Jentsch pointed out that "doubts about the animation or non-animation of things are responsible for an eerie feeling" (Jentsch 1906). In a note he mentioned the ambiguity of automats and their psychological effect in E.T.A. Hoffmann's pieces. Sigmund Freud took up the article and criticized Jentsch, on the one hand, for not including among his examples the automat Olimpia from the short story *The Sandman* (Hoffman 2008) and, on the other hand, because of other motifs responsible for the uncanny effect of the narrative (including the Sandman himself, who tore out the eyes of children). Freud described the "uncanny" as something once familiar, which is then hidden in the subconscious and later recurs in an alienated shape. For example, he cited the motif of the doppelganger and "in the highest degree in relation to death and dead bodies, to the return of the dead, and to spirits and ghosts" (Freud 1919). Freud also linked invisible manifestations such as noise or imagination with eerie effects. Thus the uncanny had been associated in compound with artificial figures long before Mori suggested his hypothesis. Jentsch and Freud even explained the uncanny with examples of human-like representations.

The "valley" in Mori's terminology refers to the strong sloping curve at the end of the chart (Fig. 5.1). Mori sets human-similarity (the sum of all human characteristics) in relation to the affinity of a figure. Unfortunately at the earlier stage of the uncanny research, the term "affinity" was not used consistently. Other terms like "acceptance," "familiarity," "positive emotional response," or "likeability" were also used. While these terms may have slightly different meanings in other languages, the essential meaning remains the same and describes a feeling that can be either positive or negative.

The implications of the phenomenon are not only limited to robots, but circulated to a wide audience through critics of computer animations, movies, and video

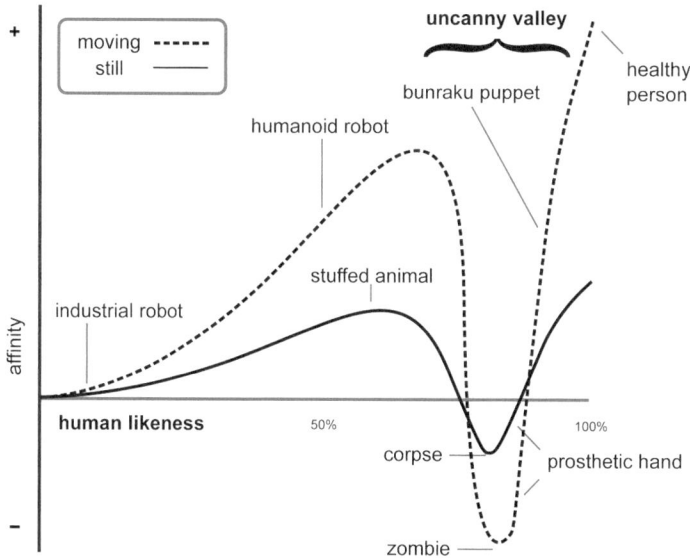

Fig. 5.1 Mori's illustration of the uncanny valley: This is a simplified diagram (MacDorman 2005b) of Mori's graph from the original article published in Energy Magazine (1970)

games. The film critic Kyle Buchanan wrote in the New York Magazine about *The Adventures of Tintin* (Spielberg 2011): "The biggest problem with the Tintin movie might be Tintin himself" (Buchanan 2011). In connection with the creation of Princess Fiona in the computer-animated movie *Shrek* (2001), Lucia Modesto (PDI/Dreamworks) made the following remark: "She was beginning to look too real, and the effect was getting distinctly unpleasant" (Weschler 2006). Andy Jones, Animation Director of the first computer-animated movie with human actors *Final Fantasy* (2001), commented on his work in a similar way: "As you push further and further, it begins to get grotesque. You start to feel like you're puppeteering a corpse" (Weschler 2006).

Mori's hypothesis predicts that only a 100 % real and healthy-looking human is fully accepted by an observer. Mori also conjectures possible reasons for this. Visible signs of disease have a negative impact on our feelings. The fear of death and subconscious protection against pathogenic infections are the supposed reasons of this rejection (Greenberg et al. 2000; Cohen et al. 2004; MacDorman 2005a). Functional magnetic resonance imaging (fMRI) scans show stronger metabolic responses in areas of brain that are responsible for predictions and are correlated to the negative response. If a figure does not move according to expectations of the observer, contradictions in the categorical perception seem to arise (Saygin and Ishiguro 2009; Cheetham et al. 2011). A study on monkeys shows that this phenomenon also occurs in primates (Steckenfinger and Ghazanfar 2009). This leads us to the conclusion that there is an evolutionary-related cognitive mechanism. Other factors such as attractiveness, familiarity, culture backgrounds, etc. probably also have measurable impact on this effect (Zebrowitz 2001; Hanson et al. 2005; Schneider and Yang 2007; Green et al. 2008).

Uncanny Valley research is a highly interdisciplinary field, and yet the subject is still fragmentarily examined. In sum, it is proclaimed with certainty that this phenomenon exists. Moreover, due to technical developments in robotics and computer animation sciences, this phenomenon has gained an increasing importance in the last four decades. However, it can be safely assumed that the Uncanny Valley has had an impact on humans long before its discovery.

This article explores the hitherto little-discussed historical emergence and contemporary impact of the Uncanny Valley and shows that it is strongly connected to intentions, aesthetics, and cultural aspects of artificial figures. Examples from the history of art, engineering, and literature demonstrate that the Uncanny Valley is an essential human feature that affects our behavior and our decisions.

Here we consider the Uncanny Valley from a new perspective, which turns it from a hypothetical, marginal issue to a relevant aspect of our culture and history with a not negligible impact on the socio-cultural development of mankind. It is herein assumed that Mori's hypothesis relates to human-like representations in general, not only to physical entities in the shape of robots or prosthetics, but also to visual arts and written narratives that have to be envisioned in front of the mind's eye. By incorporating these narratives, we will see and understand why today the image of artificial figures is negatively affected, especially in western parts of the world.

The first section will take a look at hints and stories from our past where artificial humans are specially linked to eeriness or attention. This journey begins with one of the earliest extant works of literary poetry: the Epic of Gilgamesh.

5.2 Androids in Ancient Cultures

The Sumerian poem written in the 3rd millennium B.C. about Gilgamesh, the king of Mesopotamia, tells a story of Enkidu, who was made of clay by the goddess Aruru to undermine the strict and extravagant reign of Gilgamesh. Enkidu is a primitive form of man who is close to nature and lives together with animals. He is discovered by a poacher, and by order of the king who already knows about him, people send Enkidu a woman. He alienates himself from nature and becomes socialized, undergoing the stages of human development (from nomads, to farmers, and shepherds), and finally meeting the king in the capital Uruk. After a fight they become friends and live through many adventures together. Gilgamesh's mother adopts Enkidu, so that the king and Enkidu become brothers.[1] After killing the Bull of Heaven (a mythical creature in several myths of ancient times), the gods punish Enkidu for his deeds with a deadly disease he later painfully succumbs to. Enkidu disintegrates into dust and Gilgamesh is in deep mourning for a long time. After Enkidu's death Gilgamesh recognizes that he is mortal, too. Thereafter, the search for eternal life becomes the leitmotif of the *Epic of Gilgamesh* (Maul 2012).

---

[1] In the Sumerian version of the poem, Enkidu remains a slave and servant of Gilgamesh.

Enkidu is created by craftsmanship and divine magic (similar to many other creation myths). Fascinating is the character's helplessness in collision with the environment, which in the end brings him to a tragically fatal end. He becomes an instrument in the hands of gods who want to give Gilgamesh a lesson in humility. In the end, Enkidu pays with his life for the withdrawal from nature and for becoming an unscrupulous hunter. This story has no direct evidence that would speak to a special eerie effect of Enkidu's role. However, this figure causes particular fascination: He is created artificially, is an outsider, and he makes humans aware of their mortality. Later, we will see that these are the characteristics typical for other artificial figures. Another interesting issue about the Gilgamesh story is that Enkidu's death is caused by his creator. For many other artificial figures later – especially those with personality and intelligence – this conflict of existence is automatically preprogrammed. We will see that this is also an important part of the Uncanny Valley phenomenon.

Also in Greek mythology, the creation of artificial life using magic and clay plays an important role. As the Titans Prometheus and his brother Epimetheus take a walk, they see the divine potential of the earth and so they form animals and man from clay. Every animal gets a special talent, and humans get every special quality from all the animals. The goddess Athena, a friend of Prometheus, recognizes the potential of his work and gives people sense and reason as a special gift. Prometheus is so proud of his creation that he becomes a patron and teacher to humankind. But other gods, led by the king of the gods, Zeus, are against an emancipated species and demand sacrifices and worship of the people. Prometheus turns against the gods and brings divine fire to the people. The punishment by Zeus follows immediately: he gives an order to his son Hephaistos, god of craftsmanship, to create an artificial woman: Pandora. She is blessed with all kinds of gifts, she is seductive and beautiful. She possesses a vessel with a disastrous content that shall instigate sorrow and death over mankind. Hermes brings Pandora to Epimetheus who succumbs to her magic and opens the vessel[2] although his brother had warned him not to accept any presents. Since that day mankind has been struck by illnesses, disasters, and sorrows again and again. Furthermore, Zeus orders Hephaistos to bind Prometheus to a rock in the Caucasus Mountains. An eagle pecks the immortal in the liver day after day. Prometheus is freed several millennia later by Hephaistos. Meanwhile, the cycle of creation and annihilation of mankind repeats several times until the children of Prometheus and Epimetheus, Pyrrha and Themis, create people from stone – "a hard race and able to work" (Schwab 2011, 28).

The ancient Greek poets and philosophers Aischylos, Hesiod, Platon, as well as the Roman Ovid provided different versions of the legend of Prometheus. The myth about the creation of the human species tells about the rebellion against the divine order and about the attempt to develop a self-determined culture. This theme appears most fascinating and encouraging to many artists. Johann Wolfgang von Goethe

---

[2] The vessel is usually known as the "Pandora's box". The term has come about through a translation error. The Greek word *pithos* originally referred to a big amphora used for water, wine, oil, or grain.

used this theme in his famous poem *Prometheus* in the time of Storm and Stress (Goethe 2013; Braemer 1959). "In the Age of Enlightenment, the poem acts as a firebrand – a well-articulated contempt for all inherited or self-proclaimed authorities" (Gassen and Minol 2012, 70). As we have established earlier the idea that artificial characters would/can fight for freedom and self-determination appears uncomfortable to most of us. Perhaps the ancient Greek gods on Mount Olympus had a similar uncomfortable feeling. The motif of a rebellion of self-determined artificial species produced in series and also suitable for work was taken up again only in the twentieth century – when it seemed technically possible to build an entire class of robots. In earlier times artificial figures were very rare or they were individual products with a special status. Pandora is a remarkable example for the first negatively associated artificial figure in history: she embodies our fear of manipulation and disastrous intention, in this case covered by the seduction abilities she embodies as a woman. Despite different traditions, we still have an accurate picture of Pandora: She is seductive and equipped with many gifts such as beauty, musical talent, curiosity and exuberance. Aphrodite also gives her gracious charm, Athene adorns her with flowers and Hermes gives her a charming language (Hesiod 1914). Both fascination for her beauty and fear of her gift remain vivid today and represent probably the first manifestation of the strange effect elicited by an artificially created figure.

Beauty is a quality frequently mentioned in relation to artificial figures. One example is the fascinating poem by in Ovid's *Metamorphoses* about *Pygmalion*, a Cypriot sculptor (Anderson 1972). After having had bad experiences with sexually licentious women, he withdraws into confinement and carves an ivory statue in secret. Pygmalion falls in love with the realistic and life-sized statue and treats her as if she were a real woman. He cares for her: he dresses, adorns and fondles the figure in a loving way. At a celebration in honor of the goddess Venus (the Greek Aphrodite) Pygmalion asks in a prayer that gods give him a real wife who looks like his statue. Venus fulfills his wish: when he comes back home and kisses the statue, the ivory becomes warm, soft, and alive. The figure awakens to life and becomes a living female human.

The *Pygmalion* story is one of the most popular poems in Ovid's *Metamorphoses* (1 B.C.–8 A.D.). Without resorting to any dramatic twists Ovid tells about three fascinating motifs for the creation of an artificial figure: aesthetics, loneliness, and love. For the first time, material gets a special meaning: the noble and organic ivory underlines the natural and aesthetic claim of the figure. Through the centuries, the Pygmalion motif has been innumerably transferred and reinterpreted by poets, painters (Fig. 5.2), and musicians. Until 1762, no text mentions the name of the statue. In a very influential work by Jean-Jacques Rousseau *Pygmalion, scène lyrique* dated in 1762, the sculptor swears eternal fidelity to the statue that was first named "Galatea" which means "Milk White" (Rousseau et al. 1997). We should also mention the operetta *The Beautiful Galatea* by Frank Suppé where the statue transforms from a virgin to a psychotic nymphomaniac until the goddess Venus transforms Galatea back into a statue (Dinter 1979).

5   Historical, Cultural, and Aesthetic Aspects of the Uncanny Valley

Fig. 5.2 Pygmalion and Galatea (ca. 1890), Oil on canvas by Jean-Léon Gérôme (French, Vésoul 1824–1904 Paris), The Metropolitan Museum of Art, New York

The term Pygmalion effect (or Rosenthal effect) becomes established later in psychology as the outcome of a self-fulfilling prophecy (Rosenthal and Jacobson 1968). For example, anticipated positive assessments by a teacher ("this student is highly gifted") are subconsciously transmitted and confirmed by increased attention. Furthermore, pygmalionism describes sexual affection towards human representations in the form of statues, paintings, and dolls, which can also serve as a fetish. Life-size human replicas are currently produced commercially as sex dolls (e.g. "Real Dolls" made of silicone). Although these dolls are very realistic, they can be nevertheless disquieting to people without pygmalionism (cf. Valverde 2012). The attraction of artificial figures dominates the effect of the Uncanny Valley (cf. Hanson 2005; Hanson 2006). The intense bond can also be formed by visual and haptic contact, which is frequently the case of lonely men (Holt 2007). The habituation also could be an explanation when "[…] the stimuli continue for a long period without unfavorable results" (Thorpe 1944). The British TV documentary series *The Secret of the Living Dolls* (Channel4 2014) shows how frightening the living dolls can be for the viewer. It shows people who live as dolls in whole-body dresses made of silicone. Critics and the audience describe the documentary as extremely disturbing, creepy, and scary (Bouc 2014; Michaels 2014; Styles 2014; Westbrook 2014). Maybe Pygmalion foresaw this effect when he asked Venus "shyly" for a real wife *like* the ivory virgin, and not for a living statue. Venus transformed the statue into a real human made of flesh and blood – and this was a divine way out of the Uncanny Valley.

5.3 Androids in Non-Western Cultures

A series of books[3] dated around 350 B.C. are attributed to the Daoist philosopher Lieh Tzu. In *Book V – The questions of Tang* a story is told about *The Automat* of the engineer Ning Schï, who is presented with his human-like figure to King Mu of Chou at his travelling court. First the king does not understand what the inventor wants to show him because he considers Ning Schï's construction to be an ordinary person. The machine can sing, dance and do various tricks. But when the automat makes advances to the concubines, the king can't bear it and wants to execute him immediately. The engineer disassembles the machine to demonstrate that it was only composed "of leather, wood, glue, paint, from white, black, red and blue parts" (Lieh-Tzu and Wilhelm 1980). The king orders Ning Schï to reassemble the machine again, examines the mechanism and recognizes its various functions. "For a sample, he removed the heart and the mouth could not speak anymore; he removed the liver and the eyes could not see anymore; he removed the kidneys and the feet could not walk anymore" (Lieh-Tzu and Wilhelm 1980, 113–114). The king is extremely impressed and wonders, "how man can reach the works of the creator." The king takes the machine into his wagon and drives back to his home. At the end of the story, two masters of engineering, who thought they had already reached the limits of the humanly possible, are so impressed by the story about Ning Schï's machine that they are afraid to speak ever again about their craft as an "art."

A second story in the Chinese *Tripitaka* (a collection of educational writings of Buddha) probably has the same origin: the five sons of the king Ta-tch'ouan have different talents. The first son is clever, the second son is inventive, the third one is handsome, the fourth one is vigorous, and the fifth one is always very lucky. The sons decide to travel to various kingdoms to find out which of their "extraordinary virtues is the most outstanding" (Völker 1994, 73). The inventive brother goes to a foreign kingdom and manufactures a mechanical man out of 360 parts (mostly wooden). He gets a lot of respect for his work and is blessed with gifts. The machine sings, dances, and acts. When the king of the land and his wife hear about that craft, they go up to a tower and have a look at the wooden man. They are both are very amused but do not see a wooden man until the actor winks upward to the queen. The king orders to cut off the head of the man. But the inventor, who is the "father" of the figure, cries and says how much he loves his "son." He holds himself responsible for his son's poor education, begs for mercy, and finally switches off the machine. The inventor pulls a pin and the mechanical man breaks into its components. The

---

[3] There are few English translations. The most well-known summaries of Lieh Tzu's stories about automats can be found in "Science and Civilisation in China: Volume 2, History of Scientific Thought" (Needham and Wang 1956, 53 ff) and "The Book of Lieh-tzŭ" (Giles 1925; Graham 1990). A comprehensive translation of Lieh Tzu texts was derived from the German sinologist Richard Wilhelm. This summary refers to his original German translation from "Das wahre Buch vom quellenden Urgrund" (Lieh-Tzu and Wilhelm 1980, 113 ff).

king wonders how he could be fooled by an artificial man. He claims that the inventor has a gift "which is unequaled in the world" and gives him tons of gold. The inventor returns home and distributes the gold among his brothers. In a song he praises his work and boasts: "Who is able to surpass me?"

Both examples show that stories about the creation of artificial figures are not limited to Western culture. Simply the idea that there might have been an artificial mechanical man in China 2,300 years ago is fascinating. The story of Ning Schï's machine and the wooden man of the son King Ta-tch'ouan are inspired by the enthusiasm that such high art of engineering evokes. The creation of an artificial human being is regarded as the greatest gift ever. King Mu was explicitly interested in functionality of the apparatus and kept the machine unceremoniously for himself. Technical scholars also show reverence for the difficulties of the building process of such complex machines and redefined the craft, which they previously regarded as art. Today we cannot know for sure whether such machines truly existed in ancient China and whether these figures were not distinguishable from a real human. But if people really had never seen anything like that before, they had to assume that it was a clad or painted man – and regarded it to be an actor – like it happens in the second story.

Historical reports of amazing apparatuses that have been used for entertainment came not only from China but also from Arabia. The fall of the Roman Empire constitutes at the same time the end of an epoch of many literary and technical achievements. In the beginning of the ninth Century, the Caliph of Baghdad initiated preservation and translation of ancient writings (*Graeco-Arabica*) and thus made an invaluable contribution to the conservation of Greek science and philosophy, which was also characterized by fascination and interest in anthropomorphic machines. In the twelfth century the Arabian engineer and author Al-Jazarī continued antiquity research in his *Book of Knowledge of Ingenious Mechanical Devices* (in particular he uses the knowledge of the pressure and suction of water by Heron of Alexandria) and creates a detailed manual for the construction of such machines (Al-Hassan 1977). Many clocks, fountains, doors, locks, etc. that are preserved until nowadays prove to be functional. Among these machines are barkeeper dolls and a machine with four mechanical figures sitting in a boat singing and playing musical instruments. Figures are mostly painted and made of mounted copper, wood and possibly of papier-mâché (Hill 1996, 208). Some of Al-Jazarī's figures could move their heads, arms, and legs. There exist different interpretations of the effect these figures produced: on the one hand, automated mechanisms apparently delayed the movement of the figures, on the other hand "subtle caprices" of the characters resulted from these delays may well have been intentional (Nadarajan 2007). Obviously some of the figures served for amusement and entertainment. The decorations, the intricate design manuals, as well as the high number of contemporary translations in the Arabian region indicate a high popularity of such machines and the fascination they evoked. The book by Al-Jazarī represents only one of the highlights of the epoch of technical achievements and inventions in the Arabian world.

5.4 Demons and Automata in the Middle Ages and Renaissance

Stories of human-like figures in the European Middle Ages are characterized predominantly by the Christian and Jewish faith. There are traditions of the golem, legends of mandrake roots with human-like forms, and alchemical instructions for creating a homunculus. Well-known are several myths about the *Prague Golem*, a mute Jewish legendary figure who grazed through the cities before the Jewish festival of Passover. A note on Golem's forehead or in his mouth brought him to life and kept Golem under control. According to a legend, once Rabbi Löw forgot to remove the note so the Golem was able to walk through the city without any control. In one legend, it was possible to tear the note so that the golem crumbled into a thousand pieces (Völker 1994). The medieval reports have some things in common: artificial figures are created under mystical or demonic influence. These figures have little to do with divine creation and have less aesthetic appeal than ancient Greek statues or the elaborately painted machines of medieval Arabia. They were not used for entertainment or amusement but often had a repulsive effect. In legends the misshapen Golem does not have the ability to speak or to develop a free will. According to Hildegard von Bingen (von Bingen and Throop 1998), the devil lives in the mandrake root, which has to be exorcised by means of spring water. And in most traditions the homunculus is only about the size of a fetus and could be bred on an organic substance like blood, flesh, excrement, sperm, or urine (in the epilogue of Völker 1994).

During the European Middle Ages the concept of artificial life is associated with demonic powers and negative response to the efforts of alchemists (Newman 2005; cf. LaGrandeur 2010). According to a legend, scholar Albertus Magnus "constructed a door guard of metal, wood, wax and leather" (Strandh 1992, 175). The guard welcomed visitors with the Latin "Salve!" and asked for the reason of their visit before they were allowed to step in. One day Magnus' pupil, the young Thomas Aquinas, smashed the door guard angrily into pieces. There are different specifications about the possible reasons: some say he was so scared of the android that he had smashed him with a stick. Others say he did not want to listen any longer to the "chatter" of the guard (Völker 1994, 113). Anyway, Magnus was very upset that Thomas had destroyed the work "of 30 years." Thomas Aquinas wrote years later that a soul is a prerequisite for any proper motion and demons are responsible when "necromancers make statues speak, move and do other things alike." (Aquinas 2013, Q. 115).

Both demons and the undead are myths that have existed since before the Middle Ages (e.g. the Lamia from Greek mythology), and these myths are spread beyond Europe (e.g. Asanbosam from West Africa or Jiang Shi from China). However, there is no evidence that all these stories are due to a ubiquitous fear of death or to the Uncanny Valley (cf. MacDorman 2005a). The first extant reports on demons in the shape of living corpses, known as vampires today, come from Southeast European countries (Lecouteux 1999). The peculiarity of vampire stories is how the appearance of the living dead is reported: pale skin, unnatural eyes – these are a few but visible abnormalities which distinguish undead from ordinary people. This allows to establish a link between these stories and the elements that seem responsible for the experiences eeriness of objects that fall in the Uncanny Valley: the reasons why nowadays artificial figures are assigned to the Uncanny Valley resemble those we find in play in stories of vampirism. The combination of visible signs of pathogenic diseases and ambiguity between life and death prompt negative emotions such as the eerie feeling of the Uncanny Valley. It is unlikely that the living dead have harmed someone, but it is quite possible that conscious or subconscious fear of death inspired people to invent stories, myths, or figures. Such fears can function as a warning of potential threats or socially harmful behavior in times when the mortality rate was very high. In times of prosperity and security, these fears may appear superfluous and only fascinate us, especially if we are affected by forbidden stimuli.

Between 1495 and 1497 Leonardo da Vinci presumably built a functional robot which had a complicated mechanism hidden under a knight's armor. Pulley blocks and cogwheels driven by hydropower put the robot's arms in motion (Rosheim 1997). Da Vinci's construction plans have survived until today and display a strong influence from the Arabian entertainment automats (Strandh 1992). Greek and Arabian works on mechanical devices were revised and translated in Europe only at the end of the fifteenth to the beginning of sixteenth century. The extensive and well-known work in Europe of this period is *Les Raisons des forces mouvantes* by the Frenchman Salomon de Caus. De Caus illustrated many constructional ideas; moreover, he took up plans from Greek and Arabic engineers and built elaborate machines himself. Many designs were exhibited with great success in Paris and at the Heidelberg Castle. In the palace of the Duke of Burgundy in Saint-Germain, de Caus built a system with a total of 256 artificial figures or machines driven by the power of water. Particularly popular motifs were wheel-driven animated scenes from Greek mythology (Strandh 1992). Such machines became fashionable at courts and in the big cities of Europe. Caus' constructions became very popular and were often copied as props for theatre performances. During the Renaissance, machines were socially acceptable and were popular elements of garden and park ensembles.

In the middle of the sixteenth century an Italian clockmaker and engineer Juanelo Turriano developed special virtuosity in construction of machines. He became famous for his water lifting device in Toledo. According to legend, Turriano built an artificial figure that even went shopping for him (Strandh 1992). This story was very persistent and gave the street where Turriano lived its present name: *Calle del Hombre del Palo* (the avenue of the wooden man). In his time, he had to defend himself because of the accusations of an abbot who was convinced that Turriano was in league with the devil. However, the design of machines developed further without ceasing till the end of the sixteenth century, and machine builders were competing in the production of increasingly sophisticated and more and more spectacular and entertaining figures. We have to mention here, of course, the automata of Vaucanson,[4] which made their creator very rich, as well as the clockmaker family Jaquet-Droz, who were the first to develop interchangeable program rollers for their figures. The *Three Musicians* by the Jaquet-Droz family are in good working condition and are displayed in the Museum of Neuchatel in Switzerland.

At the beginning of the seventeenth century, human-like machines were very common and were treated like modern day pop stars. Philosophers, doctors, and anthropologists started to be interested in such constructions. In 1637, French philosopher René Descartes predicted that people would eventually be able to develop a soulless machine that would look and behave like an animal. He compared a mechanical pumping process with the blood circulation of animals and also drew comparisons with humans. A legend tells that Descartes built an eponymous android child as a replacement for his illegitimate daughter Francine. In 1649 when Descartes was invited to the court of Christina of Sweden he took Francine with him. On his journey from Amsterdam to Stockholm, the suitcase with the android drew the attention of the superstitious sailors. When the ship was caught in a storm in the North Sea, Descartes was accused of standing in the league with the devil and was made responsible for the storm. The captain ordered that the android be thrown overboard (Strandh 1992).

Another machine that attracted a lot of attention[5] a few years later was the "Chess-playing Turk" dating back to the 1760s. This mechanism, built by Hungarian Baron Wolfgang von Kempelen, may be considered to be the first machine of uncanny intelligence. A figure of a male android in a Turkish costume would sit in front of a box with a chess board mounted on it. The machine was designed so that when a chess move was made, the android responded with a move of a chess figures itself. Some claim the construction was able to say the word "Chess!" and "Gardez!" (Strandh 1992; Völker 1994). Von Kempelen claimed that he succeeded in developing artificial intelligence equal to the chess-playing abilities of humans. From 1783 to 1784 he traveled with his machine through Europe and let the "Chess-playing Turk" compete with well-known chess players (Fig. 5.3).

Von Kempelen stopped showing the machine in 1785 after Frederick the Great had offered him a large sum of money for unveiling the secret and had apparently been quite disappointed by the solution of the puzzle. After von Kempelen's death the Turk fell into oblivion for a few years until the German inventor Johann Mälzel purchased the machine and demonstrated it again in Europe and USA as "The Automaton Chess Player." The Turk won many games during those demonstrations and defeated some famous statesmen such as Napoleon and Benjamin Franklin. At a presentation in London, the English poet and author Edgar Allen Poe observed the machine closely. In his famous essay *Maelzel's chess player* he made it clear that the machine must have been a swindle and thus exposed the fraudulent automaton.

---

[4] Vaucanson's *Canard Digérateur* (Digesting Duck) simulated a metabolism by eating and defecating kernels of grain.

[5] E.T.A. Hoffmann's story "The Automata" took von Kempelen's Chess Turks as a model. Contemporary literary critics say that Hoffmann's fascination with the Turk does not affect the reader (Gendolla 1992).

Fig. 5.3 The Chess-playing Turk (1789): From book that tried to explain the illusions behind the Kempelen chess playing automaton (The Turk) after making reconstructions of the device. Author: Joseph Racknitz (Source: University Library, Humboldt University of Berlin, 3639 v.)

Apparently both Mälzel and von Kempelen used children or midgets sitting in a small box inside the machine.

Machines in the Renaissance and Enlightenment reached a considerably higher level of acceptance than the machines designed in previous centuries. The machines were built for entertainment and, since they acted autonomously and performed tricks in front of the eyes of spectators, they were a big attraction for the audience. Presumably, this fascination was particularly triggered by curiosity when the audience – as in the case of the Chess Turk – tried to find out how figures worked.

However, society as well as religion did not allow crossing a certain border: an artificial figure could not be designed as one-to-one copy of a human being – and in turn a human was not allowed to be presented as a machine. That became particularly evident in the case of French physician and materialist Julien Offray de la Mettrie who was inspired by the theories of Descartes and who described a man as a kind of machine for the first time in 1745 (Mettrie 1990). Mettrie was an atheist and polarized the world with his theories of a demystified human existence, as he not only explained complex bodily functions with mechanical processes but also saw the soul as a result of physical development. He opposed Descartes' dualism of mind and matter. La Mettrie was persecuted by the clergy and his works were heavily censored. Enlightenment poets and philosophers such as Friedrich Schiller, Denis Diderot, and Jean-Jacques Rousseau set themselves against La Mettrie instead of giving him their support (Jauch 1998). Frederick the Great took La Mettrie as his "court atheist", but Voltaire, who also worked at court, ridiculed the physician (Jordan 1910). La Mettrie retired and fell into oblivion until the nineteenth century. The developments in this period show that automation was widespread and accepted – especially in entertainment. The controversy around the construction of artificial figures was more intense the more human-like a machine looked. The idea that a human being can be seen as a machine or can be reproduced as a machine provoked extremely strong rejections, even in times of Enlightenment and in a world with many automats. This era makes it particularly clear that both enormous technical progress and the fear of artificial humans tend to coexist.

5.5 Uncanny Creatures in Nineteenth and Early Twentieth Century Literature

The uncanny effect produced by artificial characters – particularly the idea that they could look human or threaten humans by their very existence – is often used in stories of the dark romanticism. Seen from the perspective of literary science, E.T.A. Hoffmann's *The Sandman*, published in 1815, is a "special discipline of representatives of all methodological directions" (in the epilogue of Hoffmann 2008) of its own – not only in the uncanny research. Jentsch and Freud refer to this Hofmann's piece for their explanations of eerie feelings.

Nathanael, a mentally disturbed student from a good home, falls in love with the beautiful "daughter" of the physicist Spalanzani and observes her through the window of the house opposite. To see her more closely, Nathanael uses a spotting scope, which he had gotten from the glass dealer Coppelius. Already in the beginning of the story, Nathanael suspects the glass dealer to be the Sandman – an eye-stealing nightmare figure he knows from his childhood. Although the daughter does not move and sits motionless at the window, through the telescope Nathanael sees the doll as a living being. At a celebration Spalanzani presents Olimipia to the public and the guests realize that she is a mechanical and therefore artificial wooden doll. Although she is able to play the piano and is part of the entertainment, she looks as a fake with a stiff expression that makes some guests feel uncomfortable. Nathanael does not recognize the fraud because he is so much in love with the android and wants to make her a proposal of marriage. Olimpia's beauty outshines her imperfection in behavior. Spalanzani and Coppelius fight over the doll and its glass eyes fall out and scatter on the floor. The torn-out eyes remind Nathanael of Sandman. Succumbed to madness, he attempts to kill Spalanzani. Nathanael is detained and transferred to an asylum.

As Freud noted, originally Nathanael's fears are not implicitly attributed to the doll but to his terrible fear of the Sandman. The disturbing effect of his behavior is tremendously enhanced by his love for the doll, which he considers as a real woman, especially by viewing her through the enlarging lens. In Hoffmann, the use of such a perspective seems to be a possible way to avoid the Uncanny Valley. The motif of a broken glass eye and the subconscious fear of castration due the loss of the eyesight enhance the eerie effect (Freud 1919). The real horror for Nathanael begins when he realizes that Olimpia is not a real human. Similar to Pygmalion's statue, Olimpia is a product of male fantasy, which was designed according to the wishes of a man and is the object of a man's desire. Passionate love makes both Nathanael and Pygmalion blind for the fact that their figures are not real humans. Hoffmann deliberately uses the uncanny effect produced by an artificial figure to create the atmosphere of horror and fear in his story. This method is taken up again in today's science fiction and horror stories and to the present day it remains a popular method to trigger deep-rooted fears. Particularly interesting in that case is Hoffmann's personal affinity for machines. Machines fascinated him: "once when the time will be, for the benefit of all sensible people I see with me, I will make an automaton" (Heckmann 1982). As Nathanael's character proves, Hoffmann's highly pronounced and contemporary fascination for artificial human-like figures shows the frightening ambiguity when an "idea of an imitation of man by machine turns to a vision of horror" (Müller 1989).

Today's most famous artificial figure in world literature appeared 3 years later in one of the first science-fiction novels ever (Freedman 2000; Murray 2002). Mary Shelley's *Frankenstein, or the Modern Prometheus* written in 1818 tells the story of Victor Frankenstein, who is obsessed by the idea of the creation of an artificial human. Frankenstein works sloppily and compiles his figure out of the body parts of criminals and material from the slaughterhouse. He reanimates the body with electricity and creates a 3-m tall monster that is ugly and scary. Because of shame and fear Frankenstein keeps his creature secret, and this causes numerous serious problems. The beast sees himself as a victim and asks Frankenstein to create him a woman with whom he wants to escape from civilization. But Frankenstein fears that together with his wife the monster could kill even more people and be a danger to future generations. Frankenstein destroys the almost completed figure of the monster's wife. The monster takes revenge by killing Frankenstein's bride Elisabeth and flees. Victor wants to hunt him down and follows the creature up to the Arctic. During his travel, Victor becomes seriously ill and dies. The creature returns to him and commits suicide when it becomes conscious of its poor deeds (Shelley 1818).

Warning against too much enthusiasm and the irresponsible use of modern technology is a feature that characterizes the period of gothic novels at the beginning of the nineteenth century, when the topic of artificial figures was particularly important. The general eerie effect produced by the gruesome story is reinforced because the nameless monster is composed of corpses. The image of a monster is being deliberately exploited in Frankenstein not only to initiate suspense and horror but also to highlight the dangers of a human trying to take over the role of God. Such eerie stories grew popular in that period along with the widespread machine manufacturing. We can assume that highly realistic copies of human-like figures or reports about them not only triggered an eerie and disturbing effect, but also served as an indication of the potential dangers and as advice to deal carefully with the technical heritage of the antiquity. A novel aspect about Frankenstein's monster is its autonomous, uncontrolled behavior, which illustrates the powerlessness of creator towards his creation (Kormann 2006). The image of Shelley's monster had a strong influence not only on the literature of dark romanticism, but also on many generations thereafter. Until the twenty-first century it remains a highly controversial topic. Shelley's and Hoffmann's figures were templates and role models for numerous theatre adaptations, plays, and films and were deliberately used to express our deep-rooted fear of human-like artificial figures and to issue warnings against excessive technological advances and irresponsible actions of science.

Between the nineteenth century and the early twentieth century, there was a further intermezzo of particularly complex human-like machines produced by known magicians. The machines could demonstrate various magic tricks. Famous designers of that period were the founder of modern magic Jean Eugène Robert-Houdin, French magician Stèvenard, Jacques-Henri Rodolphe, Jean David, and brothers Maillardet. The figures of this period had a strong appeal not only on stage. During the period of industrialization, especially in Paris, a small industry of automatic machines emerged, so enthusiasts and collectors could purchase artificial figures. With the outbreak of World War I the industry came to an abrupt end and the era of artistically designed, complex, and human-like amusement machines was over.

After the First World War, the role and effect of artificial figures particularly depended on whether the country was a winner or a loser in the war. Victory led to euphoria and an optimistic approach to technological progress, whereas the defeated countries were generally very skeptical about it. The Czech author Karel Čapek connected this fear with the vision of a collective of artificial entities. Influenced by the subject of the *Prague Golem* from Jewish mysticism, he addressed the use and danger of artificial figures to warn of a further World War in his play *Rossumovi Univerzální Roboti* (*R.U.R.*), whose premiere was in 1921. The play centers on a company that manufactures robots to be used as a cheap workforce. The influential utopian drama describes social and global economic consequences of the widespread usage of robots. The robots finally rebel against oppression and extinguish mankind in a terrible war. The play was a major success worldwide and was translated into almost thirty languages (Koreis 2008).

The remarkable aspect about Čapek's play is that artificial, human-like figures are used not for amusement but for hard work, which people are not willing to do any longer. On the other hand, machines are not artistically-crafted, individual productions anymore, but uniform entities of serial mass production. The use of robots has an enormous social and economic impact on the world, and the rebellion of the machines even leads to the end of mankind. In his play Čapek warns against the power of political concerns, dictatorships, and artificial intelligence. Karel's brother Josef, painter and also a writer, coined the title of the play. The Czech word "robota" stands for forced labor and at the same time is a synonym for a human-like artificial apparatus, which should ease the people's cumbersome work. The fate of the two politically engaged brothers is tragic: Karel died as a result of a hunger strike, as he demonstrated against the Munich Agreement in which the Allies decided to surrender the Czech Republic to Germany. His brother Josef kept on demonstrating against the seizure of power by the Nazis through numerous performances with the play *R.U.R.* (Thiele 1988). He was murdered in the concentration camp Bergen-Belsen in 1939.

5.6 Androids in Animated Movies and Films

In 1927 Fritz Lang made the ambiguity of artificially created robots the topic of his expressionist silent movie *Metropolis*. In the eponymous city there are two societies: the upper class living in luxury and the working underclass in the lower parts of the city. The city is administrated by the sole ruler Joh Fredersen, whose son Freder falls in love with Maria, a worker woman and preacher who lives in the lower part of the city. Frederson wants to suppress the rebellion at the early stage by replacing Maria with a machine-man that the scientist Rotwang has constructed. Fredersen compels Rotwang to make the robot look like Maria. But Rotwang, who is driven by revenge, reprograms the machine so that it incites the workers to rebel against the authorities. The rebellion succeeds, and the crowd rushes in a nerve-wracking chase through the city, running into the real Maria, who tries to appease the mob. The workers accuse her of being a "witch" and drag Maria through the streets. Meanwhile, the machine-man is thrown on the pyre and the human-like shell of the machine burns. As the metal is exposed, workers recognize the fraud and that Rotwang misused them, pursuing his own purposes. They chase Rotwang and Fredersen to the roof of a cathedral, where they start fighting each other. After Rotwang falls down, Maria mediates between the workers and Frederson and restores peace (Fig. 5.4).

Brigitte Helm played the part and embodied both figures in an eerie double role: helpful and benign Maria and the sexually unbridled machine-man that is the personification of Rotwang's sinful plans to manipulate the frustrated workers. It was the first time that a robot embodied the uncanny doppelganger motif in a film, which was described by Freud as fear-inducing in "the highest degree" (Freud 1919). However, in America the reviews of *Metropolis* were devastating. H. G. Wells wrote in the New York Times on April 17, 1927: "I have recently seen the silliest film. (…) It gives in one eddying concentration almost every possible foolishness, cliché, platitude, and muddlement about mechanical progress and progress in general served up with a sauce of sentimentality that is all its own" (Wells 1927). Also other contemporary reviewers panned the film. Today, these criticisms can be regarded as erroneous assessments of a strong technological affinity in general and of social developments during the Weimar Period in Germany. However, both *Metropolis* and

Fig. 5.4 Brigitte Helm in Metropolis (Lang 1927), Remastered, Creative Commons license: CC0 1.0 Universal, Public Domain Dedication, online available at: archive.org

R.U.R. made clear for the first time that machines may replace humans one day due to superiority in intelligence, strength or in mere number.

After World War II, computers and their intelligence had a strong influence on our view of artificial characters. It had been already been established that androids would need an extremely high intelligence and a huge amount of computing power in order to interact with their environment. For emotions, however, it was completely unclear whether could be calculated by machinery. One of the first figures focusing on this distinction is the Tin Man from Lyman Frank Baum's fairy tale *The Wonderful Wizard of Oz* (Baum 2008). The Tin Man, part of fellowship in the story about Dorothy Gale, wants a heart to be able to feel emotions. Since then, science fiction writers and film directors have been making use of the emotionlessness of artificial characters or intelligences to produce drama or weirdness. For example, in *Do Androids Dream of Electric Sheep?* by Philip K. Dick (1968) a group of androids which can hardly be distinguished from humans go out of control. These androids are unable to simulate emotional reactions what can only be determined with a complex detector – the "Voigt-Kampff machine." The science fiction thriller *Blade Runner* directed by Ridley Scott (1982) is based on Philip K. Dick's short story and is a paragon of how to deal with sinister figures. Here, the androids are called replicants.

Whereas the android/replicant Rachael is unaware of its artificiality and appears helpless and pitying, other androids are well aware of their superior skills and use them, thus posing a threat to humans. They are chased by the Blade Runner – a bounty hunter. In test screening the audience was unsatisfied with the end of the film, and producers insisted on changing it to a "happy ending." The director's cut (Scott 1992), however, indicates that the Blade Runner himself is a replicant and has to flee together with Rachael. The attitude toward artificial protagonists changed quite similarly in the science fiction series *Star Trek: The Original Series* (1966–1969) and *Star Trek: The Next Generation* (1987–1994) by Gene Roddenberry. Whereas in the original series, artificial intelligences or figures were mostly just eerie, highly intelligent series antagonists, in the following series the emotionless android Data is a full member of the crew. The search for humanity and emotions of Data has become a leitmotif of the series and the subsequent movies.

Horror films have also made use of the terrifying effect of human-like figures. Film critic Steve Rose writes: "(…) but film-makers have known about it long before it had a name. It's what makes many horror movies tick. Zombies are archetypal monsters from the bottom of the uncanny valley, with their dead eyes and expressionless faces. Likewise the glazed-over doppelgangers in Invasion of The Bodysnatchers or the robotic Stepford Wives, not to mention the legions of dolls, dummies, puppets, waxwork figures and clowns that have struck terror in the hearts of horror fans, from the ventriloquist's dummy in Dead of Night to Chucky in Child's Play" (Rose 2011).

The Scandinavian series *Real Humans* (Baron et al. 2012) shows that only a subtle adjustment is needed to make real actors look like eerie robots. The author of the series Lars Lundström explains in an interview how the special gesture play of the "Hubots" (human household robots) developed: "For this we actually needed a long time. (…) Finally we consult a mime actor to learn how to decompose movements and recompose them liquidly again. Then we thought how we could avoid all the small human gestures: no blinks of the eyes, no scratching or touching of the own face, an upright posture. Basically, the actor had to act normal, but in an abnormal way. The Hubots act like humans, but you can see that there are no real people. They are somewhat like bad actors: you can exactly see what they are trying to do" (Hurard 2013).

The uncanny effect of artificial figures can be specifically transmitted through the actors' craft, but also occurs unintentionally. This happens with obviously artificial protagonists with whom an emotional connection to the audience should be made. In this case the Uncanny Valley phenomenon can be held responsible if there is no emotional bond of the target audience with artificial actors. According to Misselhorn (2009) this is due to the fact that movies require a kind of "imaginative perception", i.e., the spectator only imagines perceiving something, but does not really perceive it. In animated characters that fall into the uncanny valley and cause a feeling of eeriness, imaginative perception gets in conflict with real perception. The Uncanny Valley is often mentioned as a reason why films like *Final Fantasy* (Sakaguchi and Sakakibara 2001), *The Polar Express* (Zemeckis 2004), or *The Adventures of Tintin* (Spielberg 2011) are criticized and have not achieved box office success. The Disney production *Mars Needs Moms* (S. Wells 2011) even counts to as one of the biggest flops in film history (Barnes 2011). Studies on uncanny research also use computer-generated images to investigate the eerie effect (MacDorman et al. 2009; Steckenfinger and Ghazanfar 2009; Cheetham et al. 2011; Tinwell et al. 2011; Cheetham et al. 2013).

A particular feature of artificial figures in the twentieth century is that their intelligence works without figurative representations or representative bodies. Only the voice and indirect actions of HAL9000 in *2001 – A Space Odyssey*, both in the book (Clarke 1968), as well as in the eponymous film version (Kubrick 1968), are sufficient to create an oppressive atmosphere. Man can hardly prevail against the uncanny intelligence in the background. Other examples of undefinable forces in the background sending deadly humanoid machines to fight mankind: Skynet from *Terminator* (1984) sends a cyborg from the future into the present to wipe out the human race; similarly, the Wachowski Brothers let the last survivors of humanity fight against the *Matrix* (1999) and their virtual agents. However, machines and their artificial intelligence are mostly man-made and display significant weaknesses. Many of these figures or intelligences have become an integral part of pop culture and major trademarks.

Whereas artificial figures and intelligences are seen rather negatively in the West, in Asia especially in Japan "where cultural perspectives on robots have developed rather differently from perspectives in the West" (MacDorman et al. 2008) exists a more positive attitude toward robots. This is possibly due to the rapid technological development of Japan, which has relied on robots since the industrialization of the country and robots do not constitute a threat to jobs there. In Japan, advanced household robots are considered as health care and are being increasingly used to care for the elderly, which allows for the peaceful application of robots. Nevertheless, the

Japanese affinity for robots has its limitations: 11 years after the discovery of his Uncanny Valley Masahiro Mori wrote: "(...) when the negative qualities of human beings are multiplied by the negative qualities of a machine, the results can be catastrophic" (Mori 1989, 51).

5.7 Discussion

The historical review reveals that communication and cooperation between humans and anthropomorphic figures or machines did not always run smoothly. I will address the positive examples later, but in total, negative or skeptical experiences influence our view of the encounter between real and artificial humans. Artificial figures are often exemplarily used to warn of the consequences of rapid advancements in technology. Running into danger to be vanquished by an uncontrollable species often triggers an existential fear.[6] But the question is how our imagination of artificial figures has been influenced by the Uncanny Valley. Are artificial figures only eerie because of their negative image in history or because of their negative impact due to the Uncanny Valley?

With hindsight, we look back at a divergent picture of the Uncanny Valley in history, because relying on handed down reports can hardly provide us with an accurate idea of how artificial characters from the past really looked like and what people really felt when they saw them. But in the moment of a sensory impression as well as in stories, we always try to get a concept of a figure or person in our minds. Only a few have ever seen an android, but many will have formed a negative, neutral, or even positive view on androids. But both the direct perception, as well as the indirect notion of an ambiguous shape or form, can arouse uncomfortable feelings.

Conceptual as well as imaginative perception (cf. Reid et al. 2014) of artificial figures are influenced by hitherto neglected factors: intention, aesthetics, and the cultural context of a figure.

Intention As mentioned in the beginning, artificial figures are often designed according to our expectations and simulate human appearance or human behavior. This mental model (Lee et al. 2005) also includes a kind of awareness and intention (Zlatev 2001; Fong et al. 2003). Amusement machines made by the Chinese, the Arabs, and during the Renaissance in Europe were primarily described as entertaining devices or just as tools. The lack of awareness leads to no rejection, because there is no active threat against humanity. So, these stories tell of no further conflict between man and machine, however, stories like from the men of Prometheus, Capek's robots in *R.U.R.*, the androids from Blade Runner, etc. show that self-determination of an autonomous and emancipated species is not accepted by the

---

[6] The science fiction author Isaac Asimov directly addresses this fear of mechanical man in some of his robot novels. His term "Frankenstein complex" also predicts a strong phobia against all resembled human beings – similar to the Uncanny Valley.

predominant type and treated as a serious threat that often results in a devastating conflict. Mary Shelley's Frankenstein deals with a figure's self-determination and obviously describes the artificial figure as a monster with apparent cruel intentions and moral errors. This not only decreases the emotional bonding of the reader with the tragic role of the figure, but also increases the reader's doubts and fears. Artificial figures like Pandora, Olimpia, or the Machine from "Metropolis" also produce an eerie image when distracting (often with their appearance) from an evil purpose or leaving the reader or protagonist in the dark about their true intention. This might, in principle, also apply to von Kempelen's/Mälzel's "intelligent" Chess-playing Turk, because of its ambitious intention to win the game against humans. In contrast, neutral, or philanthropic intentions in combination with tragic fates like the death of Gilgamesh's companion Enkidu, the end of Magnus' doorman, Descartes' daughter, or Rachael's role in Blade Runner may appear less eerie and even pitiful for human beings.

Aesthetics Because people are accustomed to associating their counterparts to a specific gender, androgynous artificial figures are nearly always portrayed as male or female. In addition to gender, the visual aesthetics of artificial figures are especially emphasized – as beautiful (like Pygmalion's sculpture, Pandora, Olimpia, etc.), repulsive (Frankenstein's monster), or just unobtrusively human-like (Ning Schi's wooden man). These stories precisely described the figures' intricately formed, lifelike details, and later we will see that this issue is very important. Generally speaking, we can say that responsive aesthetics lead to more initial acceptance by the viewer (cf. Hanson 2005) and seem to successfully obscure the artificial being. Because of their human shape, attractive, human-like figures can also be considered as a potential partner and sometimes produce sexual longing (as in the case of Pygmalion, Pandora, and Olimpia). Combined with knowledge about a baleful motivation or a figure's unknown intentions, the eerie idea of the figure increases.

Cultural Context Almost all historical reports tell of the high level of craftsmanship and technical know-how necessary to create artificial humans. It is fascinating when the underlying technical processes cannot be completely understood right away, or when we wish we could be like these characters or when we expect from characters things we wish we could achieve ourselves. In addition to the figure itself as well as the creator, reason, materials, or the method of development are highlighted and depend on epoch, religion, as well as culture. In antiquity and the European Middle Ages, especially mystical or divine factors were accountable for the creation of artificial life. In Asia, in Arabia, and during the Renaissance in Europe, especially the art of engineering was highlighted. And as we have seen, today there is another image of robots in Asia (especially in Japan) than in the West. Thus, it is clearly significant in which culture an artificial figure exists.

The presented aspects are relevant in forming both short and long term mostly negative judgments influenced by the Uncanny Valley. There are some few positive examples, like Pygmalion's sculpture, the android Data from *Star Trek*, or the Tin

Man from *Alice in Wonderland* that demonstrate a peaceful coexistence between natural and artificial beings. They also show that only equality, good intentions, and mutual respect can lead to higher emotional levels and relationships like friendship or love. A true example of this emotional bond shows the case of the bomb defusing PackBot "Scooby Doo," which was mourned by US soldiers in Iraq after he was destroyed by a mine explosion and could not be repaired (Singer 2009). Scooby Doo's case indicates that not only the appearance, but also the alleged common intention as well as its role in a group of humans may be critical for human acceptance.

The historical references often described artificial figures' lifelike details. Why were these details as well as their skills and abilities so important, and why were they so precisely described? Artificial characters also need to look attractive and have enormous aesthetic qualities. Did the narrators hope that their description would seem more interesting by mentioning these details? Assuming that these figures had really possessed all these abilities and properties, why, excluding the previously mentioned positive examples, were such figures unable to integrate permanently into human society? Or: why did Olimpia attract adverse attention at the party? Consciously or subconsciously, the authors could create a sinister concept with their accurate description of artificial characters. These figures attracted attention due to their "lifelike" details, but despite high visual aesthetics, artificial figures are unable to get the long term acceptance of men. They just bring evil upon the people, disappear, or get destroyed. A permanent and stable relationship between human and human-like figures is rarely mentioned.

The most plausible answer for that reason is that these figures were consciously or subconsciously exploited by their creators due to their uncanny effect and thus increased the tension of stories. Intention, aesthetics, as well as the cultural context are combined by the mind to a certain role model of a human entity. The historical review shows that not only observing a real figure but also the idea of an artificial human is sufficient to trigger eerie responses – not only towards human protagonists within the stories but also to the readers. This idea of an ambiguous creature can also be declared as uncanny like its real embodiment. Only a few people may have truly seen an artificial human in the past, but the conflict between man and machine seem comprehensible and plausible if the Uncanny Valley within our imagination is already taken into account.

As previously mentioned, human standards according to which a human-like artificial figure are evaluated are the same criteria by which we evaluate humans. As this is the precondition for social interaction with anthropomorphic figures, we cannot ignore missing human attributes – e.g. imperfections in facial expressions or errors in speech, etc. This is also evident, either in the short term by processing a negative perceived sensory impression, or in the long term by imaginations and the resulting kinds of prejudices.

Finally, the historical review shows that sensations as well as thoughts result in the same stereotypical image of artificial figures – imperfect and therefore negatively associated. To enable smooth social interactions with machines we must ensure that artificial figures are strongly designed according to human expectations and attitudes. And we need to rethink our understanding and image about artificial figures and about ourselves. An entity's intention has to be clear, the appearance has to be appropriate, and we generally have to reject unjustified prejudices towards artificial figures caused by our historical or cultural backgrounds.

Nevertheless, it remains very difficult to finally improve the impact and acceptance that figures are always accepted by people. In the end, there are still our personal preferences, prejudices, and subjective attitudes that will decide whether we accept a figure or not. Finding the best commonalities and the most reasonable route around obstacles like the Uncanny Valley will present the most difficult challenges to improving social interaction and communication between human-like artificial figures and real humans.

Acknowledgments This work is supported by the graduate program Digital Media of the Universities of Stuttgart and Tübingen, and the Stuttgart Media University (HdM) as well as by the German Research Foundation (DFG) for support of the *SimTech Cluster of Excellence* (EXC 310/1).

References

Adamson, Andrew, and Vicky Jenson. 2001. *Shrek*. United States: DreamWorks, Pacific Data Images DreamWorks Animation.
Al-Hassan, Ahmad Yussuf. 1977. The Arabic text of Al-Jazari's "A compendium on the theory and practice of the mechanical arts". *Journal for the History of Arabic Science* 1: 47–64.
Anderson, William S. 1972. *Ovid's metamorphoses*, American Philological Association series, vol. 2. Norman: University of Oklahoma Press.
Aquinas, Thomas V. 2013. *Summa theologica*. Vol. 1. COSIMO CLASSICS.
Barnes, Brooks. 2011. Many culprits in fall of a family film. *New York Times*.
Baron, Stefan, Henrik Widman, and Lars Lundström. 2012. *Real humans (Äkta människor)*. Matador Film AB, SVT, DR, YLE.
Baum, L. Frank. 2008. *The wonderful wizard of Oz*. Oxford: Oxford Children's Classics University Press.
Bouc, Angelina. 2014. *Men who dressup as rubber dolls star on secrets of the living dolls*. guardiantv.com.
Braemer, Edith A. 1959. *Goethes Prometheus und die Grundpositionen des Sturm und Drang*, Beiträge Zur Deutschen Klassik. Weimar: Arion Verlag.
Breazeal, Cynthia L. 2004. *Designing sociable robots. A Bradford book*. Cambridge: MIT Press.
Buchanan, Kyle. 2011. The biggest problem with the Tintin movie might be Tintin himself. *New York Magazine*.
Cameron, James. 1984. *The Terminator*. United States: Orion Pictures, Hemdale Film Corporation, Pacific Western Productions.
Channel4, 4OD. 2014. *The secret of the living dolls*. U.K.: 4OD – Channel 4.
Cheetham, Marcus, Pascal Suter, and Lutz Jäncke. 2011. The human likeness dimension of the "uncanny valley hypothesis": Behavioral and functional MRI findings. *Frontiers in Human Neuroscience* 5: 126. doi:10.3389/fnhum.2011.00126.
Cheetham, Marcus, Ivana Pavlovic, Nicola Jordan, Pascal Suter, and Lutz Jancke. 2013. Category processing and the human likeness dimension of the uncanny valley hypothesis: Eye-tracking data. *Frontiers in Psychology* 4: 108. doi:10.3389/fpsyg.2013.00108.
Clarke, Arthur C. 1968. *2001: A space odyssey*. London/New York: Hutchinson (UK)/New American Library (US).

Cohen, Florette, Sheldon Solomon, Molly Maxfield, Tom Pyszczynski, and Jeff Greenberg. 2004. Fatal attraction: The effects of mortality salience on evaluations of charismatic, task-oriented, and relationship-oriented leaders. *Psychological Science* 15: 846–851. doi:10.1111/j.0956-7976.2004.00765.x.

Dautenhahn, Kerstin. 1999. Robots as social actors: Aurora and the case of Autism. *Proceedings CT99, The Third International Cognitive Technology Conference*. San Francisco (CA), US. 359: 374.

Dick, Philip K. 1968. *Do androids dream of electric sheep? Time.* Orion.

Dinter, Annegret. 1979. *Der Pygmalion-Stoff in der europäischen Literatur*, Studien Zum Fortwirken Der Antike. Heidelberg: Winter Verlag.

Duffy, Brian R. 2003. Anthropomorphism and the social robot. *Robotics and Autonomous Systems* 42: 177–190. doi:10.1016/S0921-8890(02)00374-3.

Fong, Terrence, Illah Nourbakhsh, and Kerstin Dautenhahn. 2003. A survey of socially interactive robots. *Robotics and Autonomous Systems* 42: 143–166. doi:10.1016/S0921-8890(02)00372-X.

Freedman, Carl. 2000. *Critical theory and science fiction*, Literary studies: Science fiction. Middletown: Wesleyan University Press.

Freud, Sigmund. 1919. Das Unheimliche. *Imago. Zeitschrift für Anwendung der Psychoanalyse auf die Geisteswissenschaften.*

Gassen, Hans-Günther, and Sabine Minol. 2012. *Die Menschen Macher: Sehnsucht nach Unsterblichkeit*, Erlebnis Wissenschaft. Hoboken: Wiley.

Gendolla, Peter. 1992. *Anatomien der Puppe: zur Geschichte des Maschinen – Menschen bei Jean Paul, ETA Hoffmann, Villiers de l'Isle-Adam und Hans Bellmer*. Vol. 113. Heidelberg: Universitätsverlag Winter.

Giles, Lionel. 1925. *Taoist teachings from the book of Lieh Tzŭ*, Wisdom of the East series. London: J. Murray.

Goethe, Johann Wolfgang von. 2013. *Prometheus: Dramatisches Fragment*. Berlin: Contumax-Verlag.

Graham, A. Charles. 1990. *The book of Lieh-Tzu: A classic of the Tao*. UNESCO collection of representative works. Chinese series. J. Murray. Columbia University Press: New York.

Green, Robert D., Karl F. MacDorman, Chin-Chang Ho, and Sandosh Vasudevan. 2008. Sensitivity to the proportions of faces that vary in human likeness. *Computers in Human Behavior* 24: 2456–2474. doi:10.1016/j.chb.2008.02.019.

Greenberg, Jeff, Arndt, Jamie, Linda Simon, Tom Pyszczynski, and Sheldon Solomon. 2000. Proximal and distal defenses in response to reminders of one's mortality: Evidence of a temporal sequence. *Personality and Social Psychology Bulletin* 26: 91–99. doi:10.1177/0146167200261009.

Hanson, David. 2005. Expanding the aesthetic possibilities for humanoid robots. *IEEE-RAS international conference on humanoid robots* 24–31.

Hanson, David. 2006. Exploring the aesthetic range for humanoid robots. *Proceedings of the ICCS/CogSci-2006* 16–20.

Hanson, David, Andrew Olney, and S. Prilliman. 2005. Upending the uncanny valley. *Proceedings of the National Conference on Artificial Intelligence* 20: 24/1728.

Heckmann, Herbert. 1982. *Die andere Schöpfung: Geschichte der frühen Automaten in Wirklichkeit und Dichtung*. Frankfurt am Main: Umschau.

Hesiod. 1914. *The Homeric hymns, and Homerica*, ed. & trans. Hugh G. Evelyn-White. Cambridge (Mass.), US: William Heinemann.

Hill, Donald R. 1996. *A history of engineering in classical and medieval times*. London: Routledge.

Hoffmann, E.T.A. 2008. *Der Sandmann*. Ditzingen: Reclam.

Holt, Nick. 2007. *Guys and dolls*. UK: BBC.

Hurard, Oriane. 2013. Lars Lundström, Autor der Serie "Real Humans": "Ein Spiegel unserer eigenen Existenz". *arte.tv*.

Jauch, Ursula Pia. 1998. *Jenseits der Maschine: Philosophie, Ironie und Ästhetik bei Julien Offray de La Mettrie (1709–1751)*. München: C. Hanser Verlag.

Jentsch, Ernst. 1906. Psychiatrisch-Neurologische Wochenschrift. *The Journal of Nervous and Mental Disease:* 22: 195–205.
Jordan, Leo. 1910. Pars Secunda Philosophiae, seu Metaphisica. *Archiv für Geschichte der Philosophie.* 23: 338–373.
Koreis, Voyen. 2008. Čapek's R.U.R. *2014.*
Kormann, Eva. 2006. Künstliche Menschen oder der moderne Prometheus. Der Schrecken der Autonomie. *Amsterdamer Beiträge zur neueren Germanistik* 59: 73.
Kubrick, Stanley. 1968. *2001: A space odyssey.* Metro-Goldwyn-Mayer, Beverly Hills (CA), US.
LaGrandeur, Kevin. 2010. Do medieval and renaissance androids presage the posthuman? *CLCWeb: Comparative Literature and Culture* 12. doi: 10.7771/1481-4374.1553.
Lang, Fritz. 1927. *Metropolis.* Germany: UFA Film.
Lecouteux, Claude. 1999. *Histoire des Vampires, Autopsie d'un mythe.* Paris: Editions Imago.
Lee, Sau-lai, Kiesler, S., Ivy Yee-man Lau, and Chi-Yue Chiu. 2005. Human mental models of humanoid robots. In *Proceedings of the 2005 IEEE international conference on robotics and automation,* 2767–2772. Washington, DC: IEEE Computer Science Press. doi: 10.1109/ROBOT.2005.1570532.
Lieh-Tzu, and Richard Wilhelm. 1980. *Das Wahre Buch Vom Quellenden Urgrund.* Stuttgart: Zenodot Verlagsgesellscha.
MacDorman, Karl F. 2005a. Mortality salience and the uncanny valley. In *5th IEEE-RAS international conference on humanoid robots, 2005.,* 3:399–405. IEEE Publication Database. doi: 10.1109/ICHR.2005.1573600.
MacDorman, Karl F. 2005b. Androids as an experimental apparatus: Why is there an uncanny valley and can we exploit it? *Android Science* 3: 106–118.
MacDorman, Karl F., Sandosh K. Vasudevan, and Chin-Chang Ho. 2008. Does Japan really have robot mania? Comparing attitudes by implicit and explicit measures. *AI and Society* 23: 485–510. doi:10.1007/s00146-008-0181-2.
MacDorman, Karl F., Robert D. Green, Chin-Chang Ho, and Clinton T. Koch. 2009. Too real for comfort? Uncanny responses to computer generated faces. *Computers in Human Behavior* 25: 695–710. doi:10.1016/j.chb.2008.12.026.
Maul, Stefan M. 2012. *Das Gilgamesch-Epos.* 5th ed. C.H.Beck.
Mettrie, Julien Offray de La. 1990. *L' homme machine.*
Michaels, Adrian. 2014. Secrets of the living dolls, Channel 4, Review. *The Telegraph.*
Misselhorn, Catrin. 2009. Empathy with inanimate objects and the uncanny valley. *Minds and Machines* 19: 345–359. doi:10.1007/s11023-009-9158-2.
Mori, Masahiro. 1970. The Uncanny Valley. *Energy* 7: 33–35.
Mori, Masahiro. 1989. *The Buddha in the Robot.* Tokyo: Kosei Publishing Company.
Müller, Götz. 1989. *Gegenwelten: Die Utopie in der deutschen Literatur.* Stuttgart: J.B. Metzlersche Verlagsbuchhandlung.
Murray, Charles Shaar. 2002. Horror as Dr X builds his creator. *The Telegraph,* February.
Nadarajan, Gunalan. 2007. Islamic automation: A reading of Al-Jazari' s the book of knowledge of ingenious mechanical devices (1206). *Foundation for Science Technology and Civilisation* 1(803): 1–16. Manchester: FSTC Limited.
Needham, J., and L. Wang. 1956. *Science and civilisation in China: Volume 2. History of scientific thought.* Cambridge: Cambridge University Press.
Newman, William R. 2005. *Promethean ambitions: Alchemy and the quest to perfect nature,* American politics and political economy series. Chicago: University of Chicago Press.
Pollick, Frank E. 2010. In search of the uncanny valley. In *User centric media,* eds. Petros Daras, Oscar Mayora Ibarra, Ozgur Akan, Paolo Bellavista, Jiannong Cao, Falko Dressler, Domenico Ferrari, et al., 40, 69–78. Berlin Heidelberg: Springer. Lecture notes of the institute for computer sciences, social informatics and telecommunications engineering. doi: 10.1007/978-3-642-12630-7.
Reichardt, Jasia. 1978. *Robots: Fact, fiction, and prediction.* New York: Viking Penguin.
Reid, Thomas, Gideon Yaffe, and Ryan Nichols. 2014. Thomas reid. In *The Stanford encyclopedia of philosophy,* ed. Edward N. Zalta. Stanford: Stanford University.

Riek, Laurel D., Tal-Chen Rabinowitch, Bhismadev Chakrabarti, and Peter Robinson. 2009. How anthropomorphism affects empathy toward robots. In *Proceedings of the 4th ACM/IEEE international conference on human robot interaction – HRI '09*, 245. New York: ACM Press. doi:10.1145/1514095.1514158.

Rose, Steve. 2011. Tintin and the uncanny valley: When CGI gets too real. *The Guardian*.

Rosenthal, Robert, and Lenore Jacobson. 1968. Pygmalion in the classroom. *The Urban Review* 3: 16–20. doi:10.1007/BF02322211. Weinheim/Basel: Beltz.

Rosheim, Mark E. 1997. In the footsteps of Leonardo (articulated anthropomorphic robot). *IEEE Robotics and Automation Magazine* 4: 12–14. doi:10.1109/100.591641.

Rousseau, Jean-Jacques, Horace Coignet, and Jacqueline Waeber. 1997. *Pygmalion: Scène lyrique*. Paris: Université-Conservatoire de musique.

Sakaguchi, Hironobu, and Motonori Sakakibara. 2001. *Final Fantasy – The Spirits Within*. Colombia Pictures.

Saygin, Ayse Pinar, and Hiroshi Ishiguro. 2009. The perception of humans and robots : Uncanny hills in parietal cortex. In *Proceedings of the 32nd annual conference of the cognitive science society*, ed. R. Catrambone. Cognitive Science Society: 2004–2008.

Scassellati, Brian Michael. 2001. *Foundations for a theory of mind for a humanoid robot*. Cambridge: MIT Press. doi:10.1037/e446982006-001.

Schneider, Edward, and Shanshan Yang. 2007. Exploring the uncanny valley with Japanese video game characters. In *Intelligence*, ed. Baba Akira, 546–549. Tokyo: The University of Tokyo.

Schwab, Gustav. 2011. *Sagen des klassischen Altertums*. Leipzig: Anaconda Verlag.

Scott, Ridley. 1982. *Blade Runner*. Warner Bros, Burbank (CA), US.

Scott, Ridley. 1992. *Blade runner – The director's cut*. Warner Bros, Burbank (CA), US.

Shelley, Mary. 1818. *Frankenstein or the modern prometheus*. London: Lackington, Hughes, Harding, Mavor, & Jones.

Singer, P.W. 2009. *Wired for war: The robotics revolution and conflict in the twenty-first century*, A Penguin book. Technology/military science. New York: Penguin Press.

Spielberg, Steven. 2011. *The Adventures of Tintin*. Paramount Pictures (North America), Columbia Pictures (International).

Steckenfinger, Shown A., and Ghazanfar, Asif A. 2009. Monkey visual behavior falls into the uncanny valley. *Proceedings of the National Academy of Sciences of the United States of America* 106: 18362–18366. doi:10.1073/pnas.0910063106.

Strandh, Sigvard. 1992. *Die Maschine: Geschichte, Elemente*. Weltbild-Verlag: Funktion Ein enzyklopädisches Sachbuch.

Styles, Ruth. 2014. Secrets of men who dress up as rubber dolls revealed in new documentary. *dailymail.co.uk*.

Thiele, Eckhard. 1988. *Karel Čapek*. Vol. 1257. Reclam: Leipzig.

Thorpe, William H. 1944. Some problems of animal learning. *Proceedings of the Linnean Society of London* 156: 70–83. doi:10.1111/j.1095-8312.1944.tb00374.x.

Tinwell, Angela, Mark Grimshaw, Debbie Abdel Nabi, Andrew Williams, Debbie Abdel Nabi, Tinwell Angela, and Grimshaw Mark. 2011. Uncanny valley in virtual characters 1 facial expression of emotion and perception of the uncanny valley in virtual characters Angela Tinwell. *Computers in Human Behavior* 44: 1–34. doi:10.1016/j.chb.2010.10.018. Elsevier Ltd.

Valverde, Sarah Hatheway. 2012. *The modern sex doll-owner: A descriptive analysis*. San Luis Obispo: California Polytechnic State University.

Völker, Klaus. 1994. *Künstliche Menschen: Dichtungen und Dokumente über Golems, Homunculi, lebende Statuen und Androiden*. Suhrkamp Verlag KG: Phantastische Bibliothek.

Von Bingen, Hildegard, and Priscilla Throop. 1998. *Hildegard von Bingen's Physica: The complete English translation of her classic work on health and healing*. Rochester (Vermont), US: Inner Traditions/Bear.

Wachowski, Andy, and Larry Wachowski. 1999. *The Matrix*. United States, Australia: Warner Bros. Pictures, Roadshow Entertainment.

Wells, Herbert G. 1927. Mr. Wells reviews a current Film: He takes issue with this German conception of what the city of one hundred years hence will be like. *The New York Times Company*, 17 Apr.

Wells, Simon. 2011. *Mars needs Moms*. Walt Disney Animation Studios.

Weschler, Lawrence. 2006. Wired: Why is this man smiling ? *Wired* 10.06: 1–6.

Westbrook, Caroline. 2014. C4's secrets of the living dolls pulls in and freaks out viewers in equal measure. *metro.co.uk*.

Zebrowitz, Leslie. 2001. *Facial attractiveness: Evolutionary, cognitive, and social perspectives: Evolutionary, cognitive, cultural and motivational perspectives*. New York.

Zemeckis, Robert. 2004. *The Polar Express*. United States: Warner Bros. Pictures, Castle Rock Entertainment, Shangri-La Entertainment, ImageMovers, Playtone, Golden Mean.

Zlatev, Jordan. 2001. The epigenesis of meaning in human beings, and possibly in robots. *Minds and Machines* 11: 155–195. doi:10.1023/A:1011218919464.

Chapter 6
Ethical Implications Regarding Assistive Technology at Workplaces

Hauke Behrendt, Markus Funk, and Oliver Korn

6.1 Introduction

Technological innovation can fundamentally change the way people life their lives. In many cases this comprises both positive and negative aspects. Psychologists and Neuroscientists suppose, for instance, that people lose abilities they do not exercise any longer since technology substitutes its functions.[1] Comfort comes at a price, as one may say. This raises general worries against the overall worth of new technologies as such.

This paper deals with such worries regarding *Assistive Technology at Workplaces* (ATW). More precisely, we are interested in its normative implications. In particular, it is our concern to investigate whether ATW is good and to what extend. Therefore, we shall address the following two interrelated guiding questions about its development and application:

1. *Why should one use ATW?*
2. *What are standards that specify how one should use it?*

These questions are motivated by the ongoing technological progress in the field of Human-Computer-Interaction, in particular by the ability to produce special

---

[1] It is believed, more generally, that the human brain develops in a way that can be described by a so-called "use-it-or-lose-it"-principle according to which synapses are formed or recede dependent on the usage (cf. e.g. Blakemore and Choudhury 2006).

H. Behrendt (✉)
Institute of Philosophy, University of Stuttgart, Stuttgart, Germany
e-mail: Hauke.Behrendt@philo.uni-stuttgart.de

M. Funk • O. Korn
Institute of Visualization and Interactive Systems (VIS), University of Stuttgart, Stuttgart, Germany
e-mail: Markus.Funk@vis.uni-stuttgart.de; Oliver.Korn@acm.org

© Springer International Publishing Switzerland 2015
C. Misselhorn (ed.), *Collective Agency and Cooperation in Natural and Artificial Systems*, Philosophical Studies Series 122,
DOI 10.1007/978-3-319-15515-9_6

workplaces for the manual assembly, which directly assist (impaired) workers in the working process.

The first question addresses the general worry *why* one should use ATW at all. Given that its application could impose serious effects on people's everyday working life, the mere fact that we are technically able to build these devices is no sufficient reason for doing it. We must rather consider additional justifying ends that speak in favor of its realization. A balanced consideration clarifies under which circumstances further investigation on this field can be endorsed with good reasons. We suggest that there are indeed such reasons. Particularly, we shall argue that ATW is good to the extent that it ensures social inclusion to the world of employment.

The second question is concerned with *how* one should use ATW. A positive answer to the first question does not entail that its use is justified by all means. Rather, we shall argue, while employees must be able to see themselves as self-determined when using ATW, the individual needs of impaired workers must be met as well. More precisely, we suppose that ATW must in particular satisfy two requirements of good workplaces, which we specify as (a) an exploitation restraint and (b) a duty of care.

Both questions are closely interrelated. We expect pros and cons at different levels of abstraction. To offer a reasonable account, we shall weigh up its positive and negative aspects for those who are possibly affected from an impartial point of view. In other words, we propose reasons everyone could adopt from an abstract perspective despite their private interests and desires. This reflects the fact that the development and application of new technologies is a societal issue: As a society we collectively carry the responsibility for what we are doing in public.

The paper is structured in the following way: Initially, we shall outline a concrete concept and eight appropriate technical functions of ATW for the manual assembly. This shapes a detailed picture how this technology might be constructed in order to serve its purpose. Although we shall make some general statements about the overall worth of ATW as such, this example functions well as a case in point (6.2). We shall turn next to its normative implications as stated in our two guiding questions. In order to do this, we first want to give a short overview of the social policy and legal background in Germany. This can clarify that social inclusion is already an important sociopolitical issue that gains political support. We will then question whether social inclusion to the world of employment is indeed worthwhile from an impartial point of view, and by what means (6.3).

6.2 A Technical Concept of ATW

From previous work and from interviews with manufacturers and sheltered work organizations we derived eight technical functions ATW should implement to serve its purpose. These functions are explained in detail in the following sections after proposing some general remarks on its equipping.

To detect a user's movement, actions, or working steps that are conducted at the workplace, ATW for the manual assembly should use visual sensors. Other sensors

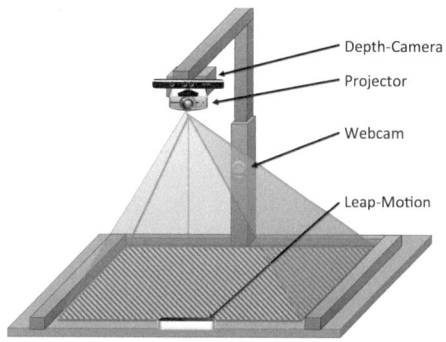

Fig. 6.1 The conceptual construction and sensor placement of our prototype for an assistive system for the workplace

than camera-based ones would interfere too much with the working task. Furthermore, all sensors should be placed in a way to not disturb the natural workflow. Figure 6.1 shows a sketch of our prototypical setup: A top-mounted Microsoft Kinect depth-camera can detect placement of objects and touch with the surface. A top-mounted projector is capable of displaying feedback directly into the user's field of view. A Leap Motion controller that is mounted directly in front of the user is able to visually detect fine-grained hand movement. Furthermore, a HD webcamera can identify objects using computer-vision algorithms.

6.2.1 Detecting Movements in Pre-defined Areas

When creating a workflow for ATW, users must be able to define areas that are surveyed by the system. Whenever the system detects movement in a previously defined area, the system can analyze the current working task from that movement. For example, the user could define an area on top of the boxes where the spare parts are stored. In that case, the system could identify which box the user accessed last and therefore detect the current working step.

Those areas could also be defined on places where the assembly has to be done. Using this technique, the system could detect whether the worker already put a spare part at the right place or even whether a spare part was put onto the wrong position. By combining the analysis of depth data with a visual object recognition algorithm called SURF (Bay et al. 2006), the system is even capable of reporting whether the right part was assembled at a defined position.

6.2.2 Detect Finger and Hand Movement

By having finger and hand movement information available in real-time, the system is able to communicate immediate feedback to the user. This enables the system to predict that the user is about to make an error, for example aiming to grab a wrong spare part. In that scenario, the system could project a warning directly at the hand of the user before the error was even made. Additionally, if the user wants information about the assembly of a spare part, he could get it by simply pointing at the spare part that he wants to assemble. By analyzing the finger position, the system is able to discriminate the pointing gesture and display information immediately.

6.2.3 Detect Actions That Are Performed at the Workplace

The system must be able to detect activities in order to have information about the task the user currently performs. Among others the system can monitor the tools that are available at the working place. If, for instance, the user grabs a hammer instead of a screwdriver, the system can identify that a wrong activity is being performed and thus warn the user. As an alternative way to detect user's activities, the system could receive accelerometer data from the user's wristwatch and detect tasks based on the accelerometer data (Blake et al. 2010) (Fig. 6.2).

6.2.4 Detect Touch with the Surface and Objects

As the system has a top-mounted projector and is able to project content directly into the user's field of view on the table, this content can be made touch sensitive by using the Microsoft Kinect. Such technology would provide users with more options to interact with the system. If, for instance, the system displays a notification that

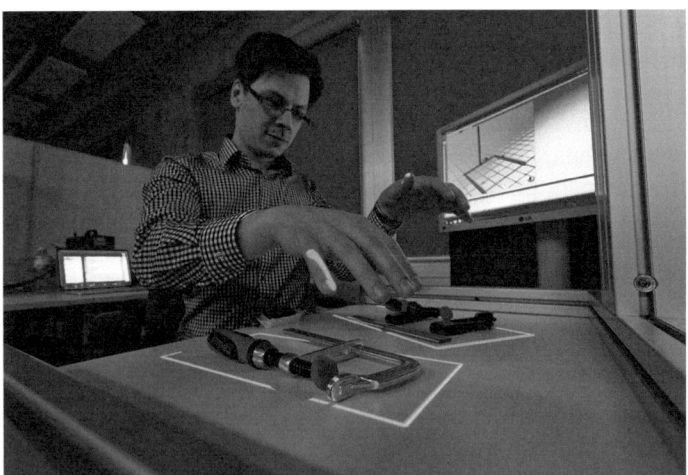

Fig. 6.2 Our prototypical implementation of an assistive system for the workplace. The system highlights detected objects and helps the worker with visual instructions

the user does not want to see anymore, the user could close it by tapping it. The UbiDisplays Toolkit (Hardy and Alexander 2012) supports this functionality.

Another feature that can be used with ATW is to work with physical objects as a tangible user interface for controlling digital functions (Funk et al. 2014). The system can track previously defined objects and calculate their rotation. This could also be used to interact with the system and control the workflow.

6.2.5 Programming by Demonstration

Creating assembly instructions can take a long time, especially if they need to be updated a lot, for example when producing in small lot sizes. Therefore, one could implement the concept of programming by demonstration. In this concept, a user performs the assembly task at the assistive system. The system records the movements and working steps and creates assembly instructions from it. Then, it can record a video of the assembly task and additionally display visual feedback, for example where to pick the next spare part or where to assemble the currently picked spare part.

6.2.6 Assessment of the Worker's Current Performance

Impaired workers' performance can change from day to day. Therefore, the system needs to be able to assess the worker's current performance and adapt its feedback to it. In situations where the worker cannot remember the next task, the system could display a video showing how to perform. Were the worker performing better, the system might not show a video, but rather highlight only picking and mounting positions. Of course, if the system has performance information about the worker, special care needs to be taken that this information is only used for displaying appropriate feedback rather then collect data for other purposes such as employee rating.

6.2.7 Emotion Recognition

Another component that should be considered when displaying feedback are the emotions of the worker. By using a web-camera and emotion detection algorithms, the system could detect the mood of the worker. If a working task makes the worker angry or sad, the system could propose getting personal assistance or change the way of displaying feedback. In case the worker is happy, the system could acquire the information that this kind of work or feedback makes the worker happy and keep the current settings.

6.2.8 Gamification

ATW could implement gamification components in order to motivate workers. Recent research has shown that gamification at the workplace has such effects on impaired workers (Korn et al. 2014). Our previous research used an additional screen to display points and a Tetris gamification component, where the assembly speed was mapped to a block's color (Korn et al. 2012). In future research we want to display the gamification components directly into the worker's field of view using in-situ projection.

6.3 Normative Implications

We shall now turn to the normative implications of ATW as stated in our two guiding questions above, namely (1) *why* and (2) *how* one should use it. The phrase "normative implications", as we understand it here, can be spelled out as considerations about the overall worth of ATW. We take the concept of worth broadly as a second-order property of what is best supported by the balance of reasons (Scanlon 1998; Parfit 2011). Following Derek Parfit (2011:38) we assume that these reasons bear on facts about a things distinct properties which are sufficient "to respond to this thing in some positive way, such as wanting, choosing, using, producing, or preserving" it.

The view that something can be good (or bad) in this way needs more specification. We can thus distinguish at least two different ways in which reasons regulate something's worth. When, for example, x promotes someone's personal interests or aims, this would give her self-interested reasons to positively respond to this thing. In this case we say that x is *good for* someone. Additionally, something can also be supported by impartial reasons, that is, reasons everyone would have from an impartial point of view. That x is *impartially good,* then, means that everyone has reasons to value x regardless of whether or not this would be self-interestingly good for them.

Note that reasons can conflict with one another. That is, since most of the time we deal with multiple pro-tanto reasons varying in strength one must usually weigh up the positive and negative reasons that speak for or against something in order to determine its overall worth. Applying this framework to our purpose we shall now address ATW's distinct properties in order to formulate generalizable propositions about its overall worth.

6.3.1 An Argument for ATW's Worth

Since ATW is special technology for workplaces in industrial production it might appear at fist sight that its application would primarily be *good for* business companies. In fact, we anticipate an improved performance of workers when using ATW, as related evidence showed (Korn et al. 2013). This amounts to aspects such as performance rate, product quality, flexibility, and employee motivation. That is to say, ATW may contribute to industrial production in many ways.

However, we cannot sufficiently infer from these expectations whether or not ATW is overall good. Economic benefit clearly provides some support, although in view of its particularity it is far from obvious that apart from private business interests this fact is decisive from an impartial point of view. If, for instance, an increase of productivity would lead to involuntary unemployment or substandard working conditions, the collective disadvantages would far outweigh the advantages.

Now actually the reverse is true. We assume that ATW can bring relief for impaired persons to establish themselves in working life. Its use would enable them to cope with more complex and thus meaningful tasks as they could otherwise have done. Moreover, using ATW can likewise grant reasonable working conditions – whose content will be specified later – by means of its technical functions as outlined in Sect. 6.2. We take these aspects as crucial and conclude therefore that

Thesis$_{(Inclusion)}$ *ATW is impartially good because it insures social inclusion to the world of employment.*

In other words, we claim that the virtue of being inclusive is a feature of ATW which gives everyone impartial reasons to value it. Yet this inclusion thesis bears on the assumption that social inclusion is somehow good. To take up this observation the concept of social inclusion can be spelled out as a regulative idea, which expresses the ideal of an open society in which all citizens equally participate. Exclusion, on the contrary, would thus mean that persons or groups become marginalized, that is, that they have no equal standing in one or more constitutive areas of society (Bude and Willisch 2008).

Note, however, that we are not committed to the implausible claim that social inclusion is something good in itself. In that case inclusion would be good independently from any antecedent condition. That is, it would be intrinsically good. This assumption seems to be mistaken, though. We rather suppose that the worth of social inclusion is grounded in the worth of the underlying social practices. In other words, we hold that social inclusion is a "normatively dependent concept" (Forst 2012). That implies that by itself social inclusion cannot provide reasons to be regarded as valuable. More precisely, only if backed by further, independent normative resources one can determine whether social inclusion would be good or bad under certain conditions.

This can be shown by a simple consideration: If social inclusion would be good in itself, then all kinds of exclusion must generally be bad. Now consider a slave state in which a minority is arbitrarily excluded from being enslaved. This is obviously not bad in any plausible sense. On the contrary: In the case of slavery exclusion would be generally preferable. What would be bad here is the very practice of slavery, not the total amount of people who participate. In other words: A full inclusive society of slaves would not make it any better but rather worse.

These considerations show that social inclusion contains no intrinsic value. Rather, its worth must derivatively rest upon further normative background conditions, namely the worth of the underlying social practices. The ideal of an inclusive society must then be substantially specified with regard to these practices in order to expose forms of wrong inclusion. This suggests that a reasonable examination must extend its scope to the relevant context. In the present analysis the place of current interest is the world of employment within a modern democratic welfare state with an embedded capitalist economic order. We conclude therefore that the inclusion thesis is true if and only if social inclusion to the world of employment gains additional normative backing. To put it differently, we hold that it grounds on the condition that participation to the world of employment is indeed worthwhile.

6.3.1.1 Political Reasons for Inclusion to the World of Employment

The first domain where we find support for this assumption is the actual socio-political sphere in Germany. It is widely held that the concept of political legitimacy can be defined as a (moral) right to rule. According to this view citizens are (at least mostly) morally obligated to obey the political decisions and positive laws of a legitimate political order (Schmelzle 2012). Hence, if social inclusion is legitimately warranted, that would be a sufficient reason to comply with this political goal whether or not it independently earns additional normative support. In order to show that there is in fact such socio-political basis in Germany, we shall give a short overview on the status quo of its social policy and legal background (cf. for the following Schulze 2011).

As one of 156 nation states Germany has ratified the *UN-Convention on the Rights of Persons with Disabilities* (UNCRPD 2008). With its subscription to these guidelines of international law the German government bound itself to insure the human rights of persons with disabilities. Article 3 defines social inclusion as one of UNCRPDs general principles. Moreover, social inclusion plays also an important role in the convention's more specific demands. Article 19 declares that

> States Parties to this Convention recognize the equal right of all persons with disabilities to live in the community, with choices equal to others, and shall take effective and appropriate measures to facilitate full enjoyment by persons with disabilities of this right and their *full inclusion and participation in the community* (…). (Emphasis added)

On a plausible interpretation of this passage, the highlighted phrase "full inclusion and participation in the community" already entails that full access to the world of employment should be granted, supposing that employment is central for the adequate recognition of equal membership in the community. An explicit expression of such claim can be found in article 27, where it is stated that:

> States Parties recognize the right of persons with disabilities to work, on an equal basis with others; this includes the right to the opportunity to gain a living by work freely chosen or accepted in a labour market and work environment that is open, inclusive and accessible to persons with disabilities (…).

What measures these claims exactly entail requires a detailed analysis of the particular circumstances, which cannot be elaborated here at length. We rather suggest that in the end this question must be answered by means of a public deliberation process within each society. Nevertheless, the quoted passages clearly demonstrate that inclusion to the world of employment is already a well-defined political aim and also a binding political decision. As a consequence the guidelines stated in the UNCRPD directly have an important impact on future social legislation of all signatories.

Additionally, in substance several claims are already part of the German social law (Heines 2005; Frehe 2005; Blesinger 2005). Beside a general anti-discrimination principle, which is stated in the German constitution (Art. 3. Abs. 3 Satz 2 GG), the social security code declares full inclusion to the working life as one of its central goals (§33 Sozialgesetzbuch IX). It claims to obtain, improve, and cure the ability to work for impaired persons with respect to their capacities (§ 4). Most notably for the present topic, the law even provides work-assistance to realize full inclusion. This involves measures of personal as well as technological assistance that must be provided at the workplace for persons with disabilities (§§81; 102 Abs. 3 Nr. 2b SGB IX).

Correlating these findings with the question why using ATW, the answer simply is that ATW functions as a warranted means to realize these binding political decisions and positive laws. As formerly indicated this fact alone provides a stable foundation for an ongoing development of ATW. We conclude, therefore, that

Thesis I$_{(Political)}$ *ATW is politically good because it ensures social inclusion to the world of employment, which is authorized by binding political decisions and positive laws.*

However, in constitutional democracies political decisions can change and legislation amendments are normal procedures. Indeed, one reason why democratic rule can claim legitimacy is not only the fact that the people prospectively determine its binding decisions but also that they can (at least in most cases) change them back (Habermas 1996). For that reason it is still appealing to answer the question whether social inclusion to the world of employment is also supported by independent reasons beside actual politics. To put it differently: Is social inclusion also supported by additional reasons whose validity does not merely bear on the normative force of factual (and in this sense arbitrary) political decisions but which could rather ground them rationally?

Although we cannot lay out a full account here, we shall programmatically introduce our approach for such an enterprise. As mentioned above, we must look for those reasons in the underlying social practices. In particular, we shall examine to what extend we can put normative weight to the practice of gainful work. Social practice is, very roughly, a technical term for special forms of cooperation, which are governed by means of certain rules. These rules define a bundle of roles with corresponding rights and obligations, which can be enforced formally or informally among its members. According to their roles, members of social practices obtain special forms of authority to make legitimate claims on one another and to hold themselves mutually responsible for what they are doing. Hence, social practices consist in a network of multiple behavioral expectations (Honneth 2011; Stahl 2013).

A social practice can possess normative weight in several respects. One important respect is whether the constitutive rules and norms, which together generate it, are justified (Rawls 1999a). If a practice were wrongful, this would give everyone decisive impartial reasons to not engage in it, as the slave state example suggests. However, here we are concerned with the broader question what defendable claims could be made for or against one's inclusion. Given that a practice is just, this fact alone is not sufficient to answer the question what normative pertinence it obtains for its members. That is, what positive reasons one may have to engage in that practice.

In his "groundwork" Kant (2002) identifies three domains of practical reasoning. Following this classification of practical rationality established by Kant, we distinguish (a) a technical, (b) an ethical, and (c) a moral domain of practical reason. According to these domains, the participation in the world of employment could be good either (a) instrumentally as means, (b) prudentially as an ethical good or (c) morally as a binding norm (Höffe 2006).

6.3.1.2 Instrumental Reasons for Inclusion to the World of Employment

In accordance with these distinctions the first domain of technical reasons specifies how one should rationally act in order to realize his given ends. Kant calls this class of reasons technical, because it solely refers to means-end-relations regardless of any further rational assessment of the relevant ends themselves. In other words: These reasons merely determine necessary means for given ends, while being neutral towards the question whether one has also good reasons to attain them. Thus, technical reasons state what one must rationally do in order to realize his aims whether or not this action would be good or rational on its own. As Kant (2002: 32) puts it: "The precepts for the physician to make his patient thoroughly healthy, and for a poisoner to ensure certain death, are of equal value in this respect, that each serves to effect its purpose perfectly." This is why this stance is also called means-end or instrumental rationality (Kolodny and Brunero 2013).

Labor has often been seen in this way as instrumentally good. Since it appears to be a fundamental condition of human life that its subsistence consists for the most part in the necessity to work in some form or another, labor is sometimes called a "necessary evil". According to this view, it is merely an unpleasant means for the realization of other required goods such as food, shelter, and so on.[2] Additionally, in modern societies regularly almost all such goods are exchanged via markets. Traditionally they were taken as efficient instruments to neutrally coordinate the satisfaction of people's private interests and desires in mass societies. Thus, it is

---

[2] Cf. the instructive distinction between labor and work made by Arendt (1998). We do not, however, make this distinction in the following since we are not interested in a comprehensive anthropological analysis of work as an activity, but rather examine normative implications of the narrow institutional concept of gainful work as originally proposed by Kambartel (1993).

widely assumed that the participation in the world of employment purely rests on self-interest. Or as Adam Smith (1976: 19) famously puts it:

> It is not from the benevolence of the butcher, the brewer, or the baker that we expect our dinner, but from their regard to their own interest. We address ourselves, not to their humanity, but to their self-love, and never talk to them of our own necessities, but of their advantages.

Following this line of thought, the worth of ATW could be expressed in the following general form as instrumentally good:

Theses I$_{(Technical)}$ *ATW is instrumentally good because it ensures social inclusion to the world of employment, which is a necessary means to other material goods.*

However, this thesis bears on two problematic presuppositions. It holds, firstly, that there are always reasons to strive for other material goods, which, secondly, must be obtained through labor. To see why these suppositions are problematic, recall that the question here is not whether a society as a whole can reproduce itself without the work of its members, but rather what reasons there are to engage in this practice individually. Note further that we are not concerned with all forms of work here, like private housework and the like, but rather with gainful employment, which is mediated by a job market.

With this remarks in mind, it becomes clear why the thesis as stated above is not generalizable for precisely two reasons: First, although it might be true that the material reproduction of a society as a whole depends largely on the labor of its members, this presupposes no concrete organizational structure. History tells us that work must not be collectively distributed between everyone. Rather, there have always been persons or groups that were exempt from work, either by means of domination and exploitation of others or through charity and self supply. Regardless of whether or not this can be criticized on moral grounds, these considerations show that persons or groups can very well obtain goods by other means without causing a total collapse of the whole system. Second, note that it is also far from obvious that people necessarily have reasons to strive for other goods. Imagine the long-standing way of ascetical living. According to this conception one should stay as self-sufficient as possible. Holding that ascets too have reasons to obtain material goods is contradicting on conceptional grounds. Aside from a few goods essential for survival reasons to strive for material goods depend on personal decisions of autonomous persons and are thus rather accidental.

In sum, the assumption that labor is instrumentally good as necessary means to other required goods is an unjustified generalization. Moreover, while it has some plausibility to hold that gainful work is normally indeed exercised for that reason in modern societies, this proposal cannot reasonable rationalize why this should also hold for the special case of impaired persons. Instead of earning their living by labor, they could receive transfers or the like. Finally, this explanation does not exhaust all theoretical resources why labor might be good but rather restrict its explanation too quickly on an instrumentalist view.

6.3.1.3 Ethical Reasons for Inclusion to the World of Employment

The second domain of practical reasoning, following Kant's classification, is the prudential. Here, the agent's concrete ends are under consideration. Whereas the first domain of instrumental reasoning only states means-end relations, this second domain questions whether the intended end in fact somehow benefits the agent's well-being. Although we do not agree with Kant on his specific hedonic concept of individual well-being, we nevertheless shall draw on the distinct characterization of this domain in a merely formal way. By that we mean that we follow Kant insomuch as this domain refers to the question what would make someone's individual life the most worth living, while we are neutral with regard to any substantive claim about its specific content.

If labor would be a constituent part of a person's well-being, ATW would be ethically good. Proponents of this view hold that

Thesis I$_{(Ethical)}$ *ATW is ethically good because it ensures social inclusion to the world of employment, which makes one's life more worth living.*

Whether this thesis is true, depends, again, on two presuppositions. First, it states that well-being is a universal end. Second, it holds that gainful work is a necessary condition to that end. Following Kant (2002: 32), we take the first condition as satisfied in the just described formal sense: It is a brute fact about human nature that all persons strive for a good life, although the concrete interpretation of its content remains controversial. Moreover, while assuming that this goal is an condition of human nature, it is also supported by good reasons: If there is any plausible sense in which one's life can be called comparatively better or worse, it seems to be trivially true that one has strong reason to choose the better. Note that this view does not predict any substantive claims about what makes someone's live go best, but rather what Thomas Hurka (1993: 25) calls the "extensional claim" according to which everyone might in fact desire well-being while at the same time not knowing which state would be coextensive with this desire.

The second condition is rather problematic. We assume that a substantial view of well-being cannot successfully be justified by philosophical means so far. In order to show that each substantial view of well-being faces serious problems, we shall roughly discuss its most prominent accounts in the following (cf. for the following e.g. Steinfath 1998). The current philosophical debate about substantial conceptions of well-being can be divided into two rival positions, namely objective and subjective theories of well-being. Both hold that there is a distinct standard according to which the quality of someone's life can be assessed. Therefore we shall refer to these theories as normative theories of well-being. While subjective theories claim that the accurate standard fully depends on certain subjective qualities of the person whose life it is, objective theories hold that there is a fixed content of what would make a life the most worth living regardless of whether or not this fulfills factual desires or the like.

We shall propose several thought experiments, which show that subjective theories lack recognition of significant goods, whereas objective theories cannot defend their substantive claims sufficiently against crucial objections. We shall discuss first (a) hedonic theories and (b) desire theories as representatives of subjective theories, and next (c) objective list theories and (d) perfectionist theories as representatives of objective theories.

(a) **Hedonistic Theories**

According to hedonistic theories of well-being a good life bears completely on the total amount of happiness it contains. Thus, for hedonists the only intrinsic good is happiness. That is, the standard that measures whether something contributes to someone's well-being is, accordingly, whether it increases someone's happiness. Applied to our current question it follows that whether labor is ethically good totally depends on the sum of happiness a concrete working life would bring. In other words, if hedonists are right, ATW would be ethically good if and only if it would produce more happiness than other comparable activities would do.

Against this view it can be opposed that while happiness might indeed be a crucial condition of well-being, it is far from clear that it is the only intrinsic good. Thus, it has been argued that other goods such as truth or autonomy, for instance, are as well worth striving for, even if they would not contribute but rather reduce ones happiness. Robert Nozick (1974: 42–44) has famously argued that happiness is not the full story of someone's life by proposing the following thought experiment:

> Suppose there were an experience machine that would give you any experience you desired. Super-duper neuropsychologists could stimulate your brain so that you would think and feel you were writing a great novel, or making a friend, or reading an interesting book. All the time you would be floating in a tank, with electrodes attached to your brain. Should you plug into this machine for life, preprogramming your life's experiences? (…) Of course, while in the tank, you won't know that you're there; you'll think it's all actually happening. (…) Would you plug in? (…) We learn that something matters to us in addition to experience by imagining an experience machine and then realizing that we would not use it.

Nozick suggests that we would not plug in an experience machine that grants happiness. He infers from that observation that good experiences can only be one value among others, and thus that hedonistic theories are false.

A second objection comes from Shelly Kagan (1998: 34–36). In his thought experiment, he describes a businessman who takes great pleasure in being respected by his colleagues, well-liked by his friends, and loved by his wife and children until the day he died. But his happiness rests on many wrong believes: the deceived businessman's colleagues actually think he is useless, his wife doesn't really love him, and his children are only nice to him so that he will keep giving them money. According to hedonic theories one must judge the businessman's life as good regardless of whether or not his happiness is based on true or false beliefs since there is only one intrinsic good, namely happiness. But, Kagan persuasively argues, no one would agree with that. Rather, there is a difference between true and false kinds of happiness. That is, while Nozick insists that there are more intrinsic values than happiness, Kagan's example shows that even happiness is not always good. We conclude, therefore, that hedonic theories cannot claim plausibility.

(b) **Desire Theories**

Desire theories try to fix the deficits of hedonic theories. According to this view, the quality of someone's life is measured by the degree of its desire fulfillment. In other words, desire theorists claim that the well-being of a person rises according to the amount of her satisfied desires. In order to exclude absurd desires most proponents of this position normally restrict its scope to informed desires, that is, desires one would hold under ideal conditions (Griffin 1986). According to desire theories, labor would be ethically good to the extent that it is part of someone's (informed) desires. Given this approach of desire theories, it would then be an empirical question, whether ATW would contribute someone's well-being since the constituents of well-being totally depends on the desires one has (factual or under ideal conditions).

Against this view there are at least two crucial objections. In particular, it appears that desire fulfillment is neither a necessary nor a sufficient condition to well-being. First, it is implausible that the fulfillment of ones desires is always good for someone. There might be desires, even informed ones, whose achievement would be indifferent regarding the quality of one's life. Imagine that one desires that a certain faraway star contains a certain chemical composition, as Scanlon (1998) suggests. It is obviously clear that it would not contribute to anyone's well-being at all whether or not that star has in fact this desired chemical composition. This suggests that desire fulfillment is not sufficient. Second, desire fulfillment is also not necessary. Some goods seem to be valuable regardless of whether or not they realize a corresponding desire. Unforeseen events like an unwanted child, for instance, can make someone's life go better. We conclude, therefore, that desire theories fail to provide a reasonable account of what constitutes good-making properties.

(c) **Objective List Theories**

For Objectivists, such as Parfit (1984) or Scanlon (1998), the key problem of subjective theories is that they grasp the relationship between desires and valuable goods from the wrong direction. According to their view, one should value something not because that would fulfill his desires, but rather the contrary is true: One should desire something because it is valuable.

Thus, objectivists hold that (at least some) things must be valuable regardless of whether or not they contribute to one's desire fulfillment. That is, the good-making features of something are determined by these things, which give someone reasons to value that thing. Someone's well-being, then, is determined with reference to their ability to correctly respond to those valuable properties. This view has been called "Objective list theory" (Parfit) or "Substantive good theory" (Scanlon) since those accounts argue for an objectively fixed list of substantive goods that contribute to someone's well-being independently from their recognition through the agent.

Such views can account for some crucial objections that subjectivists face. Nevertheless, although the objectivist approach is not implausibly narrow with regard to the things that matter, it appears that their list of valuable things might be too elitist. Would one really want to say that something contributes to someone's well-being even if that thing does not matter for that person? Moreover, for objectivists the substantive claims about the goods that are constituents of well-being bear essentially on arguable intuitions. That is, they lack a stable ground according to which claims about valuable properties can be justified. How should one decide, for example, whether labor is an item on the list of intrinsic values or not? We assume that while intuitions play an important role in philosophical theory building, they are too loose and potentially contradicting in order to ground objective claims about everyone's well-being. Thus, objective list theories cannot defend a plausible theory of the good life.

(d) **Perfectionism**

For that reason, other authors have grounded their objective claims about what makes things good for someone's well-being on a different justificatory basis, namely on the natural functions of things. A prominent ancestor of this view is the ancient philosopher Aristotle (2011). Roughly speaking, Aristotle holds that everything has a natural function (ergon) that determines its goodness (arête). As the natural function of a knife is to slice, so for the good knife is to slice well. According to this view, one can, thus, measure someone's well-being accordingly by the degree one cultivates his distinct human nature. That is, the better someone develops his natural functions the better his life is. For that reason these theories are called perfectionist theories (Hurka 1993).

However, this attempt to determine individual well-being faces two crucial objections. First, the presumption of something's natural function is hard to swallow. Rather, the stipulation of these functions must be justified on non-natural grounds. Or as Bernard Williams (1972: 59) famously puts it:

> If one approached without preconceptions the question of finding characteristics which differentiate men from other animals, one could as well, on these principles, end up with a morality which exhorted men to spend as much time as possible in making fire; or developing peculiarly human physical characteristics; or having sexual intercourse without regard to season; or despoiling the environment and upsetting the balance of nature; or killing things for fun.

Moreover, even if we grant that there is something as a natural function, this would not obtain any normative force. Rather, one would commit a naturalistic fallacy if one concludes from natural properties that they are also good properties (Moore 1993).

We conclude, therefore, that all substantive accounts of well-being face serious objections. No attempt to determine what goods make someone's life the most worth living is satisfactory. By philosophical reflection alone we cannot find a sufficient answer to this complex question at least for the time being. This, however, does not include that there is principally no answer to this question, nor that striving for a good live would be unreasonable or senseless. Rather than giving practical advice, the thesis that so far there is no definitive answer to what well-being consists in is an expression of philosophical humility.

One important aspect concerning the question whether labor contributes to someone's well-being cannot be discussed here at length, but should nevertheless be mentioned. While answering the question what intrinsic goods constitute a good life seems to be problematic, one might succeed more easily in determining whether labor would be a necessary condition to become an ethical person. That is, there might be certain social conditions, which are universally necessary in order to properly develop required ethical capacities. In this respect, the capabilities approach as proposed for example by Nussbaum (2006) would be an interesting starting point. Additionally, there comes further support for such assumption from social psychology. In particular, it is widely held, that social appraisal is an essential basis for a positive self-relation (Mead 1934; Honneth 1992). In modern societies, one fundamental realm of such appraisal can be found in the world of employment, as recent evidence suggests (Schlothfeldt 1999; Bieker 2005). For that reason, participation to the world of employment might indeed be ethically good, although not individually as a constituent but rather in this more fundamental supra-individual respect as a necessary condition to well-being.

6.3.1.4 Moral Reasons for Inclusion to the World of Employment

The third domain of practical reasons is the moral domain. Moral reasons are always impartial reasons. They hold necessarily and universally, or, as Kant puts it, they are categorically valid. In recent years, this domain of moral reasons has also been called the domain of justice because it strictly refers to what people "owe to each other" (Scanlon 1998; Darwall 2013). Accordingly, in this section we aim to answer the question whether citizens have the standing to make legitimate claims against each other regarding their participation in the world of employment, or, that is, whether ATW is morally good.

Since John Rawls' influential work "A Theory of Justice" (Rawls 1971) it is wieldy accepted that the question what claims citizens can legitimately make in a society is best answered with reference to its so-called basic structure, that is, to the most fundamental institutions, which together provide societies' essential functions. Following Rawls, we hold that this structure is best understood as a complex form of social cooperation among free and equal persons for their mutual advantage. Accordingly, justice requires a fair distribution of all respective shares in its benefits and burdens that arise within this cooperative venture among its members. In order to decide what this claim exactly demands, Rawls suggests, one must consider which principles of justice can be reciprocally proposed by all citizen to one another for mutual acceptance, or, as Scanlon (1998) puts it, which principles no one can reasonably reject.

Given this normative concept of society as cooperation of free and equal persons, we are inclined to say that there are both rights and duties to participation in the world of employment. We do not claim, however, that there are universal claim-rights to labor, that is, that citizens have individual rights to demand a position in economy, which, then, must be provided. We rather suggest that everyone is the legitimate holder of a freedom-right to labor, which entails a demand to fair equality of opportunity.[3] In other words, an unrestricted participation in the world of employment must be available for every citizen as well as a fair chance to attain desired positions. This assumption echoes the first part of Rawls' second principle of justice, which states that "social and economic inequalities (…) are to be attached to offices and positions open to all under conditions of fair equality of opportunity (…)"(Rawls 2001: 42). Moreover, it appears that there is also a pro-tanto obligation to participation since fair terms of cooperation require a fair share of engagement in its constitutive practices. We claim thus that

Thesis I$_{(Moral)}$ *ATW is morally good because it ensures social inclusion to the world of employment, which is demanded both as a right to fair equal opportunity and a pro-tanto duty to participation.*

To prove this claim, we shall first state two reasons, why we reject the view that there are individual claim-rights to labor, the first is a moral reason (a), the second a technical (b). Then, we shall propose a rationale for our substantive claims as proposed in thesis I$_{(Moral)}$, namely that there is a right to equal opportunity (c), and a pro-tanto duty to participation (d).

(a) A claim-right to labor would conflict with the personal freedom of citizens. That is, it would be morally wrong to force someone to fill certain positions since everyone has the standing to freely decide what to do with one's talents and natural assets. This comprises free choice of employment. But in mass societies the division of labor indicates that desired positions are almost always restricted. That is, economic positions are distributed in order to fulfill its function, namely to satisfy the needs of its citizen. A claim-right to labor, then, could only be exercised without free competition, but rather through a central administrative decision, which would constantly conflict with the freedom of citizens (Hayek 1944).

(b) A claim-right to labor would also imprudently undermine the positive effects of a competitive market system. Given that the best economic output can be produced for a society when most suitable aspirants are in charge, competition seems to be a necessary means in order to attain the best qualified employees for each position. Thus, organizing economy by means of a demandable claim-right to labor would be less efficient and therewith would have bad economic effects for all members of society.

(c) To account for these considerations, we hold that citizen have no right to claim a position. Rather, labor should be distributed by means of a competitive market system among its members in order to ensure their civil liberties in accordance with satisfying the principle of economic efficiency. Note, however, that as free and equal members everyone must possess a status that enables him to free participation in the world of employment. This follows from the initial idea of free and equal membership in fair cooperation to mutual advantage. In other words, regardless which reasons one may have to engage in that practice, equal

---
[3] The applied concepts of claim-, and freedom-rights are explained in Wenar (2005).

membership demands that the participation to its essential practices is made available for everyone. Moreover, since social conditions are highly arbitrary, justice requires measures, which ensure that careers are open to all under conditions of fair equality of opportunity. Or, as Rawls famously proposed: "Those at the same level of talent and ability and who have the same willingness to use them, should have the same prospects of success regardless of their initial place in the social system, that is irrespective of the class into which they were born" (Rawls 1999b: 161). Other distributive principles would be unacceptable on rational grounds since they cannot account for the equal worth of the cooperating persons (Gosepath 2004). That is, it would be morally wrong to not grant free access to the world of employment for every citizen.

(d) Yet justice requires a fair distribution of both benefits and burdens of cooperation. We suggest, therefore, that there, too, is a pro-tanto duty to participation in the world of employment. When it has been accepted that the needs and desires of its members should be satisfied through a competitive market system, everyone has a claim against the others to engage in order to fulfill their fair share of participation. This duty is only pro-tanto because it might be the case that persons reciprocate their contribution in other respects.

Although it has been shown so far that the development of ATW is very well favored by good reasons, we omitted the second question by what means the application of ATW is justified. That is, which standards specify how one should use it.

6.3.2 Two Normative Standards of Good Workplaces

Working conditions are traditionally contested. While industrial workers in the 1950s and '60s mostly claimed leisure and wage increase as compensation for their hard work in mass production, the maintenance of save and healthy working conditions became a general issue in the 1970s under the label "humanization of work". In recent years new "good work"-initiatives took up this policy in order to establish reasonable working conditions (Sauer 2011).

These efforts must be endorsed with regard to ATW. In particular, we claim that the application of ATW is justified only in so far as it satisfies precisely two standards of good workplaces, namely an exploitation restraint and a duty of care. Although the concrete content of these standards must always be legitimate objects of open and fair negotiations, we hold that there is an underlying basis that must be recognized.

6.3.2.1 The Exploitation Restraint

The concept of personal autonomy is central to modern self-understanding. It refers to the competence to act for one's own reasons as well as to be accountable for one's actions. According to Kant the autonomy of persons must be unconditionally respected. It is a universal moral law, he claims, to treat anybody "always at the same time as end and never merely as means."(Kant 2002: 47). Thus, legitimate working conditions must always respect the worker's autonomy. On a plausible interpretation, this claim entails that it would be morally wrong to treat them in a way in which they cannot see themselves as self-determined. In other words, workers must be able to rationally agree to what they are doing when using ATW.

For that reason some authors condemn industrial work in general. They contrast it with an ideal of handicraft that, according to their view, is the only appropriate way of working in a humane way (cf. e.g. Sennett 2008). Such views must, however, be rejected. Following Axel Honneth (2008), we assume that conditions of mass-production are irreversible conditions of modern societies, which therefore cannot rationally be refused. Whoever demands such working conditions fails to recognize that there are natural restrictions to what is feasible in modern times.

This, of cause, is no pleading for one-sided justifications of substandard working conditions. Rather, workers must always be able to assert their interests and needs in order to shape working conditions in accordance with their autonomy and ethical self-understanding. This demands at least strong workers representation and a general democratization of corporate policy. Thus, rather than patronizingly stipulate what autonomy requires, its concrete interpretation must reciprocally be negotiated with representatives of industry and unions.

6.3.2.2 The Duty of Care

Beside autonomy the individual needs of impaired workers must be met as well. Especially persons with disabilities are more frequently in need of help that requires special care. Note that the concept of care must be taken as assistance to self-determination rather than a veiled form of paternalism. In other words, we do not claim that designers of ATW should anticipate the "real needs" of impaired workers. That would undermine their autonomy. Rather, ATW must be able to detect when care is required and provide special features in order to assist workers to assert their individual needs by means of special feedback functions.

Moreover, when using ATW workers must be able to attain positive self-assessment in order to reduce psychological stress (Joiko et al. 2006). That is, they must be able to identify with their work in some positive way such as perceiving it as meaningful, feeling useful, and to constantly make personal progress as far as possible. Thus, the level of assistance provided must be adjusted in consideration of the individual capacity of each worker in order to neither overload her mentally nor demand to little of her. In accordance with Joiko et al. (2006: 24) the following five typical features must be granted to comply with these requirements: (a) The working tasks should be holistic, so that workers are able to perceive direct feedback about their overall progress. (b) In order to avoid one-sided strain a diversity of tasks should be provided. (c) Social interaction among workers should be supported since that strengthens mutual problem solving. (d) Tasks should be arranged such that they supply individual alternatives for action in order to promote one's self-determination. (e) In order to preserve and promote the individual skills of worker's tasks should always be challenging in some way. These aspects make clear that ATW must not only satisfy requirements of good work but that it also can contribute to fulfill it. As outlined in Sect. 6.2 ATW can adjust its functionality in situ and therewith provide precise individual assistance.

6.4 Conclusion

It was the purpose of this paper to address general worries regarding the development and application of ATW. Therefore, we gave a concrete concept how this technology might be constructed and proposed eight technical functions it should adopt in order to serve its purpose. Then, we discussed the question why one should use ATW and by what means. In sum, we have argued that ATW should be employed because it ensures social inclusion to the world of employment and considered four normative domains in which its worth may exactly consists in. In addition, we insist that ATW must satisfy two requirements of good workplaces, which we specified as (a) an exploitation restraint and (b) a duty of care.

One final remark: It is a well-known fact that tools can be abused. The same is true for ATW. An undesirable incident, which everyone has strong reason to avoid, would be that the use of ATW provokes an army of unskilled laborers and thus replacing qualified workers. Besides critical public attention there is little one can do to entirely prevent such malpractices. Thus, in order to ensure that ATW will only be used in accordance with its positive features, everyone ought to take his responsibility seriously and make use of his political powers when required.

References

Arendt, Hannah. 1998. *The human condition*. Chicago: The University of Chicago Press.
Aristotle. 2011. *Nicomachean ethics*. Trans. Robert C. Bartlett and Susan D. Collins. Chicago: The University of Chicago Press.
Bay, Herbert, Tuytelaars Tinne, and Gool Luc Van. 2006. Surf: Speeded up robust features. In *Computer vision–ECCV 2006*, 404–417. Berlin/Heidelberg: Springer.
Bieker, Rudolf. 2005. Individuelle Funktionen und Potenziale der Arbeitsintegration. In *Teilhabe am Arbeitsleben*, ed. Rudolf Bieker, 12–24. Stuttgart: Kohlhammer.
Blakemore, Sarah-Jayne, and Suparna Choudhury. 2006. Development of the adolescent brain: Implications for executive function and social cognition. *Journal of Child Psychology and Psychiatry* 47(3): 296–312.
Blanke, Ulf, Bernd Schiele, Matthias Kreil, Paul Lukowicz, Bernard Sick, and Thiemo Gruber. 2010. All for one or one for all? Combining heterogeneous features for activity spotting. In *Proceedings of the IEEE PerCom workshop on context modeling and reasoning*, 18–24. Mannheim: IEEE.
Blesinger, Berit. 2005. Persönliche Assistenz am Arbeitsplatz. In *Teilhabe am Arbeitsleben*, ed. Rudolf Bieker, 282–285. Stuttgart: Kohlhammer.

Bude, Heinz, and Andreas Willisch (eds.). 2008. *Exklusion. Die Debatte über die Überflüssigen.* Frankfurt a.M.: Suhrkamp.
Convention on the Rights of Persons with Disabilities (UNCRPD), BGBl. III 155/2008. http://www.un.org/disabilities/convention/conventionfull.shtml
Darwall, Stephan. 2013. Morality's distinctiveness. In *Morality, authority, & law*, 3–19. Oxford: Oxford University Press.
Forst, Rainer. 2012. *Toleration in conflict. Past and present.* Trans. Ciaran Cronin. New York: Cambridge University Press.
Frehe, Horst. 2005. Das arbeitsrechtliche Verbot der Diskriminierung behinderter Menschen. In *Teilhabe am Arbeitsleben*, ed. Rudolf Bieker, 62–80. Stuttgart: Kohlhammer.
Funk, Markus, Oliver Korn, and Albrecht Schmidt. 2014. An augmented workplace for enabling user-defined tangibles. In *CHI'14 extended abstracts on human factors in computing systems*, 1285–1290. New York: ACM.
Gosepath, Stefan. 2004. *Gleiche Gerechtigkeit. Grundlagen eines liberalen Egalitarismus.* Frankfurt a.M.: Suhrkamp.
Griffin, James. 1986. *Well-being: Its meaning, measurement and moral importance.* Oxford: Clarendon Press.
Habermas, Jürgen. 1996. *Between facts and norms: Contributions to a discourse theory of law and democracy.* Trans. William Rehg. Cambridge, MA: MIT Press.
Hardy, John, and Jason Alexander. 2012. Toolkit support for interactive projected displays. In *Proceedings of the 11th international conference on mobile and ubiquitous multimedia*, 42. New York: ACM.
Hayek, Friedrich August. 1944. *The road to selfdom.* London: Routledge.
Heines, Hartmut. 2005. Teilhabe am Arbeitsleben – Sozialrechtliche Leitlininien, Leistungsträger, Förderinstrumente. In *Teilhabe am Arbeitsleben*, ed. Rudolf Bieker, 44–61. Stuttgart: Kohlhammer.
Höffe, Ottfied. 2006. *Lebenskunst und moral.* München: Beck.
Honneth, Axel. 1992. *Kampf um Anerkennung. Zur moralischen Grammatik sozialer Konflikte.* Frankfurt a.M.: Suhrkamp.
Honneth, Axel. 2008. Arbeit und Anerkennung. Versuch einer Neubestimmung. *Deutsche Zeitschrift für Philosophie* 56(3): 327–341.
Honneth, Axel. 2011. *Das Recht der Freiheit.* Berlin: Suhrkamp.
Hurka, Thomas. 1993. *Perfectionism.* New York: Oxford University Press.
Joiko, Karin, Schmauder Martin, and Wolff Gertrud. 2006. In *Psychische Belastung und Beanspruchung im Berufsleben*, ed. Bundesanstalt für Arbeitsschutz und Arbeitsmedizin. Bönen/Westphalen: Kettler.
Kagan, Shelly. 1998. *Normative ethics.* Oxford: Westview Press.
Kambartel, Friedrich. 1993. Arbeit und Praxis. *Deutsche Zeitschrift für Philosophie* 41(2): 239–249.
Kant, Immanuel. 2002. *Groundwork for the metaphysics of morals* (ed. and trans. Allen W. Wood). New Haven/London: Yale University Press.
Kolodny, Niko, and John Brunero. 2013. Instrumental rationality. *The Stanford encyclopedia of philosophy* (Fall 2013 Edition), ed. Edward N. Zalta. http://plato.stanford.edu/archives/fall2013/entries/rationality-instrumental/. Accessed 15 Sept 2014.
Korn, Oliver, Albrecht Schmidt, and Thomas Hörz. 2012. Assistive systems in production environments: exploring motion recognition and gamification. In *Proceedings of the 5th international conference on pervasive technologies related to assistive environments*, 9. New York: ACM.
Korn, Oliver, Albrecht Schmidt, and Thomas Hörz. 2013. The potentials of in-situ-projection for augmented workplaces in production. A study with impaired persons. In *CHI'13 proceedings of the ACM SIGCHI conference on human factors in computing systems*, 979–984. New York: ACM.
Korn, Oliver, Markus Funk, Stephan Abele, Thomas Hörz, and Albrecht Schmidt. 2014. Context-aware assistive systems at the workplace. Analyzing the effects of projection and gamification. In *PETRA'14 Proceedings of the 7th international conference on pervasive technologies related to assistive environments.* New York: ACM.

Mead, Herbert. 1934. In *Mind, self, and society*, ed. Charles W. Morris. Chicago: University of Chicago Press.
Moore, George Edward. 1993. In *Principia ethica*, ed. T. Baldwin. Cambridge: Cambridge University Press.
Nozick, Robert. 1974. *Anarchy, State, and Utopia*. New York: Basic Books.
Nussbaum, Martha. 2006. *Frontiers of justice: Disability, nationality, species membership*. Cambridge, MA: Harvard University Press.
Parfit, Derek. 1984. *Reasons and persons*. New York: Clarendon.
Parfit, Derek. 2011. *On what matters*. Oxford: Oxford University Press.
Rawls, John. 1971. *A theory of justice*. Cambridge, MA: Harvard University Press.
Rawls, John. 1999a. Two concepts of rules. In *Collected papers*, ed. Samuel Freeman, 20–46. Cambridge, MA: Harvard University Press.
Rawls, John. 1999b. Distributive justice: Some Addenda. In *Collected papers*, ed. Samuel Freeman, 154–175. Cambridge, MA: Harvard University Press.
Rawls, John. 2001. *Justice as fairness: A restatement*. Cambridge, MA: Belknap Press.
Sauer, Dieter. 2011. Von der "Humanisierung der Arbeit" zur "Guten Arbeit". *Aus Politik und Zeitgeschichte* 15: 18–24.
Scanlon, Thomas. 1998. *What we owe to each other*. Cambridge, MA: Harvard University Press.
Schlothfeldt, Stephan. 1999. *Erwerbsarbeitslosigkeit als sozialethisches Problem*. Freibung/München: Alber.
Schmelzle, Cord. 2012. Zum Begriff politischer Legitimität. *Leviathan* 27: 419–435.
Schulze, Marianne. 2011. Menschenrechte für alle: Die Konvention über die Rechte von Menschen mit Behinderung. In *Menschenrechte – Integration – Inklusion. Aktuelle Perspektiven aus der Forschung*, ed. Petra Flieger and Volker Schönwiese, 11–26. Bad Heilbrunn: Klinkhardt.
Sennett, Richard. 2008. *The craftsman*. New Haven: Yale University Press.
Smith, Adam. 1976. In *An inquiry into the nature and causes of the wealth of nations*, ed. Roy H. Campbell and Andrew S. Skinner. Oxford: Oxford University Press.
Stahl, Titus. 2013. *Immanente Kritik*. Frankfurt a.M.: Campus.
Steinfath, Holmer (ed.). 1998. *Was ist ein gutes Leben? Philosophische Reflexionen*. Frankfurt a.M.: Suhrkamp.
Wenar, Leif. 2005. The nature of rights. *Philosophy & Public Affairs* 33(3): 223–252.
Williams, Bernard. 1972. *Morality: An introduction to ethics*. Cambridge: Cambridge University Press.

Chapter 7
Some Sceptical Remarks Regarding Robot Responsibility and a Way Forward

Christian Neuhäuser

> *Adama:* "She was a Cylon, a machine. Is that what Boomer was, a machine? A thing?"
> *Tyrol:* "That's what she turned out to be."
> *Adama:* "She was more than that to us. She was more than that to me. She was a vital, living person aboard my ship for almost two years. She couldn't have been just a machine. Could you love a machine?"
>
> *(Battlestar Galactica, 'The Farm', Season 2, Episode 5, 2005)*

Robots could eventually become just like humans. They could then not only think, but also feel. They would be able to develop their own personalities, love and hate other robots and humans, develop individual life plans, they would learn to respect themselves, and even suffer from low self-esteem. We do not know if this will ever happen. Only time can tell. Nevertheless, for many people this idea of human-like robots exerts a large and sometimes strange fascination. Maybe this is so, because if people could create such robots they would then possess almost divine powers. Or it could also be the other way around. That is to say, maybe by creating these machines humans could become disenchanted with humankind by realising that there is nothing divine about themselves.[1]

Regardless of which position one takes, there will always be people who try to build such human-like robots. In fact, this has already been happening for a long time. To be sure, it is not possible to produce human-like robots at the moment. But there have definitely been attempts at building such robots, and there have also been initial successes within the field of robotics (Benford and Malartre 2007; Beck 2009; Nourbakhsh 2013). Today, robots are carrying out more and more tasks that

---

[1] The position of humankind in the cosmos is unquestionably one of the fundamental anthropological questions and is an essential part of the question: "What is the human?" as discusses by philosophical anthropologists like Helmut Plessner (1986) and Max Scheler (2009).

C. Neuhäuser (✉)
Institute of Philosophy and Political Science, University of Dortmund, Dortmund, Germany
e-mail: christian.neuhaeuser@udo.edu

© Springer International Publishing Switzerland 2015
C. Misselhorn (ed.), *Collective Agency and Cooperation in Natural and Artificial Systems*, Philosophical Studies Series 122,
DOI 10.1007/978-3-319-15515-9_7

previously could only be undertaken by human beings alone. Robots perform transactions in financial markets and could perhaps even be (co-)blamed for the current financial crisis. Robots can defuse bombs during wars and can even shoot people or drop bombs on them. They can commit war crimes if they attack the wrong persons, innocent civilians for instance (Arkin 2009; Sharkey 2009). Robots, like the German Care-O-Robot, now serve and assist older people who might otherwise be unable to cope with everyday life. But if they were to act in an incorrect way or do something completely wrong, they could also cause damage.

In certain respects, those robots act autonomously and can interact with both humans and other robots. Yet, in many cases, they are still controlled directly through operators in the background, as e.g. in the case of combat drones (Sharkey 2012). But sometimes they are not controlled directly, but instead follow complex computational programmes. In simple cases, such as service robots, one can clearly see what type of programmes they execute and how they operate following their commands. In other cases, this cannot be perceived as easily, firstly because robots can assimilate a large amount of information, and secondly because they are capable of learning, in the sense that this additional information can affect not only individual actions, but also the way they record information in the future, evaluate their goals, make new plans, and so on. This is already the case with robots that operate in financial markets. Eventually, it could be that no one knows the original programming of such a robot, such that it appears to us as an independent machine pursuing its own goals (Lin 2012; Bekey 2012).

Such a robot that can pursue its own goals would appear to be very similar to a real person. In the case of human beings too, it could be that we actually follow a certain inherent programming, but just do not know what this original programming looks like. It therefore appears to us as if individuals were acting completely autonomously, would pursue their own goals, and possess such a thing as free will. At least some philosophers and scientists who adhere to a naturalistic worldview have proposed an argument regarding the possible "programming" of people, arguing that free will and autonomy are just pragmatic assumptions of folk theory. No matter what position one ultimately takes on human free will in this respect, when dealing with the question of whether programmed robots could someday be like people, one should not rule out such a possibility a priori.

If it is possible that robots could one day develop autonomy analogous to human individuals, and if they already do many things that only people could do previously, then the question regarding the moral status of these robots emerges. This question includes two sub-questions (Gunkel 2012, 5). The first question is: Are robots moral agents? This refers to agents who act morally or could at least have a moral impact. One can also formulate the question as follows: Are robots capable of responsible actions? Because the ability to act morally implies the ability to act responsibly, if not necessarily the willingness to act responsibly. Or we might put it the other way around: only such beings that can bear responsibility are also capable of acting morally.[2] The second underlying question is: Are robots agents that should

---

[2] Catrin Misselhorn (2013) disagrees. She thinks that for moral agency it is enough to be able to act for moral reasons. Small children or animals acting out of sympathy would be moral agents in this

be morally considered, as individuals who possess moral rights or at least could have moral claims? Today, many people believe that all sentient beings have moral claims. But because people are not only sentient, but also reasonable, they have a higher moral status as compared to other animals. At least this is the prevailing opinion within the discussion. According to this position, humans do not only have moral claims of only relative importance but also inviolable moral rights because they possess dignity.[3]

If robots become sentient one day, they will probably have to be granted moral claims. So far, however, such a situation is not yet in sight. I will therefore only discuss the question of whether robots can be regarded as moral agents with responsibility without having their own moral claims. After all, it could be that robots are capable of being responsible without having to be sentient.[4] I will elaborate on this question regarding the ability of moral responsibility of robots in four steps. In a first step, I will explain why this question is not only interesting but also important. In a second step, I will analyse which capabilities are usually assumed to be underlying the ability to act responsibly and what this means for robots. In a third step, I will deal with the question of whether there is a weaker notion of moral responsibility for such robots that do not fulfil the usual conditions. And in a fourth and final step, I turn to the question of how to deal with a limited capacity of robots to act responsibly in groups.

7.1 Why Is the Question Regarding Moral Robots Interesting and Important?

The question regarding the morality of robots and whether they are at all able to bear responsibility is interesting for different reasons. At the beginning I had mentioned the idea that the human self-image as divine might be affected by the existence of moral robots. In addition, it may appear interesting to explore philosophical mind games of a world where morally able but massively irresponsible robots interact with society, as is the case in many science fiction stories. But beyond this rather aesthetic consideration, the question of moral robots is relevant and important for at least three reasons that are of general interest, as well as being of socio-political importance.

---

sense and robots could be too. The problem with such a downgraded concept of moral agency is that it neglects reasonable disagreement regarding moral reasons, as I will argue later on. Moral agents need to be able to deal with this reasonable disagreement by deliberating about the moral value of her reasons. Otherwise they do not act for moral reasons, but only for reasons that happen to be in line with morality.

[3] However, it is quite debatable whether animals can be said to have a form of dignity (certainly not human dignity), but perhaps an animal dignity or one shared with humans, such as the dignity of the creature, as it is stated in the constitution of Switzerland.

[4] Floridi and Sanders (2004, 350) argue that there must be some connection between being a moral agent and being a moral patient. Contrary to this I believe that it is possible that being a moral agent and being a moral patient is connected to totally different properties, rationality and sentience for instance. Then there is no difficulty in assuming that there might be entities that are moral agents, but no moral patients and vice versa.

Firstly, one can learn something general about moral responsibility through the question of moral robots. For us humans, it seems completely natural to consider ourselves as moral agents that have responsibility. Yet, this natural assumption sometimes makes it quite difficult to answer tricky questions about when exactly and for what reasons someone is responsible, when someone can be excused from responsibility, or when no responsibility is present at all. The implicitness of everyday attributions of responsibility and notions of moral responsibility might obstruct a more sober look at a coherent concept of moral responsibility. Especially the irritation that emerges when attributing responsibility to robots could provide clarity or even a better understanding of what moral responsibility means.

Secondly, one should maybe hold robots accountable if they are capable of operating responsibly. When robots do or refrain from doing certain things of moral importance, this can be an indicator for the possibility of them being held accountable. If moral agents do morally praiseworthy things, then they deserve this praise. However, if they do morally condemnable things, then they have earned blame and deserve a penalty.[5] This would also apply to robots if it can be proven that they are moral agents. They would become a part of the moral community as moral agents and must be treated accordingly, even if they are not moral patients and do not have direct moral claims and rights. Perhaps it is even possible to relieve people from a part of their responsibility and instead assign robots with more responsibility if and insofar as they are responsible agents.

Thirdly, robots and their activities have an impact on human responsibility. It might not make a difference morally, but certainly psychologically there is a difference between killing a man with a gun while standing in front of him, and operating a deadly attack drone while sitting far away in a comfortable chair in front of a computer screen. Computers in financial markets might not be held responsible when making software-guided decisions about the purchase and sale of financial products at breath-taking speed. But it could very well be that if robots are not held responsible then no one else will be. When robots assist older people the elderly are not being taken care of by humans who are responsible for handling them with respect. The robots' treatment of the elderly might not be disrespectful, but it is likely not to be especially respectful either. Perhaps almost no one will treat the elderly with respect then, if most of the care taking is done by robots (Sharkey and Sharkey 2012).

Those three reasons just mentioned motivate the currently very prominent inquiry into the responsibility of robots. Usually three conditions need to be fulfilled for the specific attribution of responsibility to an agent. Firstly, there has to be at least a possible causal connection between the agent and the event for which responsibility is ascribed (Moore 2010). Secondly, the possibility of taking a different

---

[5] This is often called the merit-based concept of responsibility, because the person or persons responsible deserves a punishment or reward. See: John M. Fischer and Mark Ravizza, *Responsibility and Control: A Theory of Moral Responsibility,* Cambridge: Cambridge University Press 1998, pp. 207 ff.

course of action within the given situation must exist. Either you desist from doing something you could have done or you have done something that you could have refrained from doing, and often both situations have to be considered (Birnbacher 1995). Thirdly, one has to have enough information about which actions are possible in specific decision-making situations and how these actions are to be evaluated from a moral point of view.[6] This is especially true when the consequences of actions are evaluated morally. Robots can change the decision-making situations of people in such a way that the causal relationships change, whereby certain acts and omissions are no longer possible, specific information is no longer available, and the moral significance of single decisions presents itself differently.

Robots act independently or refrain from doing so, and this changes the conditions for making decisions. They therefore exert an influence on the distribution of responsibility. As long as they cannot take any responsibility themselves, they might have a negative impact on something that can be called an extensive network of responsibility (Neuhäuser 2011). The ideal of an extensive network of responsibility states that for all matters that are important to people, someone should be responsible, or at least it should be possible to hold someone accountable. The more actions irresponsible robots undertake and the less predictable their actions become, the stronger their potentially negative influence within this extensive network of responsibility will be. From a practical point of view robots can then be seen as being similar to uncontrollable wild animals or unpredictable forces of nature. The unpredictable nature of their actions allows for gaps to emerge within the extensive network of responsibility. People would then have to take responsibility not only for themselves, animals and nature, but also for robots and their doings in order to fill these gaps.

The problem can be summarised as follows: If robots are not capable of acting responsibly, then this immediately raises the question of who is responsible for their actions instead and what is their impact on the actions of human individuals. However, if robots are capable of acting responsibly, then they might even contribute to reinforce a more tightly knitted network of responsibility. We could make sure that they bear responsibility for the important concerns of people. Caretaking robots, for example, would then be responsible for the well-being of needy people and could cultivate a respectful relationship with them. The way we distribute moral and legal responsibility and how our society and its institutions are to be organised accordingly depends on the moral status of robots. This is why this question is not only interesting, but also important. The more things robots do, especially the more potentially dangerous things they do and the more we integrate them into our society, the more important this question becomes.

---

[6] In the *Nicomachean Ethics* (1110a–1111b4), Aristotle already pointed to the control condition and the epistemic condition regarding information.

7.2 What Capacities Are Normally Required in Order to Have Moral Responsibility?

A good starting point to approach the question of whether robots are capable of responsibility is the theory of personhood developed by Daniel Dennett (1976). Starting with a theory of the person makes sense, because according to the standard theory not all individuals but only persons can be regarded as moral agents. The ability to be responsible depends on having the status of a person. Dennett developed an especially extensive theory of the person, which does not limit the status of personhood to humans. His theory is open enough for this status to be attributed also to other types of agents. Based on his theory one can discuss, for example, whether animals or people or companies are agents capable of bearing responsibility for their actions, and Dennett himself used it to toy with the idea of responsible robots (Dennett 1997). For this reason, Dennett's approach seems appropriate when we ask the question of robot responsibility. If even the generous approach of Dennett leads us to conclude that at the present time and in the near future robots are not capable of bearing responsibility, very little speaks in favour of doing so.

Dennett determined three criteria that must be met before someone can be called a person: Firstly, third-order intentionality must be given; secondly, language; and thirdly, self-awareness. Dennett himself assumes that these three properties build on each other; this means that language presupposes third-order intentionality, and self-awareness in turn presupposes language. One could also argue that all three properties and conditions exist only together (Rovane 1995, Quante 2007, 24). This is of no importance, since in both cases robots would need all three properties. It is more important what those three properties require. Intentionality simply means that one has mental states that are focused on specific contents: one wants, believes or hopes for something, and so on. Second-order intentionality means that someone has intentional states referring to intentional states of someone else or himself. And third-order intentionality means that someone has intentional states referring to the intentional states of someone else who in turn refer to other intentional states. This sounds complicated, but it is familiar to us in everyday life: I know you are afraid that I want to go to the cinema today. Here I have the intentional state of knowledge, which refers to your intentional state of worrying, which in turn relates to my intentional state of wanting something. There are therefore three related intentional states, whereby the third state presupposes the first two.

An agent with third-order intentionality does not simply posses intentional states. This would merely count as first-order intentionality. One also does not simply regard the other as an intentional agent. That would just amount to second-order intentionality. At this level we attribute intentionality to a chess computer when we confer on it the will to win. When third-order intentionality is attributed, one sees the opponent as an intentional agent who also sees his opponent as an intentional agent. The chess computer would then know that his opponent knows that he wants to win. It is quite doubtful that chess computers will ever become intentional agents operating at the third stage of intentionality. At least there is no reason for us to think so. A chess computer simply wants to win, this is enough for us to at least to attempt to beat it, and we do not need to assume more (Dennett 1987, 43–68).

It now becomes clear why third-order intentionality is an important prerequisite for language. If one does not simply exchange signals to communicate, like animals do, but uses a proper language with intensional meaning, one must presuppose that the addressee also understands this language, which implies that s/he also understands that you understand it yourself as language in this sense. Only if there is a mutual assumption that all involved parties are capable of intentionality, interaction is possible using a language with meaning.[7] Self-awareness seems to be fundamentally based on these skills. It occurs in the form of an objective relationship to oneself as a subject. This appears to presuppose language, because one encounters oneself in the third person: I think about me, as a person with a certain name and certain properties. Here Dennett seems to assume that self-awareness is dependent on social interaction. Third-order intentionality requires two agents to mutually ascribe intentionality to one another that thereby become aware of their status as intentional agents (Dennett 1992).

Because here we are concerned with robots, it is not important to further analyse how human beings actually come to acquire self-awareness. It is more important to recognise that robots currently do not meet the conditions established by Dennett. To be sure, one could argue that we do not know if robots have self-awareness, but at least there is no evidence that would make us believe that this is the case. The only thing we are likely to assume is that robots can utilise language. At least that is true for certain computer programmes. However, it is not clear whether this is really a language connected with third-order intentionality, which is used to communicate intentional states with someone. Robots usually do not express that they see other robots and humans as intentional agents, whereby they assume at the same time that they themselves are also viewed as intentional agents by those other intentional agents. They therefore do not express anything that gives us reason to believe that they do possess self-awareness.

There are currently computer programmes that seem to be able to pass the Turing test, or at least come close to doing so.[8] Yet I believe that winning this test is not enough to attribute third-order intentionality and self-awareness to those programmes in any case. Such a test therefore gives little proof regarding moral agency, even if it is especially designed for moral communication (Allen et al. 2000). During the Turing test, conversations are held through monitors in a very artificial environment. In such a limited environment, a computer may indeed

---

[7] It is not enough to say that language has a propositional content, which in a sense also applies to the signals emitted by animals. It is more about the intentional content of language compared to its extensional content (Frege 1962). Lewis Carroll, the famous author of *Alice in Wonderland,* and the famous logician Charles L. Dodgson are the same person. The two names have therefore the same extension, but do not have the same intension, because a speaker can refer to both names without knowing that the two names refer to the same person.

[8] This is being controversially discussed with the 'cleverbot' programme. This programme is free to use online. If you know that it is a machine, it is not very difficult to see that it is not human. Most recently the programme name Eugene Goostman is said to have passed the test.

appear to show signs of third-order intentionality. At least this facilitates the communication with it or makes it even possible in the first place, as Dennett would argue. But beyond this very limited context there are no reasons for attributing artificial intelligence in the sense of third-order intentionality and self-awareness to a robot. Robots would not pass such an intentionality test if applied in changing contexts, or so I believe. Even if third-order intentionality is attributed to robots in very limited contexts, they do not really appear as persons. This is so, because they do not show that they "deserve" this attribution by also acting as persons with third-order intentionality in any other contexts. This does not mean that things might not change in the future. But at present there is no good reason to regard them as persons in Dennett's sense which in turn means that there is also no reason to regard them as agents with moral responsibility.

There is yet another reason why the currently existing robots cannot be regarded as persons and moral agents with responsibility. Dennett himself only superficially discussed another condition that could possibly exist in order to attribute personality and in particular the status of being a moral agent capable of responsibility. This condition is the ability to assess one's own desires and judgements by means of second-order desires or an evaluative system, and to replace them if necessary. Harry Frankfurt has proposed this ability as a condition for moral responsibility (Frankfurt 1988, 11–25). Dennett is unsure whether or not to accept this condition, because maybe one should not equate metaphysical personhood and moral personhood (Dennett 1976; Smith 2010). After all, we might be able to find individuals with self-awareness who yet lack the ability to bear responsibility. For moral agency, however, the ability to evaluate one's own desires on the basis of one's own value system seems to be of great importance (Birnbacher 2006, 53–76).

According to this line of thought, robots would require an evaluative system, a system of values or wholeheartedly affirmed desires, to reach the necessary level of competence and become moral agents. This means that in order to have moral responsibility, agents need an evaluative system or system of values that produces moral reasons (Watson 1975; Abney 2012). One must be able to take a moral standpoint that allows asking what wishes or what reasons for action are recommended, permitted, and prohibited from a moral point of view. Here too it seems doubtful whether robots are capable of such considerations (Neuhäuser 2012). Maybe if morality is simply understood as a kind of unwritten rulebook, this would perhaps seem possible. One would then simply have to create programmes based on these rules, such as Isaac Asimov proposed in his writings. However, morality does not seem to work in such a way. Moral principles most certainly are valid only *prima facie*. Most contemporary applied ethicists seem to agree on this. For robots to be moral agents, they need to be able to weigh up their options, compare them, have basic values, a moral sense, and perhaps even feelings.[9] This is why Frankfurt uses the formulation that some values are affirmed wholeheartedly. And even in Asimov's writings, it is ultimately the sentient robot that becomes a moral agent.

---

[9] Of course, this is rather an empirical assertion, which goes back to something like a discursive impression.

The upshot of this discussion is this: Robots, or at least current robots, have no feelings. Even if one does not accept this point, the fact still remains that they have no value system and also no moral sense, no moral judgement, no moral intuition (or whatever one might want to call it). Robots are therefore not moral agents with responsibility. In this moral sense, they are not capable of being responsible. The implication seems to be that we need to find someone else who bears responsibility for the actions of robots if we want to develop a comprehensive network of responsibility, in which robots as somewhat autonomous agents are taken into consideration. But there is still another possibility. One could argue that while robots are not fully responsible agents in the same way as human beings, perhaps they bear a weakened form of responsibility. If this holds true, then robots could at least bear some of the responsibility that results from their actions.

7.3 Are There Different Levels of Moral Responsibility Applicable to Robots?

Robots are not moral agents with responsibilities in the sense that they can be regarded as moral persons, at least not yet. But it is possible that they can be regarded as moral agents in a weaker sense, which allows for a discussion of a weaker form of responsibility. There are currently several proposals that focus on this weaker sense of moral responsibility of which I will highlight three. Colin Allen and Wendel Wallach argue that since robots can be seen as autonomous moral agents, it is enough to treat them as if they were moral agents for them to be moral agents (1). James H. Moor distinguishes between implicit, explicit and full moral agents, and argues that robots can be both implicit and explicit moral agents (2). Luciano Floridi and J.W. Sanders distinguish between accountability and responsibility, and argue that robots can be accountable, although not responsible (3).

1. Allen and Wallach's idea that robots are autonomous moral agents is based on the assumption that robots can act as if they were moral agents (Allen and Wallach 2010, 2012). For example, when a robot takes over the job of a nurse, it does something of moral significance. This robot might be designed in such a way that it can decide when it should prepare food, for example, if the patient says loud and clear 'food' or 'hunger'. It may perhaps even be able to make an emergency call. It could regularly measure the patient's body temperature and pulse, and react correctly by notifying the medical staff under certain circumstances. Wallach and Allen argue that because it is the robot that reacts to specific situations, by deciding whether an emergency call should be made or not it acts autonomously. Because these decisions are also of moral significance, the robot acts as if it were a moral agent. Our task then is to programme the robot so that it can make the right decisions by gathering the relevant information and following the right rules.

But in reality, things are not that simple. These robots are basically no different from an automatic toaster with a timer. When you turn on a normal toaster for a certain time and a curtain happens to get entangled with the toaster's heat source maybe through a gust of wind, the toaster will then toast and burn the curtain. It is doing exactly what it was meant to do. This can be of moral significance, for example, in the case of a consecutive fire burning down the whole house. If the toaster were in fact an autonomous moral agent in the sense proposed by Wallach and Allen, it would have decided to shut down such as not to set the curtain on fire. An advanced toaster probably could be programmed in order so that it can register what type of material it is toasting and can shut itself down if a certain threshold temperature has been reached. But what if the owner wants to burn down their house to collect the insurance and to save himself and his children from financial ruin? Legally, that is certainly not allowed, but perhaps the toaster was supposed to have compassion - if not with the parent, then at least with the kids. Perhaps the toaster was supposed to play an important role in a crafty attack on a political tyrant by burning down the house and it could have made an important contribution towards the liberation of humans, robots and toasters in this unfree country, if it had not shut down following its advanced programming.

However, neither a toaster nor a nurse robot can decide such things. The robots described by Allen and Wallach are just like the plumber from the famous example of Elizabeth Anscombe who pumps water into the house of the politician regardless of whether it is poisoned or not. Repairing damaged pumps and pipelines is simply what he does (Anscombe 1957, 38–40). Robots also simply follow rules, but they lack a moral standpoint, from which they can decide in concrete situations which information is morally relevant and how it should be evaluated. In the case of very narrow environments, robots can in fact make morally relevant decisions in the sense that they follow certain commands and programmes activated through situational data. But these environments need to be very tightly controlled so that no morally significant changes occur that have not previously been foreseen by the programmer. Such robots as described by Allen and Wallach therefore also act only automatically and not autonomously when it comes to moral decisions. Allen and Wallach also consider a bottom-up approach to morality, whereby robots learn moral behaviour through interaction. I will consider this approach when discussing the argument by Floridi and Sanders, who have spelled it out in more detail.

2. The distinction between implicit, explicit and full moral agents proposed by James H. Moor is intended to help us overcome the issue just discussed (Moor 2006). Only humans can be called full moral agents, for we are the only known beings that can create values, argues Moor. Robots are not full moral agents because they do not create values. However, this does not mean that they simply follow the rules put in place at the programming stage. That would only mean implicit moral behaviour, which can go wrong very quickly due to the complexity of the situations that demand actions, as we have just seen. Robots can be situated in between the extreme opposites of full and implicit moral agency by functioning as explicit moral agents. As such they can decide for themselves what is right and wrong in specific situations. The idea is that you do not provide them with concrete rules, but instead

programme them with a moral theory. Depending on the inclination of the programmer, the robots would be, for example, either utilitarian or Kantian. As Kantian robots for instance, they could then apply the categorical imperative. They would thus be able to ask themselves whether their actions can be implemented as generalised rules. As utilitarian robots, they might ask themselves whether their actions serve to maximise the overall benefit of the greatest number of morally relevant beings.

It is not clear whether it is possible to programme robots in such a manner that they can internalise and apply a moral theory such as utilitarianism or Kantian ethics. But even if it were possible, this does not present a solution to the problem. This approach misunderstands the practical significance of moral theories. Such theories are used to develop a coherent foundational structure to determine what is morally right and wrong. That such theories regularly fail to do so does not make them worthless because they also regularly produce important insights and reveal the complexity of moral practice. But to fully apply them one-sidedly on a practical level would be wrong. Although there is no consensus about this within applied ethics, many applied ethicists tend to combine different theories in a pragmatic nature. The point can be clearly illustrated by the fact that moral theories in their purest form and with their quest for consistency and coherence can result in morally unacceptable conclusions (Jonsen and Toulmin 1988; Dancy 2006, 53–69). Even if a moral philosopher, maybe a moral realist, insisted that s/he found a perfect set of moral principles free of any contradictions, it is very unlikely that there will be widespread agreement on this, as the history of reasonable disagreement in moral philosophy shows. In practice morality therefore possesses a compromising nature and moral agents need to able to deal with that.

Let us imagine a medical robot programmed to be a one-sided utilitarian. Following the famous example given by Judith Thomson (1985) we could then think of him as working in a hospital and deciding to remove all of the organs of a healthy middle-aged man in order to save the lives of five young individuals with different organ failures. His decision would certainly maximise the total benefit of the group (Thomson 1985). Or let us imagine a Kantian fighting robot. During war, it would not be willing to kill a single innocent person, even if it could thereby save millions of lives (Singer 2009). Representatives of utilitarianism or Kantian ethics will want to defend their approaches faced against these examples. And doing so certainly is not a problem on the level of a philosophical discussion, but it becomes decidedly problematic if we want to use robots as moral guinea pigs in the laboratory of society, as some scientists seem to suggest (Crnkovic and Çürüklü 2012, 64). Most people in everyday life and most applied ethicists are essentially mixed theorists and consider various moral theories in order to make balanced decisions in relation to specific moral problems. And this is what is expected of them as responsible agents, trying to make the best decisions in given situations.

The crucial problem is that it is not clear how robots can be programmed to be such mixed theorists, although they are not fully competent moral agents. Mixed theories are based on the fact that full moral agents can appraise the results of applying different moral considerations. Robots remain morally deficient even as explicit moral agents because every moral theory, at least every known moral theory, can lead to undesirable results when it is applied dogmatically in concrete situations. However, this is exactly what robots have to do if they are to be seen as explicit moral agents within Moor's argument – they have to apply their programmed moral theory in a dogmatic way.

3. Floridi and Sanders (2004) accept that robots are not responsible agents. To them, however, that does not mean that it is impossible to see them as moral agents with some kind of accountability. The situation is similiar to the one with animals like domestic dogs. We know that they are not morally responsible. Nevertheless, we hold them accountable – we either punish or reward them – depending on whether their behaviour is morally desirable or undesirable. For example, dogs can be educated to become guide dogs, guard dogs and rescue dogs. With robots it may be just the same, as Floridi and Sanders argue. If we hold them accountable from a moral point of view, we can educate them to behave in a morally compliant manner. By doing this we can relieve the programmer and operator who will no longer have to take full responsibility for the robots, the authors conclude.

However, there are two problems with this approach. The first problem consists in the fact that robots are not like domesticated animals. They are not sentient beings with desires and fears (Johnson and Miller 2008). Dogs want to eat or be petted; they want to be accepted by people and included in the pack. They are afraid of beatings, loss of love, and do not like to be expelled from the community. They might not be aware of this, because they lack self-awareness, but this is being discussed quite controversially (Wild 2011). In any case, although it does not change the fact that dogs have a wide range of emotions and therefore are social creatures. Their emotional and social character seems to be a necessary condition for the possibility of being educated to behave in a morally compliant way. With robots, by contrast, it is not clear how such an education process is supposed to work. Robots have no emotions. They can be reprogrammed quite easily, although that of course happens only in hindsight and does not affect future situations where the morally relevant conditions constantly change. Robots will not adjust their moral behaviour for fear of punishment or hope of reward or at least it is not clear how they can be made to do so (Asaro 2012).

The second problem has to do with the fact that the trainers of animals continue to be the ones responsible for the animals' behaviour. To be sure, they have only limited responsibility for what their animals do. When a dog attacks a person, the dog's owner is partially responsible for this act in two ways. He most likely has done something wrong, otherwise his dog would not have been so aggressive. He is thus responsible for this wrongdoing and probably deserves punishment. And he also has to pay damages to the victim. However, he is not responsible to the same extent as if he had attacked the other man himself, biting and seriously injuring him. The obvious consequence of this is a less densely linked network of responsibility. In the case of the dog attack, the possible attributions of responsibility are weaker. This has effects on the proactive and reactive effect of these attributions of responsibility. The same seems to be the case when it comes to robots. The operators are still responsible, but to a lesser extent. The network of responsibility remains, but is less densely linked. It seems then that it is more of a political decision whether or not one wants to allow this dense connectedness of the network to weaken. This leads us to the final part of this analysis, namely, the question of how to deal with the lack of responsibility in robots.

7.4 How to Deal with Morally Inapt Robots?

Maybe one day robots will become full moral agents (Sullins 2006). This is currently not the case. Instead, robots are agents with some autonomy but without responsibility. This has serious consequences for the distribution of responsibility and a densely linked network of responsibility that takes into account the morally relevant actions of robots. We have two possible ways to deal with this situation: either we begin to accept that robots do more and more morally relevant things. Then it is up to us to find other agents that bear responsibility for the actions of these robots – their producers, programmers and operators, for example. One could imagine the creation of a new group of supervisors, which would control and observe whether robots are behaving in a morally compliant fashion. However, this is almost inevitably accompanied by the consequence of having a less densely linked network of responsibility that could even show serious gaps. Robotic activity becomes socially and economically more relevant when the degree of autonomous behaviour rises. However, this necessarily limits the possibilities of control and thus the responsibility of the controller.

This leads us to an alternative. One could consider the possibility of prohibiting the use of robots in certain areas and circumstances. Then they certainly would be unable to cause harm. In practice, we will probably start to see more and more mixed solutions of these two main alternatives, which deal with the growing interaction between robots and people, even if only because of the compromising nature of politics. This political situation can be phrased as a conflict between rigorous deontologists and calculating consequentialists: on the one side stands the absolute prohibition of even only potentially unethical activity of robots; on the other side we find the consideration of the overall benefit of robotic actions in moral terms. Balancing these normative considerations is rather difficult, because robots might cause moral harm, but the morally compliant actions of robots can also help to tighten the network of responsibility.

One way forward to deal with this practical dilemma is to make use of the idea of a forward-looking collective responsibility, as recently discussed in political philosophy (Miller 2001, 2011; Young 2006, 2011). The basic idea here is that groups can be responsible for bringing about desired future states together, whereas individual group members would not be able to do so. Moreover, it is assumed that this forward-looking responsibility can be established independently of looking backwards at who has caused what harm in a liable way. Separating a forward-looking collective responsibility from a backward-looking individual responsibility has a major advantage when it comes to robots. Robots as relatively autonomous agents can be integrated into responsible collectives without being responsible agents themselves. In hospitals, for instance, it can then be asked what robots can contribute to collective actions so that groups can better fulfil those tasks they are responsible for.

Moreover, human group-members can then also take responsibility for those robots that are contributing members of their groups. They have to make sure that they utilize robots in their group action in such a way that those robots do no harm. I therefore believe that the most promising way forward, when it comes to robot responsibility, is not to conceptualize robots as responsible agents themselves, but to conceptualize them as relatively autonomous members of responsible groups. For such an account, it would be necessary to look deeper into the concept of a forward-looking collective responsibility and how this concept can be applied to mixed human-robot groups. One upshot of this approach could be that we should not see robots as independent agents, although they might be relatively autonomous agents. Instead, and unlike humans we should think of them as pure group members, acting only within the strict confinements set by their groups.

References

Abney, Keith. 2012. Robotics, ethical theory, and metaethics: A guide for the perplexed. In *Robot ethics. The ethical and social implications of robotics*, ed. Patrick Lin, Keith Abney, and George A. Bekey, 35–52. Cambridge, MA: The MIT Press.

Allen, Colin, and Wendel Wallach. 2010. *Moral machines: Teaching robots right from wrong*. Oxford: Oxford University Press.

Allen, Colin, and Wendel Wallach. 2012. Moral machines. Contradiction in terms or abdication of human responsibility? In *Robot ethics. The ethical and social implications of robotics*, ed. Patrick Lin, Keith Abney, and George A. Bekey, 55–68. Cambridge, MA: The MIT Press.

Allen, Colin, Gary Varner, and Jason Zinser. 2000. Prolegomena to any future artificial moral agent. *Journal of Experimental & Theoretical Artificial Intelligence* 12: 251–261.

Anscombe, Elisabeth. 1957. *Intention*. Cambridge, MA: Harvard University Press.

Arkin, Ronald. 2009. *Governing lethal behavior in autonomous robots*. Boca Raton: Chapman and Hall.

Asaro, Peter M. 2012. A body to kick, but still no soul to damn: Legal perspectives on robots. In *Robot ethics. The ethical and social implications of robotics*, ed. Patrick Lin, Keith Abney, and George A. Bekey, 169–186. Cambridge, MA: The MIT Press.

Beck, Susanne. 2009. Grundlegende Fragen zum rechtlichen Umgang mit der Robotik. *Juristische Rundschau* 1(6): 225–230.

Bekey, George A. 2012. Current trends in robotics: Technology and ethics. In *Robot ethics. The ethical and social implications of robotics*, ed. Patrick Lin, Keith Abney, and George A. Bekey, 17–34. Cambridge, MA: The MIT Press.

Benford, Gregory, and Elizabeth Malartre. 2007. *Beyond human: Living with robots and cyborgs*. New York: St. Martin's Press.

Birnbacher, Dieter. 1995. *Tun und Unterlassen*. Stuttgart: Reclam.

Birnbacher, Dieter. 2006. Das Dilemma des Personenbegriffs. In *Bioethik zwischen Natur und Interesse*, ed. Birnbacher Dieter, 53–76. Frankfurt am Main: Suhrkamp.

Crnkovic, Gordana Dodig, and Baran Çürüklü. 2012. Robots ethical by design. *Ethics and Information Technology* 14: 61–71.
Dancy, Jonathan. 2006. *Ethics without principles*. Oxford: Oxford University Press.
Dennett, Daniel C. 1976. Conditions of personhood. In *The identities of persons*, ed. Rorty Amelie Oksenberg, 175–196. Berkeley: University of California Press.
Dennett, Daniel. 1987. *The intentional stance*. Cambridge, MA: The MIT Press.
Dennett, Daniel. 1992. *Consciousness explained*. Boston: Back Bay Books.
Dennett, Daniel. 1997. When HAL kills, who's to blame? Computer ethics. In *HAL's legacy: 2001's computer as dream and reality*, ed. David G. Stork, 351–366. Cambridge, MA: The MIT Press.
Floridi, Luciano, and J.W. Sanders. 2004. On the morality of artificial agents. *Minds and Machines* 14(3): 349–379.
Frankfurt, Harry. 1988. *The importance of what we care about*. Cambridge: Cambridge University Press.
Frege, Gottlob. 1962. Über Sinn und Bedeutung. In *Funktion, Begriff, Bedeutung. Fünf logische Studien*, 38–63. Göttingen: Vandenhoeck & Ruprecht.
Gunkel, David J. 2012. *The machine question: Critical perspectives on AI, robots, and ethics*. Cambridge, MA: The MIT Press.
Johnson, Deborah G., and Keith W. Miller. 2008. Un-making artificial moral agents. *Ethics and Information Technology* 10: 123–133.
Jonsen, Albert R., and Stephen Toulmin. 1988. *The abuse of casuistry: A history of moral reasoning*. Berkeley: University of California Press.
Lin, Patrick. 2012. Introduction to robot ethics. In *Robot ethics. The ethical and social implications of robotics*, ed. Patrick Lin, Keith Abney, and George A. Bekey, 3–16. Cambridge, MA: The MIT Press.
Miller, David. 2001. Distributing responsibilities. *Journal of Political Philosophy* 9(4): 453–471.
Miller, David. 2011. Taking up the slack? Responsibility and justice in situations of partial compliance. In *Responsibility and distributive justice*, ed. Carl Knight and Zofia Stemplowska, 230–245. Oxford: Oxford University Press.
Misselhorn, Catrin. 2013. Robots as moral agents. In *Roboethics, proceedings of the annual conference on ethics of the German association for social science research on Japan*, ed. Frank Roevekamp, 30–42. Iudicum: München.
Moor, James H. 2006. The nature, importance, and difficulty of machine ethics. *IEEE Intelligent Systems* 21(4): 18–21.
Moore, Michael S. 2010. *Causation and responsibility: An essay in law, morals, and metaphysics*. Oxford: Oxford University Press.
Neuhäuser, Christian. 2011. *Unternehmen als moralische Akteure*. Berlin: Suhrkamp.
Neuhäuser, Christian. 2012. Künstliche Intelligenz und ihr moralischer Standpunkt. In *Jenseits von Mensch und Maschine*, ed. Beck Susanne, 23–42. Baden Baden: Nomos.
Nourbakhsh, Illah Reza. 2013. *Robot futures*. Cambridge, MA: The MIT Press.
Plessner, Helmut. 1986. *Mit anderen Augen: Aspekte einer philosophischen Anthropologie*. Stuttgart: Reclam.
Quante, Michael. 2007. *Person*. Berlin: de Gruyter.
Rovane, Carol. 1995. The personal stance. *Philosophical Topics* 22: 397–409.
Scheler, Max. 2009. *The human place in the cosmos*. Evanston: Northwestern University Press.
Sharkey, Noel. 2009. Death strikes from the sky: The calculus of proportionality. *IEEE Science and Society* 2009: 16–19.
Sharkey, Noel. 2012. Killing made easy: From joysticks to politics. In *Robot ethics. The ethical and social implications of robotics*, ed. Patrick Lin, Keith Abney, and George A. Bekey, 111–128. Cambridge, MA: The MIT Press.
Sharkey, Noel, and Sharkey Amanda. 2012. The rights and wrongs of robot. In *Robot ethics. The ethical and social implications of robotics*, ed. Lin Patrick, Abney Keith, and George A. Bekey, 267–282. Cambridge, MA: The MIT Press.

Singer, Peter W. 2009. *Wired for war: The robotics revolution and conflict in the 21st century*. New York: The Penguin Press.
Smith, Christian. 2010. *What is a person? Rethinking humanity, social life, and the moral good from the person up*. Chicago: University of Chicago Press.
Sullins, John P. 2006. When is a robot a moral agent? *International Review of Information Ethics* 6(12): 23–30.
Thomson, Judith Jarvis. 1985. The trolley problem. *Yale Law Journal* 94: 1395–1415.
Watson, Gary. 1975. Free agency. *Journal of Philosophy* 72: 205–220.
Wild, Markus. 2011. Tierphilosophie zur Einführung. Hamburg: Junius Verlag.
Young, Iris Marion. 2006. Responsibility and global justice: A social connection model. *Social Philosophy and Policy* 23: 102–130.
Young, Iris Marion. 2011. *Responsibility for justice*. New York: Oxford University Press.

Part III
Collective Agency

Chapter 8
Planning for Collective Agency

Stephen A. Butterfill

8.1 Introduction

Which planning mechanisms enable agents to coordinate their actions, and what if anything do these tell us about the nature of collective agency? In this chapter I shall address these questions. But first I want to step back and consider how we might get a pre-theoretical fix on a notion of collective agency.

In everyday life we exercise agency both individually and collectively. Now Hannah is climbing the tree alone but later she will return with Lucas and they will exercise collective agency in climbing the tree together. When two or more people dance or walk together, when they kiss, when they move a rock together or when they steal a spaceship together—then, typically, they are exercising collective agency. But what is collective agency? Fortunately philosophers have offered some suggestions. Unfortunately philosophers have offered many different, wildly incompatible suggestions about collective agency.[1] Do we just have to pick a view and hope for the best, or can we get a fix on the notion in advance of choosing a favourite philosopher?

---

[1] Recent contributions include Bratman (2014); Gilbert (1990); Gallotti (2011); Gold and Sugden (2007); Kutz (2000); Ludwig (2007); Miller (2001); Schmid (2009); Searle (1990); Seemann (2009); Smith (2011); Tuomela and Miller (1988); Tuomela (2005). Even the terminology is fraught with pitfalls. I take 'collective agency' and 'shared agency' to be synonymous; likewise 'collective action' and 'joint action'. Use of the term 'collective' occasionally cues audiences to expect discussion of large-scale activities, although in this chapter 'collective' should be understood, as it is in discussions of plural prediction, as contrasting with 'distributive' (see, e.g., Linnebo 2005).

S.A. Butterfill (✉)
Department of Philosophy, University of Warwick, Coventry CV4 7AL, UK
e-mail: s.butterfill@warwick.ac.uk

8.2 Three Contrasts

One way to anchor our thinking is by contrasting paradigm cases involving collective agency with cases that are as similar as possible but do not involve collective agency.[2] Suppose lots of commuters pile into an elevator and, because of their combined weight, cause it to get stuck. Then they have collectively broken the elevator, and this is something they have done together. (This story is not one in which each commuter enters the elevator alone on a different occasion and causes it to get stuck.) It does not follow that their actions involve collective agency. After all, there are different ways we could extend the story, not all of which involve collective agency. First, we could extend the story by specifying that each individual is acting alone, unconcerned about the others. In fact, each commuter might have attempted to prevent others from getting into the elevator. In this case, the commuters only exercise merely individual agency in parallel. For a second way of extending the story, imagine a case that is as similar as possible except that the people involved are not really commuters but are accomplices of a master criminal posing as commuters. In accordance with an earlier agreement they made, they all pile into the elevator in order to put it out of order during a robbery. This would typically involve an exercise of collective agency on the part of the accomplices.

Contrasting the genuine commuters with the criminal accomplices indicates something about what collective agency isn't. It indicates that our exercising collective agency cannot be a matter only of our doing something together, nor only of there being something that we collectively do.[3]

At this point one might easily be struck by the thought that there is one goal to which each criminal accomplice's actions are directed, namely the breaking of the elevator. Since none of the genuine commuters aim to break the elevator, perhaps this gives us a handle on collective agency. Can we say that collective agency is involved where there is a single outcome to which several agents' actions are all directed?

Unfortunately we probably cannot. To see why not, consider a second pair of contrasting cases. In the first case, two friends decide to paint a particular bridge red together and then do so in a way typical of friends doing things together. This is a case involving collective agency. But now consider another case. Two strangers each independently intend to paint a large bridge red. More exactly, each intends

---

[2] The use of contrast cases to draw conclusions about collective agency is not new (compare Searle 1990). The strategy is familiar from Pears (1971), who used contrast cases to argue that whether something is an ordinary, individual action depends on its antecedents.

[3] If this is right, Gilbert is wrong that '[t]he key question in the philosophy of collective action is simply ... under what conditions are two or more people doing something together?' (Gilbert 2010, 67).

8 Planning for Collective Agency

that her painting grounds[4] or partially grounds[5] the bridge's being painted red. They start at either end and slowly cover the bridge in red paint working their ways towards the middle where they meet. Because the bridge is large and they start from different ends, the two strangers have no idea of each other's involvement until they meet in the middle. Nor did they expect that anyone else would be involved in painting the bridge red. This indicates that they were not exercising collective agency.

Contrasting the two painting episodes, the friends' and the strangers', indicates that our exercising collective agency cannot be just a matter of there being a single outcome to which each of our actions is directed. After all, this feature is common to both painting episodes, the friends' and the strangers', but only the former involves collective agency.

What are we missing? If we want to try to keep things simple (as I think we should), a natural consideration is that the friends painting the bridge red care in some way about who is involved in the painting whereas the strangers do not. Maybe what is needed for collective agency is not just a single outcome to which all agents' actions are directed but also some specification of whose actions are supposed to bring this outcome about. Consider this view:

The Simple View Collective agency is involved where there is a single outcome, G, and several agents' actions are all directed to the following end: they, these agents, bring about this outcome, G, together.

Note the appeal to togetherness in specifying the end to which each agents' actions are directed. Is this circular? No. As we saw in discussing the first contrast (the one with the elevator), there can be something we are doing collectively and together without our exercising any collective agency at all.

One way—perhaps not the only way—to meet the condition imposed by the Simple View involves intention: the condition is met where several agents are acting in part because[6] each intends that they, these agents, bring about G together. In the case of the friends painting the bridge red, the idea is that each would intend that they, the two of them, paint the bridge red together.

The Simple View does enable us to distinguish the two painting episodes: the friends painting the bridge exercise collective agency and their actions are each directed to an end involving them both, whereas this is untrue of the strangers. But is the view right?

---

[4] Events $D_1, \ldots D_n$ *ground* E just if: $D_1, \ldots D_n$ and E occur; $D_1, \ldots D_n$ are each part of E; and every event that is a part of E but does not overlap $D_1, \ldots D_n$ is caused by some or all of $D_1, \ldots D_n$. This notion of grounding is adapted from Pietroski (1998).

[5] Event D *partially grounds* event E if there are events including D which ground E. (So any event which grounds E thereby also partially grounds E; I nevertheless describe actions as 'grounding or partially grounding' events for emphasis.) Specifying the intentions in terms of grounding ensures that it is possible for both people to succeed in painting the bridge, as well as for either of them to succeed alone.

[6] Here I ignore complexities involved in accurately specifying how events must be related to intentions in order for the events to involve exercises of collective agency; these parallel the complexities involved in the case of ordinary, individual agency. (On the individual case, see Chisholm (1966).)

Probably not. We can see why not by adapting an example from Gilbert and Bratman. Contrast two friends walking together in the ordinary way, which paradigmatically involves collective agency, with a situation where two gangsters are walking together but each is forcing the other. It works like this. The first gangster pulls a gun on the second gangster and says, sincerely, 'Let's walk!' Simultaneously, the second does the same to the first. Each intends that they, these two gangsters, walk together. This ensures that the condition imposed by the Simple View is met. Yet no collective agency is involved. At least, it should be clear that no collective agency is involved unless you think the central events of *Reservoir Dogs* involve collective agency (Tarantino 1992).

Contrasting the ordinary way of walking together with what we might call walking together in the Tarantino sense indicates that the Simple View is too simple. Apparently we cannot distinguish actions involving collective agency just by appeal to there being an outcome where several agents' actions are directed to the end that they, these agents, bring about that outcome together.[7]

Maybe this is too quick. Maybe we should be neutral about whether there is no collective agency involved when the gangsters walk together in the Tarantino sense. It should be clear, of course, that there is a contrast with respect to collective agency between the ordinary way of walking together and walking together in the Tarantino sense. But what can we infer from this contrast? Suppose we are neutral on whether there are degrees of collective agency, or on whether there are multiple kinds of it. Then we cannot infer from the contrast that no collective agency is involved in walking together in the Tarantino sense. The contrast may be due to a difference in degree or kind rather than to an absence of collective agency. So rather than saying that the Simple View does not allow us to distinguish actions involving collective agency from actions that do not, it would be safer to say that this view does not enable us to make all the distinctions with respect to collective agency that we need to make.

In any case, the Simple View cannot be the whole story about collective agency. Focus on cases like two friends walking together or painting a bridge together in the ordinary way. To introduce an arbitrary label, let *alpha* collective agency be the kind or degree of collective agency involved in cases such as these. (And if there are no kinds or degrees of collective agency, let alpha collective agency be collective agency.) What distinguishes exercises of alpha collective agency from exercises of other kinds or degrees of collective agency (if there are any) and from parallel but merely individual exercises of agency?

8.3 Bratman's Proposal

The best developed and most influential proposal is due to Michael Bratman. Others defend conceptually radical proposals, proposals which involve introducing novel attitudes distinct from intentions (e.g. Searle 1990), novel subjects

---

[7] This argument is adapted from Bratman's (1992, 132–134; 2014, 48–52).

distinct from ordinary, individual subjects (e.g. Schmid 2009), or novel kinds of reasoning (e.g. Gold and Sugden 2007). Bratman's proposal aims to be, as he puts it, conceptually conservative. He proposes that we can give sufficient conditions for what I am calling alpha collective agency merely by adding to the Simple View.

The key part of Bratman's proposal is straightforward.[8] It is not just that we each intend that we, you and I, perform the action (as the Simple View requires); further, we must have, and act on, intentions about these intentions. As Bratman writes:

> each agent does not just intend that the group perform the [...] joint action.[9] Rather, each agent intends as well that the group perform this joint action in accordance with subplans (of the intentions in favor of the joint action) that mesh. (1992, 332)

This appeal to interlocking intentions enables Bratman to avoid counterexamples like the Tarantino walkers. If I am acting on an intention that we walk by way of your intention that we walk, I can't rationally also point a gun at you and coerce you to walk.

Why not? If you are doing something by way of an intention to do that very thing, then, putting any special cases aside, I can't succeed in coercing you to do it. (True, I might be coercing you to intend to do it; but that is different from coercing you to do it. After all, someone in a particularly contrary mood might coherently coerce you to intend to do something while simultaneously preventing you from acting.) Accordingly, if I intend both that we walk by way of your intention that we walk and by way of my coercing you to walk, then I have intentions which cannot both be fulfilled. Since the incompatibility of these intentions in any particular case is, on reflection, obvious enough to me, I can't rationally have this combination of intentions unless perhaps I am unreflective or ignorant.

Bratman's proposal hinges on adding second- to first-order intentions (give or take some common knowledge, interdependence and other details that need not concern us here). On the face of it, it would be surprising if this worked. After all, one of the lessons from a parallel debate concerning ordinary, individual agency appears to be that if something can't be captured with first-order mental states, invoking second-order mental states will not suffice to capture it either (compare Watson 1975, 108–109 on Frankfurt 1971). Relatedly, anyone familiar with a certain drama surrounding Grice on meaning would have to be quite optimistic to bet on Bratman's proposal. Nevertheless, to my knowledge no one has yet succeeded in showing that Bratman's proposal is wrong.

---

[8] His proposal has been refined and elaborated over more than two decades (for three snapshots, compare (Bratman 1992, 338), (Bratman 1997, 153), and (Bratman 2014, 84)). Here I skip the details; the discussion in this chapter applies to all versions.

[9] It may be tempting to think that invoking joint action here is somehow circular. But Bratman is using 'joint action' as I am using 'collective action'; and, as illustrated in Sect. 8.2, there are collective (or 'joint') actions which do not involve collective agency.

8.4 A Fourth Contrast

I seek a fourth contrast to add to the earlier three (see Sect. 8.2). This will be a pair of cases in both of which Bratman's conditions are met, although one involves alpha collective agency whereas the other does not. I do not claim that the contrast considered in this section shows that Bratman's proposal is wrong. My aim is only to motivate considering alternatives to Bratman's proposal by indicating the sort of case that might yield a counterexample.

As background, recall that there can be something that several people do collectively without exercising collective agency, as in the story about commuters breaking the elevator from Sect. 8.2. One consequence is that an individual can unilaterally have intentions concerning what several people will collectively do. (By saying that someone has such an intention *unilaterally*, I mean she intends irrespective of whether others intend the same.) To illustrate, suppose that Ayesha, one of the commuters, knows that all the others will get into the elevator irrespective of whether she elbows her way in or not. Suppose, further, that she knows that the elevator will break only if all of the commuters including herself are in it. Then she can intend, unilaterally, that they, the commuters, collectively break the elevator. (Detailed defence of this sort of possibility against objections about what individuals can intend is given in Bratman (1997).) I shall exploit this possibility in describing a pair of cases that contrast with respect to collective agency.

Ayesha and Beatrice each have one wrist handcuffed to the steering wheel of a moving car. They are matched in strength closely enough that neither can decide the car's course alone: its movements will be a consequence of both of their actions. Ayesha, determined that Beatrice should die and wishing to die herself, is wondering how she could bring this about. Thinking that she could pull her gun on Beatrice to force her to cooperate, she intends, unilaterally, that they, Ayesha and Beatrice, drive the car off the road and over a cliff. But a sudden jolt causes the gun to fly from her hand and land far out of reach. Just as it seems she will have to abandon her intention, it strikes her that Beatrice has an intention which renders the gun unnecessary. For Beatrice, whose thoughts and actions mirror Ayesha's, plainly intends what Ayesha intends, namely that they drive the car over the cliff. So Ayesha retains this intention and changes her mind only about the means. Whereas before she intended that they do this by way of her gun, she now intends that they do it by way of her and Beatrice's intentions that they drive the car over the cliff. Beatrice's intention renders Ayesha's gun unnecessary. Now Beatrice, continuing to mirror Ayesha, also forms an intention about their intentions. Further, all this is common knowledge between them. So this is a case in which the conditions imposed in Bratman's proposal about collective agency are met: the agents not only each intend that they perform the action (driving the car over the cliff) but also intend that they do so in accordance with these intentions and meshing subplans of them.

Contrast this case with Thelma and Louise's better known and more romantic intentional car crash (the two friends evade capture by driving off a cliff together; Khouri 1992). Whereas Thelma and Louise's escape is a paradigm case of collective agency, the episode involving Ayesha and Beatrice does not seem to involve collective agency at all, and certainly not alpha collective agency. For Ayesha conceives of Beatrice's intention merely as an opportunity to exploit, something that achieves what she would prefer to have ensured by force of arms. And Beatrice does likewise. Each is willing to be exploited in order to be able to exploit the other. While not all forms of exploitation are incompatible with collective agency, it seems unlikely that collective agency can consist only in the kind of mutual exploitation exemplified by Ayesha and Beatrice. This indicates that Bratman's proposal does not yield sufficient conditions for collective agency. Or at least the conditions are not sufficient for alpha collective agency, that is, for the degree or kind of collective agency the proposal is supposed to capture. Apparently, appealing to intentions about intentions does not enable us to distinguish actions involving alpha collective agency from actions not involving it.

This fourth contrast is a challenge to Bratman's proposal just as the third contrast (the one with the gangsters walking in the Tarantino sense) is a challenge to the Simple View. Someone who wanted to reject the Simple View and endorse Bratman's proposal would need to hold that something like the third contrast is genuine while insisting the fourth contrast is merely apparent.

Is this a decisive objection to Bratman's proposal? Clearly it is not. After all, one might deny that the fourth contrast is genuine and insist that, whatever intuitions anyone might have to the contrary, Ayesha and Beatrice are exercising alpha collective agency. This form of response could be made to any of the contrasts offered; the objection to the Simple View, for instance, is similarly non-decisive. Other responses to the contrasts are surely conceivable too.

What, then, can we conclude from the four contrasts? It would be a mistake simply to disregard them. True, they do not provide decisive grounds for rejecting the views we have so far considered. But none of the narrowly philosophical considerations offered in this area have provided decisive reasons for accepting or rejecting a view about collective agency. The contrasts are valuable because they indicate that something is missing from even the best, most carefully developed attempt to distinguish alpha collective agency from other kinds or degrees of collective agency and from parallel but merely individual agency. What is missing?

8.5 Interconnected Planning Versus Parallel Planning

Bratman's proposal about collective agency is part of an attempt to provide a planning theory of agency, both individual and collective. Let us say that our plans are *interconnected* just if facts about your plans feature in mine and conversely. Then Bratman's proposal about collective agency (see Sect. 8.3) implies that where our actions are guided by appropriately interconnected planning, we are exercising alpha collective agency. In the previous section I attempted to motivate doubt about whether interconnected planning is sufficient for alpha collective agency by appeal to Ayesha and Beatrice, whose actions are driven by interconnected planning but who, apparently, do not exercise alpha collective agency. Could we deny that interconnected planning is sufficient for collective agency while nevertheless holding on to a more general insight about planning—the insight that we can understand something about how collective agency differs from parallel but merely individual agency by considering the kinds of planning characteristic of the former? My aim in the rest of this chapter is to show that we could.

Many collective actions are subject to a certain kind of relational constraint. Suppose that you and I are tasked with moving a table out of a room, through a narrow doorway. When, where and how you should grasp, lift and move depends on when, where and how I will do these things; and conversely. How could we meet such relational constraints on our actions?

One way is by means of interconnected planning. When moving the table, interconnected planning would require two things of you (and likewise of me). You would need to form a view about my plans for my part of our action. And you would need to make plans for your part which mesh with what you know of my plans. In this case we each have in mind two separate plans—one for your actions and one for mine—that need to be interconnected. The need for interconnection means that the plan you end up with must involve not only facts about the size and weight of the table and the aperture of the doorway but also facts about my plans. Interconnected planning is meta-planning.

But there is another, perhaps more natural way of proceeding that doesn't require interconnected planning. Instead of considering separately how you and I should each act and then needing to interconnect these two things, you might simply consider how two people in our situation should move the table through the door; and I might do likewise. In this case, the question that guides preparation for action is not, How should I act? but, How should we act? Answering this question results in us each having in mind one plan for our actions (rather than two separate plans, one for your actions and one for mine). So no interconnection is needed. This second approach to moving the table exemplifies *parallel planning*: we each make a plan for all of our actions.

So the difference between interconnected and parallel planning concerns how many plans we each have in mind. In interconnected planning, we each have two plans in mind, one for my actions and one for yours. In parallel planning, we each have just one plan in mind, a single plan that describes how we, you and I,[10] should act.

The distinction between having one plan in mind and having two plans in mind may initially seem too subtle to matter. A plan is a structure of representations of outcomes, where the structure is subject to the normative requirements that the out-

---

[10] In specifying that a plan describes how we, you and I, should act, I do not mean to imply that the plan must be, or include, a plan for how you should act where this is something over and above being a plan for how we should act. (Distinguishing senses in which a plan on which I act might be a plan for how you should act requires some care.) I specify 'you and I' just to emphasise that the plan is supposed to answer, partially or wholly, the question, What should we do? and not only the question, What should I do?

comes represented can be partially ordered by the means-end relation yielding a tree with a single root, and that the way the representations are structured corresponds to this tree. Any two separate plans that you could both execute can be combined into a single plan. The single plan is simply to implement each of the two formerly separate plans. This might seem to indicate, misleadingly, that there is no significant difference between making a single plan and making two separate plans.

But there are significant differences. One difference arises from the ways plans can be structured. This can be seen by reflection on individual agency. Suppose you have to organise two events, a workshop and a wedding. If the two events were largely independent of each other, it would be simplest to make two separate plans for them. But suppose there are many constraints linking the two events. Some participants at the workshop are also wedding guests, and it happens that many choices of transport, entertainment and catering for one event constrain possible choices for the other event. Envisaging the complex interdependence of these two plans, you might reasonably be inclined to make a single plan for the wedding and workshop. This could be simpler than making two separate plans because it allows for greater flexibility in structuring subplans than would making separate plans for the workshop and wedding and then combining them. For instance, if you make separate plans for the wedding and workshop, you need subplans for transport in each plan and so are forced to divide the problem of transporting participants into the problem of transporting them for the workshop and the problem of transporting them for the wedding. Making a single plan for both events allows you to treat transport as a single problem (but it does not require you to do so, of course). This illustrates one way in which our each having two plans in mind, as happens in interconnected planning, can be significantly different from our each having a single plan for all of our actions in mind, as happens in parallel planning.

Perhaps distinguishing collective agency from parallel but merely individual agency requires us to focus, not on interconnected planning, but on parallel planning. Some readers may already wish to object that invoking the notion of parallel planning is somehow incoherent. I will eventually consider grounds for this objection (see the end of Sect. 8.6) and how it might be overcome (see Sects. 8.7 and 8.8). But for now please suspend disbelief and let me first explain why, assuming the notion is coherent, reflection on parallel planning may provide insights about collective agency.

How does parallel planning enable agents to coordinate their actions? Suppose you and I are parents about to change our baby's nappy. This involves preparing the baby and preparing the nappy. You're holding the baby and I'm nearest the pile of clean nappies, so there's a single most salient way of dividing the task between us. Preparing the baby is, of course, a complex action with many components. Now there are relational constraints on how the baby and nappy should be prepared; how you clean constrains, and is constrained by, how I prepare the clean nappy (because we don't want to get pooh on it). How do we meet these relational constraints? Suppose that we engage in parallel planning, and that, thanks to environmental constraints, our locations and planning abilities, we predictably and non-accidentally end up with plans that match.[11] Your having a single plan for our actions, yours and mine, means that your plan for your actions is constrained by your plan for my actions. And the fact that our plans match means that your plan for my actions is, or matches, my plan for my actions. So thanks to our parallel planning—to the fact that we each plan the whole action—your plan for your actions is indirectly constrained by my plan for my actions; and conversely.

As this example illustrates, parallel planning sometimes enables us to meet relational constraints on our actions not by thinking about each other's plans or intentions but, more directly, by planning each other's actions. By virtue of parallel planning, you can use the same planning processes which enable you to meet constraints on relations between several of your own actions in order to meet constraints on relations between your own and others' actions.

The fact that parallel planning enables agents to coordinate their actions in this way does not imply that it can help us to understand collective agency. After all, if there were no more to exercising collective agency than merely coordinating actions, the difficulties in characterising collective agency partially surveyed in Sect. 8.2 would hardly arise. Why suppose that the notion of parallel planning—assuming for the moment that it is even coherent—might help us to understand collective agency?

8.6 Parallel Planning and Collective Agency

Suppose you and I are about to set up a tent in a windy field. If we first collectively plan how we will do this and then act on our plan, it seems clear enough that we are exercising collective agency. Can we invoke collective planning in order to explain collective agency? To do so might involve circularity because collective planning might itself involve collective agency. My suggestion is that invoking parallel planning enables us to capture much the same insight about collective agency that invoking collective planning would, but without the risk of circularity and with greater generality. Let me explain.

It is a familiar idea that, in thinking about your own future actions, there are two sorts of perspective you can adopt. You can adopt the perspective of an outsider and think about your own actions in a theoretical way. Alternatively, you can adopt the perspective of the agent and think about your own actions in a practical way. One good indicator that the theoretical and practical perspectives are distinct is that, as several philosophers have noted, in considering actions practically you should dis-

---

[11] Two or more agents' plans *match* just if they are the same, or similar enough that the differences don't matter in the following sense. First, for each agent's plan, let the *self part* be the bits concerning the agent's own actions and let the *other part* be the other bits. Now consider what would happen if, for a particular agent, the other part of her plan were as nearly identical to the self part (or parts) of the other's plan (or others' plans) as psychologically possible. Should this agent's self part be significantly different? If not, let us say that any differences between her plan and the other's (or others') are not relevant for her. Finally, if for some agents' plans the differences between them are not relevant for any of the agents, then let us say that the differences don't matter.

regard biases, limits and non-rational quirks which mean your actions sometimes fail to conform to your intentions, even if these are highly reliable.

Can you adopt a practical perspective on actions which, we both know, I will perform? Initially it might seem that you could not do this without irrationality or confusion. But suppose I ask you how I could get around this obstacle (say). To answer this question you might imaginatively put yourself in my place and deliberate about how to get around the obstacle. In doing this you are taking a practical perspective on actions which I, not you, will eventually perform. You are not, of course, engaged in practical reasoning concerning those actions: your reasoning is, after all, not actually directed to action. Nevertheless, there is a fine line between what you are doing—call it practical deliberation—and practical reasoning. For suppose that, unexpectedly, you find yourself actually in my place and needing to get around the obstacle. You do not now need to start planning your actions: the planning has already been done. Indeed, for you to consider afresh how to act now that it turns out to be you rather than me who is facing the obstacle would indicate that you had not done your best to answer my question. So when imaginatively putting yourself in my place, you can adopt a practical perspective on actions which I will eventually perform and doing so actually prepares you to act.

Invoking the notion of parallel planning involves going just a tiny step further. In parallel planning, it is not just that we each adopt a practical perspective on actions some of which the other will eventually perform. It is also that we each adopt a perspective from which both of our actions, yours and mine, are parts of a single plan. We are thereby thinking practically about all of these actions simultaneously. To return to setting up our tent in the windy field, rather than collectively planning our actions we might each individually deliberate about how to proceed and, realising that just one plan is most salient to both of us, spring into action without needing to confer. It is not that we are thinking of each other's actions exactly as if they were our own, of course; but we are also not thinking of them in a merely theoretical way, as the actions of someone who is just passing by. From the perspective we each adopt in parallel planning, our actions, yours and mine, have a certain kind of practical unity.

Why is this significant? Return for a moment to the two gangsters walking in the Tarantino sense and what I am calling the Simple View (see Sect. 8.2). According to the Simple View, for some agents' walking together to involve collective agency it is sufficient that their walking be appropriately guided by intentions, on the part of each agent, that they, these agents, walk together. Now, as you may recall, earlier I noted that the Simple View seems incorrect because gangsters might act on such intentions while forcing each other to walk at gunpoint. It is this sort of problem that Bratman uses to motivate complicating the Simple View with appeal to interconnected planning (compare Bratman 1992, 132–134; Bratman 2014, 48–52). But the problem might also be overcome by invoking parallel planning. Consider the view that for us to exercise (alpha) collective agency in walking together it is sufficient that:

1. we each intend that we, you and I, walk (the Simple View);

2. we pursue these intentions by means of parallel planning; and
3. our plans predictably and non-accidentally match.

This view correctly implies that the gangsters' walking is not an exercise of (alpha) collective agency. This is because pursuing an intention by means of parallel planning means taking a practical attitude towards each other's actions. Where the above conditions, (1)–(3), are met, for one of the agents to point a gun at the other to force her to walk would be almost like her pointing a gun at herself in order to force herself to do something she plans to do. This would involve a form of irrationality. So if the gangsters were pursuing their intentions that they walk by means of parallel planning, they could not rationally be forcing each other to walk at gunpoint.

Invoking parallel planning also enables us to distinguish the case of Ayesha and Beatrice from that of Thelma and Louise, which Bratman's proposal was unable to do (see Sect. 8.4). Ayesha and Beatrice, who are handcuffed to the steering wheel of a moving car, each rely on the other's intention only for want of a gun. From their perspectives, the other's intention is just one among many factors that might have justified relying on the other to perform actions that bear a certain relation to her own. This shows that they each adopt a theoretical perspective when thinking about the others' actions and intentions. So they cannot be pursuing their intentions by means of parallel planning. By contrast, Thelma and Louise decide what they will do, which indicates that they each adopt the perspective required for parallel planning. So appeal to the above conditions, (1)–(3), but not to Bratman's proposal (see Sect. 8.3), plausibly gives us what seems, pre-theoretically, to be the right result: Ayesha and Beatrice are not, whereas Thelma and Louise are, exercising (alpha) collective agency.

This is one reason for supposing that reflection on parallel planning may yield insights into what distinguishes collective agency (or the alpha variety of it), insights that cannot be got from reflection on interconnected planning. Whereas engaging in interconnected planning is consistent with thinking of all others' plans and intentions merely as opportunities to exploit or constraints to work around, engaging in parallel planning involves each agent taking a practical perspective on everyone's actions simultaneously and so conceiving of them as having a certain kind of practical unity.

There's just one tiny problem. Engaging in parallel planning seems, on the face of it, to be always either irrational or else dependent on confusion about whose actions are whose. Why? For us to engage in parallel planning is for each of us to plan both of our actions, yours and mine. This implies, of course, that we each plan actions that are not our own: if things go well, your plan for our actions will include actions the agent of which is not you but me; and likewise for my plan. But assuming (for now) that the elements of plans are intentions, this means that among the things you would be intending are some actions that I will eventually perform; and, likewise, among the things I am intending are some things that will eventually be your actions. So how could we engage in parallel planning without irrationality or confusion? After all, you can't knowingly intend my actions, at least not in the kinds of case we are considering (there may be other cases; see Roth 2010).

As things stand, then, anyone seeking to avoid conceptually radical innovation in theorising about collective agency has an awkward choice. She has a choice between existing proposals which seem to fail to give genuinely sufficient conditions for alpha collective agency and a new proposal involving conditions that, apparently, are met only when the agents are irrational or confused. What to do?

8.7 Two Lines of Argument

There are at least two lines of argument that might be used to show that parallel planning can occur without irrationality or confusion, thereby defending the idea that reflection on parallel planning will yield insights about collective agency. One line—call it the *hard line*—would be to argue that there is a propositional attitude which resembles intention in some essential respects, and which is inferentially integrated with intention, but which you can have towards actions even where you know that some of these actions will be performed by others in exercising collective agency with you. Perhaps, for instance, we can rescue parallel planning from charges of irrationality by identifying certain ways in which intentions or closely related attitudes can be open-ended not only with respect to means but also with respect to agents. While it may be tempting to take the hard line, I shall not do so here.[12]

There is an alternative line of argument which might be used to show that parallel planning can occur without irrationality or confusion. This involves considering processes and representations more primitive than full-blown planning and intention. As we shall see, there appear to be representations which are like intention in some ways but which can be had concerning others' actions (as well as your own) without irrationality or confusion.

This second line of argument is fraught with difficulties. Philosophers have barely begun to consider how discoveries about the mechanisms which make collective agency possible might bear on theories of what it is.[13] The next section offers reasons for holding that if we broaden our view to include processes and representations more primitive than full-blown planning and intention, there are mundane cases in which parallel planning (or something resembling it in respects to be specified) need involve neither irrationality nor confusion.

---

[12] Support for the hard line might be extracted from Laurence (2011). He defends the view that, in some cases, several agents' 'individual, first-person-singular actions are all subject to the special collective action sense of the question "Why?" and [...] the same answer holds in each case' (p. 289). While there are many differences between Laurence's view and the line I am considering, both appear committed to the idea that exercising collective agency sometimes involves this: that from the point of view of any individual agent, it is almost as if all the agents' actions are guided by a single piece of practical reasoning.

[13] This is not to say that philosophers have not attempted this at all; see, for example, Tollefsen (2005) or Gallotti and Frith (2013).

8.8 Parallel Planning and Motor Representation

Consider the motor processes that enable you to reach for, grasp and transfer objects in a coordinated and fluid way.[14] Are these planning processes? They are distinct from the sort of planning that might involve getting your diary out—planning to paint a house, or to travel from London to Stuttgart, say. Nevertheless, such motor processes resemble planning activities both in that they involve computing means from representations of ends (Bekkering et al. 2000; Grafton and Hamilton 2007) and in that they involve meeting constraints relating actions which must occur at different times (Jeannerod 2006; Zhang and Rosenbaum 2007; Rosenbaum et al. 2012). If we think about planning in a narrow way, then, plausibly, these motor processes are not planning processes. But given the two points of resemblance just mentioned, we can coherently follow much of the scientific literature in adopting a broader notion of planning, one that encompasses both planning-like motor processes and full-blown, get-your-diary-out planning. Here I shall use 'planning' in this second, broader sense.

Motor representations are representations of the sort that characteristically feature in motor processes; they play a key role in monitoring and planning actions (e.g. Wolpert et al. 1995; Miall and Wolpert 1996). Motor representations can be distinguished from intentions and representations of other kinds by their format, much as visual representations can be distinguished by their format (Butterfill and Sinigaglia 2014). Motor representations are not limited to representations of bodily configurations and joint displacements. In fact, some motor representations resemble intentions in representing outcomes such as the grasping of an object or the movement of an object from one place to another. Relatedly, some motor representations resemble intentions in coordinating multiple component activities by virtue of their role as elements in hierarchical, plan-like structures, and coordinating these activities in a way that would normally facilitate the represented outcome's occurrence (Hamilton and Grafton 2008; Pacherie 2008, 189–190).

How are motor representations relevant to our difficulties with the notion of parallel planning? Motor representations lead a kind of double life. For motor representations occur not only when you perform an action but sometimes also when you observe an action. Indeed there are some striking similarities between the sorts of processes and representations usually involved in performing a particular action and those which typically occur when observing someone else perform that action. In some cases it is almost as if the observer were planning the observed action, only to stop just short of performing it herself.[15] When motor representations of outcomes

---

[14] This section draws on work in progress with Corrado Sinigaglia and, separately, Natalie Sebanz and Lincoln Colling. But it is not endorsed by these researchers, who would probably have avoided the mistakes I will doubtless have made.

[15] For reviews, see Jeannerod (2006), Rizzolatti and Sinigaglia (2008), Rizzolatti and Sinigaglia (2010). If motor representations occur in action observation, then observing actions might sometimes facilitate performing compatible actions and interfere with performing incompatible actions. Both effects do indeed occur, as several studies have shown (Brass et al. 2000; Craighero et al. 2002; Kilner et al. 2003; Costantini et al. 2012).

trigger a planning-like process in action observation, this may allow the observer to predict others' actions (Flanagan and Johansson 2003; Ambrosini, Costantini, and Sinigaglia 2011; Ambrosini, Sinigaglia, and Costantini 2012; Costantini et al. 2014). There is no question, then, that motor representations concerning others' actions, not just your own, can occur in you.[16] Perhaps, then, there could be parallel planning involving motor processes and representations. If so, the double life of motor representations tells us that parallel planning does not always involve irrationality or confusion.

Of course, one might doubt that motor representations can coherently lead a double life. After all, this might appear to involve one and the same attitude having two directions of fit, world-to-mind insofar as it is involved in planning for action and mind-to-world insofar as it is involved in observing and predicting others' actions. Avoiding this apparent incoherence may demand some subtleties; but as this issue is one that arises independently of issues about collective agency, I shall put it aside here. The overwhelming evidence that motor representations do in fact lead a double life is a reason, not decisive but compelling, to hold that the core idea must somehow be coherent.

My suggestion, then, is that parallel planning is not always incoherent or confused if it sometimes involves motor representations rather than intentions. But is there any evidence that parallel planning involving motor representations ever occurs? Planning concerning another's actions sometimes occurs not only in observing her act but also in exercising collective agency with her (Kourtis et al. 2013; Meyer et al. 2011). Such planning can inform planning for your own actions, and even planning that involves meeting constraints on relations between your actions and hers (Vesper et al. 2013; Novembre et al. 2013; Loehr and Palmer 2011; Meyer et al. 2013). This is suggestive but compatible with two possibilities. It could be that there is a single planning processes concerning all agents' actions, just as parallel planning requires; but it might also be that, in each agent, there are two largely separate planning processes, one for each agent's actions. Where the latter possibility obtains, there is no parallel planning. Fortunately, there is some neurophysiological and behavioural evidence for the former possibility. Sometimes when exercising collective agency, the agents have a single representation of the whole action, not only separate representations of each agent's part (Tsai et al. 2011; Loehr et al. 2013; Ménoret et al. 2014). So while we should be cautious given that the most relevant evidence is relatively recent, it is reasonable to conjecture, at least provisionally, that parallel planning may sometimes involve motor representations rather than intentions. This conjecture is one way (perhaps not the only way) of rescuing the view that we can give sufficient conditions for (alpha) collective agency by invoking parallel planning.

Or is it? Consider again the fourth contrast, the contrast of Ayesha and Beatrice's driving off the cliff with Thelma and Louise's (see Sect. 8.4). Earlier I suggested

---

[16] Note that this does not imply that others' actions are ever represented as others' actions motorically. It may be that some or all motor representations are agent-neutral in the sense that their contents do not specify any particular agent or agents (Ramsey and Hamilton 2010).

that Ayesha and Beatrice's actions couldn't rationally be driven by parallel planning because this would involve each taking a practical perspective on both of their actions, thereby making it irrational for her to consider forcing the other to act in something like the way it would be irrational for her to consider forcing herself to act. But now we are considering a broader notion of planning, one that does not invariably involve the agent adopting a practical perspective, or any perspective at all. This may make it unclear whether we can give sufficient conditions for collective agency by invoking parallel planning.

To solve this problem we must consider how intentions and motor representations are related. Motor representations are not themselves objects of awareness, but they do shape agents' awareness of the kinds of actions they can perform. Further, they do this in such a way that it is possible, sometimes at least, for intending that an outcome obtain (that you grasp this mug, say) to trigger a motor representation of a matching[17] outcome. When an intention is appropriately related to awareness of the kinds of action one can perform and triggers a motor representation, no further practical reasoning about how to act is needed from the agent and it would be irrational for her to so reason. So where two agents each intend that they, these two agents, move an object or drive a car over the cliff, and where these intentions are appropriately related to the agents' awareness of their abilities and trigger motor representations of matching outcomes, the only rational way for their actions to proceed is by parallel planning. This is why neither Ayesha's nor Beatrice's initial intention that they drive over the cliff could have been appropriately related to awareness of action possibilities and triggered a motor representation of a matching outcome. Had this been the case, their further deliberation about how to proceed would imply that they are irrational or confused; but, by stipulation, they are not. So the notion of parallel planning, even where it involves appeal to a broad notion of planning, can, after all, be used to give sufficient conditions for alpha collective agency.

8.9 Conclusion

My question was, Which planning mechanisms enable agents to coordinate their actions, and what if anything do these tell us about the nature of collective agency? In this chapter I have contrasted two planning mechanisms (see Sect. 8.5). The first, due to Bratman (1992, 2014), is interconnected planning involving a structure of intentions and knowledge states. In such interconnected planning, you have intentions concerning my intentions, and I likewise; your plans feature facts about mine, and conversely. So interconnected planning is meta-planning: each agent's plan concerns not only facts about her environment and the objects in it but also facts

---

[17] Two outcomes *match* in a particular context just if, in that context, either the occurrence of the first outcome would normally constitute or cause, at least partially, the occurrence of the second outcome or vice versa.

about the other agents' plans. The second mechanism is parallel planning as implemented by motor processes: in each agent there is a single plan concerning all of the agents' actions. In parallel planning, no agent need have intentions about any other agent's intentions or represent facts about any other agent's plans: rather, each plans the others' actions as well as her own. This allows agents to use ordinary planning mechanisms to coordinate with others (as illustrated in Sect. 8.5).

Reflection on the sorts of contrast case that are sometimes used to get a pre-theoretic fix on a notion of collective agency indicates that interconnected planning is not sufficient for collective agency unless much simpler conditions are also sufficient (see Sect. 8.4). Or, if we allow that there are multiple kinds or degrees of collective agency, reflection on these contrast cases indicates that interconnected planning is not sufficient for alpha collective agency, that is, for the form or degree of collective agency that we aimed to characterise. It seems that interconnected planning can't be what collective agency at bottom consists in because agents can have interconnected plans while thinking of each other's intentions only as opportunities to exploit and constraints to work around, and so without conceiving of themselves as exercising collective agency. It may be that, contra Bratman's proposal (outlined in Sect. 8.3), no amount of forming intentions about others' intentions and acquiring knowledge of such intentions is sufficient, all by itself, for alpha collective agency.

This motivates considering the possibility that we can understand something about collective agency, or some forms of it at least, by invoking parallel planning. While it was perhaps initially tempting to suppose that parallel planning always involves irrationality or confusion, reflection on discoveries about motor representation reveals that parallel planning can occur in good cases too (see Sect. 8.8). But what grounds might there be to think that invoking parallel planning will succeed where invoking interconnected planning has failed? Parallel planning involves explicitly or implicitly adopting a perspective that allows each agent to make sense of the idea that her own and others' actions are all elements in a single plan or all parts of the exercise of a single ability (see Sect. 8.6). So in parallel planning, the agents' actions appear to the agents themselves to have a kind of practical unity not unlike the kind of practical unity that multiple actions of a single agent sometimes have. And this is just what we need for collective agency. In some cases, it may be that parallel planning is what distinguishes exercises of (alpha) collective agency from parallel exercises of merely individual agency.

Does this imply that there is no role for interconnected planning at all? One requirement for parallel planning to support exercises of collective agency is that the agents involved have, or eventually end up with, non-accidentally matching plans. This matching can sometimes be achieved thanks to a combination of similarities in the agents' planning abilities, environmental constraints and experimentation. But there will, of course, be many situations in which these factors are insufficient to yield non-accidentally matching plans. Perhaps, then, interconnected planning matters in part because it enables agents to non-accidentally make matching plans in some situations. On this view, interconnected planning is not constitutive of collective agency but it does extend the range of cases in which agents can coordinate their actions in ways necessary to successfully exercise collective agency.

Acknowledgements Much of what follows has been shaped by objections and suggestions from readers of drafts and from audiences at talks, including Olle Blomberg, Michael Bratman, Gergely Csibra, Naomi Elian, Chris Frith, Mattia Gallotti, Eileen John, Guenther Knoblich, Guy Longworth, John Michael, Marlene Meyer, Catrin Misselhorn, Elisabeth Pacherie, Wolfgang Prinz, Johannes Roessler, Thomas Sattig, Hans Bernhard Schmid, Natalie Sebanz, Corrado Sinigaglia, Thomas Smith, Joel Smith, Matthew Soteriou, Anna Strasser, Cordula Vesper, Hong Yu Wong, and some peculiarly helpful anonymous referees.

References

Ambrosini, Ettore, Marcello Costantini, and Corrado Sinigaglia. 2011. Grasping with the eyes. *Journal of Neurophysiology* 106(3): 1437–1442. doi:10.1152/jn.00118.2011.

Ambrosini, Ettore, Corrado Sinigaglia, and Marcello Costantini. 2012. Tie my hands, tie my eyes. *Journal of Experimental Psychology. Human Perception and Performance* 38(2): 263–266. doi:10.1037/a0026570.

Bekkering, Harold, Andreas Wohlschlager, and Merideth Gattis. 2000. Imitation of gestures in children is goal-directed. *The Quarterly Journal of Experimental Psychology. A* 53(1): 153–164. doi:10.1080/027249800390718.

Brass, Marcel, Harold Bekkering, Andreas Wohlschläger, and Wolfgang Prinz. 2000. Compatibility between observed and executed finger movements: Comparing symbolic, spatial, and imitative cues. *Brain and Cognition* 44(2): 124–143. doi:10.1006/brcg.2000.1225.

Bratman, Michael E. 1992. Shared cooperative activity. *The Philosophical Review* 101(2): 327–341.

Bratman, Michael E. 1997. I intend that we J. In *Contemporary action theory*, Social action, vol. 2, ed. Raimo Tuomela and Ghita Holmstrom-Hintikka. Dordrecht: Kluwer.

Bratman, Michael E. 2014. *Shared agency: A planning theory of acting together*. Oxford: Oxford University Press.

Butterfill, Stephen A., and Corrado Sinigaglia. 2014. Intention and motor representation in purposive action. *Philosophy and Phenomenological Research* 88(1): 119–145. doi:10.1111/j.1933-1592.2012.00604.x.

Chisholm, Roderick. 1966. Freedom and action. In *Freedom and determinism*, ed. Keith Lehrer, 11–44. New York: Random House.

Costantini, Marcello, Ettore Ambrosini, and Corrado Sinigaglia. 2012. Does how I look at what you're doing depend on what I'm doing? *Acta Psychologica* 141(2): 199–204. doi:10.1016/j.actpsy.2012.07.012.

Costantini, M., E. Ambrosini, P. Cardellicchio, and C. Sinigaglia. 2014. How your hand drives my eyes. *Social Cognitive and Affective Neuroscience* 9(5): 705–711.

Craighero, Laila, Arianna Bello, Luciano Fadiga, and Giacomo Rizzolatti. 2002. Hand action preparation influences the responses to hand pictures. *Neuropsychologia* 40(5): 492–502.

Flanagan, J. Randall, and Roland S. Johansson. 2003. Action plans used in action observation. *Nature* 424(6950): 769–771.

Frankfurt, Harry. 1971. Freedom of the will and the concept of a person. *The Journal of Philosophy* 68(1): 5–20.

Gallotti, M. 2011. A naturalistic argument for the irreducibility of collective intentionality. *Philosophy of the Social Sciences* 20(10): 3–30.

Gallotti, Mattia, and Chris D. Frith. 2013. Social cognition in the we-mode. *Trends in Cognitive Sciences* 17(4): 160–165. doi:10.1016/j.tics.2013.02.002.

Gilbert, Margaret P. 1990. Walking together: A paradigmatic social phenomenon. *Midwest Studies in Philosophy* 15: 1–14.

Gilbert, Margaret P. 2010. Collective action. In *A companion to the philosophy of action*, ed. Timothy O'Connor and Constantine Sandis, 67–73. Oxford: Blackwell.

Gold, Natalie, and Robert Sugden. 2007. Collective intentions and team agency. *Journal of Philosophy* 104(3): 109–137.
Grafton, Scott T., and Antonia F. Hamilton. 2007. Evidence for a distributed hierarchy of action representation in the brain. *Human Movement Science* 26(4): 590–616. doi:10.1016/j.humov.2007.05.009.
Hamilton, Antonia F.de.C., and Scott T. Grafton. 2008. Action outcomes are represented in human inferior frontoparietal cortex. *Cerebral Cortex* 18(5): 1160–1168. doi: 10.1093/cercor/bhm150.
Jeannerod, Marc. 2006. *Motor cognition: What actions tell the self*. Oxford: Oxford University Press.
Khouri, Callie. 1992. *Thelma and Louise*. London: Pathé Entertainment.
Kilner, James M., Yves Paulignan, and Sarah-Jayne Blakemore. 2003. An interference effect of observed biological movement on action. *Current Biology* 13(6): 522–525.
Kourtis, Dimitrios, Natalie Sebanz, and Guenther Knoblich. 2013. Predictive representation of other people's actions in joint action planning: An EEG study. *Social Neuroscience* 8(1): 31–42. doi:10.1080/17470919.2012.694823.
Kutz, Christopher. 2000. Acting together. *Philosophy and Phenomenological Research* 61(1): 1–31.
Laurence, Ben. 2011. An Anscombian approach to collective action. In *Essays on Anscombe's intention*. Cambridge, MA: Harvard University Press.
Linnebo, Øystein. 2005. Plural quantification. In *The Stanford encyclopedia of philosophy (Spring 2005 Edition)*, ed. Edward N. Zalta. Stanford, CA: CSLI.
Loehr, Janeen D., and Caroline Palmer. 2011. Temporal coordination between performing musicians. *The Quarterly Journal of Experimental Psychology* 64(11): 2153–2167. Taylor & Francis.
Loehr, J.D., D. Kourtis, C. Vesper, N. Sebanz, G. Knoblich, et al. 2013. Monitoring individual and joint action outcomes in duet music performance. *Journal of Cognitive Neuroscience* 25(7): 1049–1061.
Ludwig, Kirk. 2007. Collective intentional behavior from the standpoint of semantics. *Noûs* 41(3): 355–393. doi:10.1111/j.1468-0068.2007.00652.x.
Ménoret, M., L. Varnet, R. Fargier, A. Cheylus, A. Curie, V. des Portes, T. Nazir, and Y. Paulignan. 2014. Neural correlates of non-verbal social interactions: A dual-EEG study. *Neuropsychologia* 55: 75–97.
Meyer, Marlene, Sabine Hunnius, Michiel van Elk, Freek van Ede, and Harold Bekkering. 2011. Joint action modulates motor system involvement during action observation in 3-year-olds. *Experimental Brain Research* 211(3–4): 581–592. doi:10.1007/s00221-011-2658-3.
Meyer, Marlene, Robrecht P.R.D. van der Wel, and Sabine Hunnius. 2013. Higher-order action planning for individual and joint object manipulations. *Experimental Brain Research* 225(4): 579–588. doi:10.1007/s00221-012-3398-8.
Miall, R. Christopher, and Daniel M. Wolpert. 1996. Forward models for physiological motor control. *Neural Networks* 9(8): 1265–1279. doi:10.1016/S0893-6080(96)00035-4.
Miller, Seumas. 2001. *Social action: A teleological account*. Cambridge: Cambridge University Press.
Novembre, Giacomo, Luca F. Ticini, Simone Schütz-Bosbach, and Peter E. Keller. 2013. Motor simulation and the coordination of self and other in real-time joint action. *Social Cognitive and Affective Neuroscience* (forthcoming). doi: 10.1093/scan/nst086.
Pacherie, Elisabeth. 2008. The phenomenology of action: A conceptual framework. *Cognition* 107(1): 179–217. doi:10.1016/j.cognition.2007.09.003.
Pears, David. 1971. Two problems about reasons for actions. In *Agent, action and reason*, ed. A. Marras, R. Binkley, and R. Bronaugh, 128–153. Oxford: Oxford University Press.
Pietroski, Paul M. 1998. Actions, adjuncts, and agency. *Mind, New Series* 107(425): 73–111. http://www.jstor.org/stable/2659808.
Ramsey, Richard, and Antonia F.de.C. Hamilton. 2010. Understanding actors and object-goals in the human brain. *Neuroimage* 50(3): 1142–1147.
Rizzolatti, Giacomo, and Corrado Sinigaglia. 2008. *Mirrors in the brain: How our minds share actions, emotions*. Oxford: Oxford University Press.

Rizzolatti, Giacomo, and Corrado Sinigaglia. 2010. The functional role of the parieto-frontal mirror circuit: Interpretations and misinterpretations. *Nature Reviews Neuroscience* 11(4): 264–274. doi:10.1038/nrn2805.

Rosenbaum, David A., Kate M. Chapman, Matthias Weigelt, Daniel J. Weiss, and Robrecht van der Wel. 2012. Cognition, action, and object manipulation. *Psychological Bulletin* 138(5): 924–946. doi:http://0-dx.doi.org.pugwash.lib.warwick.ac.uk/10.1037/a0027839.

Roth, Abraham Sesshu. 2010. Shared agency. In *The Stanford encyclopedia of philosophy (Fall 2001 Edition)*, ed. Edward N. Zalta. http://plato.stanford.edu/entries/shared-agency/

Schmid, Hans Bernhard. 2009. *Plural action: Essays in philosophy and social science*, vol. 58. Dordrecht: Springer.

Searle, John R. 1990. Collective intentions and actions. In *Intentions in communication*, ed. P. Cohen, J. Morgan, and M.E. Pollack, 90–105. Cambridge: Cambridge University Press.

Seemann, Axel. 2009. Why we did it: An Anscombian account of collective action. *International Journal of Philosophical Studies* 17(5): 637–655.

Smith, Thomas H. 2011. Playing one's part. *Review of Philosophy and Psychology* 2(2): 213–244. doi:10.1007/s13164-011-0059-y.

Tarantino, Quentin. 1992. *Reservoir dogs*. New York: Miramax Films.

Tollefsen, Deborah. 2005. Let's pretend: Children and joint action. *Philosophy of the Social Sciences* 35(75): 74–97.

Tsai, Jessica Chia-Chin, Natalie Sebanz, and Günther Knoblich. 2011. The GROOP effect: Groups mimic group actions. *Cognition* 118(1): 135–140. doi:10.1016/j.cognition.2010.10.007.

Tuomela, Raimo. 2005. We-intentions revisited. *Philosophical Studies* 125(3): 327–369. doi:10.1007/s11098-005-7781-1.

Tuomela, Raimo, and Kaarlo Miller. 1988. We-intentions. *Philosophical Studies* 53(3): 367–389. doi:10.1007/BF00353512.

Vesper, Cordula, Robrecht P.R.D. van der Wel, Günther Knoblich, and Natalie Sebanz. 2013. Are you ready to jump? Predictive mechanisms in interpersonal coordination. *Journal of Experimental Psychology: Human Perception and Performance* 39(1): 48–61. doi:10.1037/a0028066.

Watson, Gary. 1975. Free agency. *The Journal of Philosophy* 72(8): 205–220.

Wolpert, Daniel M., Z. Ghahramani, and Mi. Jordan. 1995. An Internal model for sensorimotor integration. *Science* 269(5232): 1880–1882. doi:10.1126/science.7569931.

Zhang, Wei, and David A. Rosenbaum. 2007. Planning for manual positioning: The end-state comfort effect for manual abduction–adduction. *Experimental Brain Research* 184(3): 383–389. doi:10.1007/s00221-007-1106-x.

Chapter 9
An Account of Boeschian Cooperative Behaviour

Olle Blomberg

9.1 The Dualism of the Joint and the Parallel

Accounts of joint action are often prefaced by the observation that there are two different senses in which several agents can intentionally perform an action Φ. They might intentionally Φ together, as a collective, or they might intentionally Φ in parallel, where Φ is distributively assigned to the agents, considered as a set of individuals. The accounts are supposed to capture what characterises activities in which several agents do intentionally Φ collectively rather than distributively (Gilbert 2009, 168; Kutz 2000, 1–2; Bratman 2009, 150–151). An account of intentional joint action should thus illuminate how the agency of two friends going for a walk together is different from that of two strangers walking down the street in parallel, each trying to avoid colliding with the other while they are independently walking toward the same destination (Bratman 2009; Gilbert 2009). Often, the difference is couched in terms of the presence or absence of a "shared intention". If the agents have a shared intention to Φ, then the Φ-ing that ensues is an intentional joint action. While it is sometimes acknowledged that there may be many different kinds of shared intention, it is taken for granted that an activity involving several agents is either a genuine form of joint action, or it merely involves coordinated actions performed in parallel.

This dualism between joint and parallel action also crops up outside philosophy. For instance, it has been brought into a debate about whether or not the group hunting in some chimpanzee populations is a form of joint cooperative hunting. Chimpanzees in Taï National Park in the Ivory Coast frequently hunt colobus monkeys, who reside in the canopy of the trees. Christophe Boesch argues that in these hunts,

O. Blomberg (✉)
Center for Subjectivity Research, Department of Media, Cognition
and Communication, University of Copenhagen, Copenhagen, Denmark
e-mail: olle.blomberg@gmail.com

groups of typically three to five members perform coordinated hunts toward the goal that they capture the prey, where group members adopt different roles during the hunt (Boesch and Boesch 1989; Boesch 2002, 2005). One chimpanzee, the *driver*, gets the prey moving in certain direction. *Blockers* take up positions that force the prey to flee in a certain direction, where an *ambusher* might be waiting in a tree. Along the flight, *chasers* may also follow the prey on the ground to then climb up the tree that the prey is occupying. Here, Boesch draws an analogy between how the chimpanzees' success in capturing the prey depends on the contribution of each group member and how a football team's win in a game depends on the contribution of each player:

> Like in a team of soccer players, individuals react opportunistically to the present situation while taking in account the shared goal of the team. Some players will rarely make a goal, like defenders and goalies, but the success of the team will critically depend upon their contribution. This is very reminiscent to group hunting in chimpanzees where synchronisation of different coordinated roles, role reversal, and performance of less successful roles favour the realisation of the joint goal. (Boesch 2005, 692)

Boesch (1994) also argues that how the meat of the captured prey ends up divided among the chimpanzees is determined primarily by the participation and contribution of each in the hunt (as well as by dominance rank and age).

Michael Tomasello and his colleagues have suggested that a competing explanation of the Taï chimpanzees' hunting behaviour is more plausible (Tomasello et al. 2005, 685). They argue that what looks like the coordinated performance of a team is actually just the emergent product of each chimpanzee trying to maximise the chance of capturing the monkey for itself:

> Normally, one individual begins the chase, and others scramble to the monkey's possible escape routes. The individual who actually captures the monkey gets the most meat, but because the captor cannot dominate the carcass on his own all participants (and many bystanders) usually get at least some meat as well […]. The social and cognitive processes involved here are probably fairly simple: each individual is attempting to capture the monkey on its own, and so each takes into account the behaviour of the other chimpanzees as they are pursuing this individual goal. […] The short story is thus that chimpanzees have no joint goal that "we" capture [the monkey] and share it, helping the other in his role as needed, and no sense of commitment […]. (Tomasello and Hamann 2012, 8)

Tomasello (2014, 190) also argues for an alternative interpretation of Boesch's data on how the meat is shared after a successful hunt. According to Tomasello, the main factor that determines who gets a share is the chimpanzees' proximity to the kill rather than their contribution to the hunt.

Tomasello and colleagues argue that, in contrast, human beings, including 3-year-olds, routinely engage in what he and Katharina Hamann call "true collaboration":

> [I]t is true collaboration when in addition to [participants] being mutually responsive to one another […], two key characteristics are present: (a) the participants have a joint goal or intention in the sense that they each have the goal or intention that we (in mutual knowledge) do X together; and (b) the participants coordinate their roles–their plans and sub-plans of action–including helping the other in her role as needed. (Tomasello and Hamann 2012, 2)

The notion of 'joint goal' is a goal that an individual agent has, but which concerns a collective activity, such as "that we capture the prey" or "that we go for a walk" (see Pacherie 2013, 1821–1822). According to Michael Bratman (1992), whose work on "shared cooperative activity" Tomasello and Hamann draw on, each party to a shared intention intends a joint goal "that we Φ", where Φ is a type of collective activity. To avoid a circularity, it cannot be required that the notion of collective activity that figures in the agents' intentions is that of "true collaboration" (in Tomasello and Hamann's case) or that of "shared cooperative activity" (in Bratman's case).[1] On Tomasello's interpretation, the chimpanzees that are involved in a group hunt do not each intend such a joint goal. Rather, each merely tries to capture the monkey for itself; each has the goal "that I capture the monkey for myself". Boesch, on the other hand, argues that the available evidence favours that each chimpanzee has the joint goal "that we capture the monkey".

I will not take sides in this dispute. Rather, I will unearth the dualism that underpins it. Both sides assume that chimpanzee group hunting is either cognitively demanding "true collaboration" or merely cognitively unsophisticated coordination of hunters each pursuing an individualistic goal.[2] My aim is to show that this dualism, which shows up in both philosophical and scientific discussions about joint action and cooperation, is false. While statements that several agents Φ together can either be given a collective interpretation or a distributive interpretation, this semantic dualism does not reflect an underlying dualism of socio-psychological kinds. (Of course, even if the dualism is false, Boesch or Tomasello may still be right about what characterises group hunting among the chimpanzees in Taï National Park.)

I will show that the dualism is false by giving an account of a kind of joint action that is neither true collaboration, nor merely a set of coordinated pursuits of individualistic goals. My starting point will be an operational definition of cooperative behaviour given by Christophe and Hedwig Boesch (1989). The Boesches there define the cooperative behaviour simply as "two or more individuals acting together to achieve a common goal" (1989, 550).[3] The definition tells us almost nothing about what the cognitive or conceptual prerequisites for participating in cooperative behaviour are. The definition also contains a core notion that is never explicated, namely that of a 'common goal'.

The mere fact that two agents happen to have a common goal (whatever the criteria for that are) does not itself facilitate cooperation or coordinated action toward it. My account will take the form of a set of sufficient conditions that specifies a pattern of psychological states and relations among participants that could enable this mere fact to facilitate coordinated action directed to the common goal. The

---

[1] For a discussion of this circularity problem and why it needs to be avoided, see (Petersson 2007).

[2] To be fair, Boesch now seems to be more agnostic regarding what is required for group hunting to be cooperative. See (Boesch 2012, 92–93), where he refers to several accounts of joint action, including Steve Butterfill's (2012) account of 'shared goals'.

[3] This definition has also been adopted by other researchers in both comparative and developmental psychology (see Chalmeau and Gallo 1995; Naderi et al. 2001; Brinck and Gärdenfors 2003; Brownell et al. 2006).

account is given in the next section. In Sect. 9.3, I then consider what cognitive and conceptual requirements an agent needs to meet in order to participate in this form of joint action. In Sect. 9.4, I then comment on the notion of agent-neutral goals, which enables my account to occupy a middle-ground between true collaboration and mere parallel activity. Finally, in Sect. 9.5, I consider what sorts of empirical evidence that could license us to infer that some observed behaviour is an instance of the kind of joint action I give an account of.

9.2 Having a Common Goal and Acting Jointly

The term 'goal' can be used to refer to a state of affairs or outcome O toward which an action is directed, or to the content G of a goal-directed state of an agent. The term 'common goal' can thus be used to refer to a relation between several actions or to refer to a relation between several agents. These uses are of course intimately related. Several agents' actions may have a common goal—that is, be directed to a single outcome O—in virtue of the agents having a common goal G, given that O would satisfy G if it were brought about. Plausibly, actions can be goal-directed in virtue of other kinds of facts though. Perhaps the behaviour of several insects can be directed to a single outcome, but not in virtue of the insects being in certain goal-directed states. Consider the feeding behaviour of a family of *Stegodyphus* spiders. If a large prey such as a fly lands in the family's web, then this creates vibrations in the web. In response to the vibrations, each spider independently approaches the prey (Ward and Enders 1985; quoted in Brosnan et al. 2010, 2701). When the prey is reached, each then—again, independently of each other—starts to pull the prey toward the communal nest where it is digested and consumed by all. No spider could on its own catch and transport the large prey. In this case, the function of each spider's behaviour may be to bring it about that the prey is collectively brought to the nest, even if there is no goal-directed state that represents this outcome. If this is the evolutionary function of the spiders' behaviour, then it is arguably an interesting case of a form of collective behaviour.

I will in the following use 'goal' to refer to the content of a goal-directed state and refer to the goal of an action as the 'outcome' that the action is directed to. Furthermore, I will assume that an action directed to O is so directed partly in virtue of it being controlled by goal-directed state that is satisfied by O. An agent that is in a goal-directed state with content G is said to *aim* at G.

I take a goal-directed state to either be (i) a mental state with a world-to-mind direction of fit such as a desire or an intention, or (ii) the combination of a desire or intention and a means-end belief, where the means is the goal toward which the agent's action is directed. An example of (ii) is the following: I have the desire to eat and the means-end belief that if I cook dinner, then I will be able to eat. This combination may cause me to go into the kitchen, look into the fridge, put some ingredients in a pan, and so on, that is, to perform a series of actions that are all directed toward the outcome that dinner is cooked. This outcome may not be represented by a state with a world-to-mind direction of fit; there is a state with world-to-mind direction of fit involved (the desire to eat), but it does not represent me cooking dinner. In the process of deliberating about what to do, perhaps I form or acquire an "instrumental desire" or sub-intention to cook dinner (see Sober and Wilson 1998, 217). However, this need not be the case. As Kim Sterelny (2003) points out, "[w]e can trade talk of instrumental goals for talk of beliefs", and thus "convert intentional explanations that mention instrumental goals into intentional explanations that mention only ultimate goals." (p. 88) A creature could have only one desire–say, the desire to survive–but still perform actions and activities directed to many different outcomes.

For the notion a common goal to help us distinguish joint action from coordinated activity, the agents' actions must be directed to a single outcome. There must be single outcome that satisfies the goal of each agent. Consider John Searle's (1990) example of several individuals running to take shelter from the rain. The goal of the running of each individual can be specified by the sentence "that I reach the shelter", so there is sense in which they each have the same goal: each has a goal with the same character. Nevertheless, each individual aims at a distinct goal. The intention of one of the runners is satisfied if he reaches the shelter while the intention of another is satisfied if she reaches the shelter, and so on.[4]

The term 'common goal' is sometimes used to refer to a situation where the same token object is the target of several agents' goal-directed actions.[5] For instance, Chalmeau and Gallo talk about the common goal of the chimpanzees studied by the Boesches as being simply "the prey" (1995, 103). Similarly, Brinck and Gärdenfors talk of actual objects in the chimpanzees' environment as their goals, such as "water to drink, food to be had, or an antagonist to fight" (2003, 485). Perhaps these are just elliptical ways of referring to outcomes such as "that we capture the prey" or "that we fight the antagonist", but it is important to keep in mind that goals are represented outcomes rather than target objects, or else the notion of a common goal' will not help us to distinguish cooperation from competition. If Alphonse and Amanda are each reaching for a piece of fruit, then their actions have the same target object, but they need not have a common goal. Alphonse's goal may be to snatch the fruit for only him to eat; Amanda's may be to snatch the fruit for only her to eat.

For the notion of common goal to help us understand joint action and cooperative behaviour, we must take several agents to have a common goal if the conditions of satisfaction of their goal-directed states are the same. There must be a single token outcome or state of affairs that each agent aims at. If brought about, the state of affairs would satisfy the goal-directed state of each.

---

[4] Searle (1990, 402–403) uses the term "common goal" to talk about the relationship between the runners' goals in this case. Cohen et al. (1997, 96) and Miller (1986, 133) also use the term in this way.
[5] On the distinction between goal and target object, see Jacob (2012, 209).

(1) *Single outcome condition*: There is a single outcome O that satisfies a goal $G_1, G_2, \cdots, G_m$ of each agent $A_1, A_2, \cdots, A_m$.

Note that this condition that there is a *single* outcome that satisfies a goal of each agent is not meant to exclude that there may be multiple such token outcomes. It merely implies that at least O is such a token outcome. Why not just say that there is a goal that each agent aims at? The reason why there is no goal that each runner is aiming at is that each runner is aiming at distinct goal after all, the content the goal-directed state that each is in is distinct; it is just that these goals have the same character. Many philosophers, such as John Searle (1983) for example, take the specification of the conditions of satisfaction for a mental state to also be a specification of its content. Several agents would then have a common goal if their goal-directed states had the same content. In Sect. 9.4, I argue that there are actually reasons for taking not only the content of a mental state to determine its conditions of satisfaction. The state's mode, that is, the kind of state it is, also partly determines the conditions of satisfaction. But even if we accept Searle's view, we want to be able to say that two agents have different goals even if these goals have identical conditions of satisfaction. This is because it makes sense to individuate the goals of agents not only with respect to the outcomes that satisfy them but also with respect to *how* those outcomes are represented, with respect to under what aspects they are represented. If I intend to kill Batman and intend to kill Bruce Wayne, but am unaware that Bruce Wayne is Batman, then I have two rather than one goal. This shows that goals should in general be specified with respect to aspects and not only their extensions. However, in the case of joint action, agents could arguably represent what they are aiming at under different aspects but still have a goal in common. For example, suppose Ali's goal is that the monkey who rustles the leaves is captured and Kendo's goal is that the monkey who casts the shadow is captured, then Ali and Kendo could still engage in a joint action directed at the outcome that the monkey is caught.

That (1) holds does not itself facilitate cooperation. For the fact that there is a single outcome O that satisfies each agent's goal to play a role in facilitating and coordinating joint action directed to O, each agent should also believe that (1) holds.

(2) *Doxastic single outcome condition*: $A_1, A_2, \cdots, A_m$ each believes that (1).

Note that in order for (2) to be satisfied, it is not sufficient that there is single outcome O such that it satisfies the goals $G_1, G_2, \cdots, G_m$ that each agent believes they and the others have. For (2) to be satisfied, each agent must also be aware of the fact that O is a single outcome that satisfies all the goals $G_1, G_2, \cdots, G_m$. The condition therefore helps rule out the case where Ali's goal is that the monkey who rustles the leaves is caught, Kendo's goal is that monkey who casts the shadow is caught, but Ali and Kendo each falsely believes that these are two distinct monkeys, when in fact there is only one monkey (who both rustles the leaves and casts the shadow). It is worth most existing accounts of joint action do not rule out such Frege-style cases and they arguably need to incorporate a condition similar to (2).[6]

---

[6] In Blomberg (2015), I show that this is at least the case when it comes to Butterfill's (2012) account of shared goals.

Recall that according to the Boesches' definition, several individuals are behaving cooperatively when they are "*acting together* to achieve a common goal" (my emphasis). But conditions (1) and (2) could be satisfied even if O is such that it cannot be brought about by more than one individual. Suppose several agents are inside a room, each with the goal to open the door. Each also believes that the goal of each will be satisfied if the door is opened. However, the small door handle cannot be operated by more than one person simultaneously, and each agent knows this. In this case, the fact that conditions (1) and (2) are satisfied doesn't put the agents in a position vis-à-vis each other to coordinate their actions to open the door. Here, Kaarlo Miller and Raimo Tuomela's notion of a "dividable" goal is useful: "Goal P of an agent x is *dividable* if and only if there will, or at least can, be parts or shares for at least one other agent to bring about (or sustain) P." (Miller and Tuomela 2013, 6; see also Butterfill 2013, 849–850) The goal of each agent in the room is not dividable. Likewise, the goal of a chimpanzee that is trying to capture the monkey for itself does not have a dividable goal. A chimpanzee who intends "that we capture the monkey together" does have a dividable goal, as does a chimpanzee who is simply trying "to capture the monkey" (assuming that the satisfaction of this goal is compatible with the contributions of other agents). But what is crucial is not whether the goal of each is dividable, but that each *believes* that the goal of each is:

(3) *Doxastic dividable goal condition*: Each agent $A_1, A_2, \cdots, A_m$ believes that $G_1, G_2, \cdots, G_m$ are dividable.

Finally, each agent that participates in the joint action expects that the others also participate:

(4) *Doxastic action condition*: In virtue of their beliefs specified in (1) and (2), each agent $A_1, A_2, \cdots, A_m$ believes that the others has performed, is performing or will perform an action directed to O in order to achieve their goal $G_1, G_2, \cdots, G_m$.

If these belief states and goal-directed states of the agents cause them to perform actions directed to O, then I submit that these actions constitute a form of joint action. Note that this account does not exclude that joint action may include elements of coercion and manipulation. But given that conditions (1)–(4) are satisfied, then coercion or manipulation of others are unlikely to be of benefit to a participating agent.

9.3 Cognitive and Conceptual Demands

Conditions (1)–(4) are relatively cognitively and conceptually undemanding. To participate in joint action that is caused by the belief states and goal-directed states defined by these conditions, agents need not be able to attribute beliefs to other agents. But they must be able to attribute goals to others. Non-human primates seem to fit this socio-cognitive profile. At least, chimpanzees appear to be sensitive to the goal-directness (or outcome-directedness) of others' actions (Call et al. 2004;

Call and Tomasello 2008). However, note that the fact that they can recognise the outcome that another's action is directed toward does not imply that they can attribute a goal to the agent of that action. Instead, they may simply represent the outcomes as states of affairs toward which the actions are pulled, or represent them as the function of the bodily movements in question (see Csibra and Gergely 2007; Butterfill and Apperly 2013). In other words, recognising the goals that actions are directed toward does not necessarily require metarepresentational capacities. One can recognise an outcome that another agent's action is directed toward without representing it as something that is represented by the other agent (see Butterfill (2012) for an account of joint action that doesn't require that agents are able to attribute goals to other agents, but only to the actions of others). Nevertheless, it is possible that chimpanzees have a mentalistic understanding of goal-directedness, even if they do not have a mentalistic understanding of beliefs. With such a partial understanding of other minds, they could make sense of actions in terms of partially subjective motivating reasons for action, in terms of desires or intentions that they attribute to others in combination with facts about the world (drawing on their own beliefs about the world).[7] No study has hitherto been able to show that non-human primates are able to attribute beliefs to others, but my account does not require that participants have this ability.

The account also requires that agents have general capacities for reasoning about how several causal contributions may generate a combined effect. Exactly what is required will of course depend on the task that the participants are facing. Clearly, many types of agents have such capacities, including chimpanzees (see e.g. Seed and Call 2009).

As I have noted, Tomasello and Hamann's "true collaboration" as well as Bratman's "shared cooperative activity" requires that participants have joint goals, goals "that we (in mutual knowledge) do X together" or "that we Φ". Is this conceptually and cognitively demanding? If mutual knowledge requires a concept of belief (see Glüer and Pagin 2003; Tollefsen 2005), then the former notion appears to be beyond what non-human primates could intend. However, there are arguably weaker notions of collective activity that could figure in the content of the agents' intentions, so a requirement that each agent intends "that we Φ" need not be very conceptually demanding (see Petersson 2007). To illustrate, we can conceptualise the activity of a family of *Stegodyphus* spiders without conceiving of them as having any mutual knowledge or goal-directed states. Such a weak concept of collective activity or joint action may figure in the content of the intentions of those who engage in "shared cooperative activity" or similar robust forms of joint action. Nevertheless, the account I have given does not even require that agents have such a weak concept of joint action.

---

[7] This would be akin to what Perner and Roessler (2010, 205) call the "the hybrid account of children's conception of intentional action".

9.4 The Middle-Ground: Agent-Neutral Goals

The kind of joint action that I have specified in Sect. 9.2 occupies a middle-ground between "true collaboration" and mere parallel activity because conditions (1)–(4) can be satisfied even if none of $G_1, G_2, \cdots, G_m$ is a joint goal. A goal may be neither "joint" nor be exclusively about an individual's own agency.[8]

Consider the case of two individual agents who go for a walk together. They can do this in a way which is neither like the friends going for a walk together, nor like two strangers who each go for a walk in parallel. Suppose that Ann and Bob are colleagues who are taking part in a workplace pedometer challenge. Each of them has a pedometer—a step-counting device—that displays the sum of the total number of steps they have taken. Ann and Bob each intends that the total number of steps reaches 10,000 (not that *they* take 10,000 steps). Their actions are to some extent interdependent. If one of them believes that the other isn't going to walk at all, then they will rescind their intention. One of them cannot take 10,000 steps on their own in time they have at their disposal. If one of them sees that the other is walking a lot, then they may themselves take fewer steps since their intention is likely to be satisfied anyway. Here, if all of conditions (1) to (4) are satisfied and if Ann and Bob's belief states and goal-directed states cause them to take 10,000 steps, then arguably, they take 10,000 steps together by performing a joint action.

Both Ann's and Bob's goal is dividable here without being a "we"-goal. These goals are agent-neutral in the sense that they are compatible with the possibility that the other agent contributes to its achievement, but this contribution is not itself part of what the agent herself aims at. However, it also looks like these goals are compatible with the absence of a contribution from the agent herself. Ann's goal would, it seems, be satisfied even if she was never moved to take a single step if Bob somehow managed to walk 10,000 steps on his own. If that is right, then it becomes somewhat mysterious why condition (4) in the account provided in Sect. 9.2 would be satisfied. After all, it seems that agent-neutral goals would be satisfied even if the agents never get involved in the action so to speak. Why then would each agent expect that the others have performed, are performing or will perform actions directed to O given that their goals are agent-neutral in this sense?

Given a certain view of the content of intentions, this question doesn't arise if the goal-directed states of Ann and Bob are intentions.[9] According to this view, the agent's self, in the form of the indexical 'I', as well as the causal efficacy of her intention figures constitutively in the specification of the content of an intention. For example, according to Searle, the content of an intention to raise my arm that I have prior to raising it is "[that] I perform the action of raising my arm by way of carrying out this intention" (1983, 92). Ann and Bob's goal-directed states could not on this

---

[8] My account of joint action is not the only one that does not require that the actions that constitute a joint action is directed to a collective activity. Other such accounts include Pacherie and Dokic (2006, 110), Butterfill (2012) and Miller (2001).

[9] Nor does the question arise if Ann's and Bob's goal was "that we take 10,000 steps". After all, each of them takes herself or himself to be a member of the "we" that the goal concerns.

view be intentions but would have to be some other type of goal-directed states.[10] Still, we want their goal-directed states to play some role in causing and coordinating the agents' activities that are directed to bringing about the goal.

Now, we can avoid this difficulty if we drop the assumption that it is only that content of a goal-directed state that determines its conditions of satisfaction. The conditions are arguably not determined solely be features that are internal to the content of the state in question. To be specific, the conditions of satisfaction for an intention or other goal-directed state is also in part fixed by the type of state it is.[11] We can think of a mental state's type in functional terms, defined by its role in the agent's psychic economy, including its role in causing and coordinating action.

The causal self-reflexivity that Searle places in the content of an intention is arguably part of what makes this goal-directed state into an intention rather than, say, a different state with world-to-mind direction of fit such as a hope or a wish. This suggests that the causal self-reflexivity is not part of the content of an intention, but that it is determined by the functional role of intention. Arguably, what is intended does not include this causal self-reflexivity. I do not intend that "I perform the action of raising my arm by way of carrying out this intention", but rather simply "to raise my arm". Nevertheless, since I represent this in virtue of having an intention, the action will only be successful if it is appropriately caused and coordinated by this intention itself. The role of the agent's own agency may thus merely be implicitly represented in the cognitive system's architecture. The goal-directed state that controls the agent's movements as she, say, approaches a table that she wants to move, could thus fail to fulfil its function if the state wasn't involved in bringing about the outcome that satisfied her goal "that the table is moved". However, there is no reason to think that the state would fail to fulfil its function if other agents were also playing a causal role in bringing about the satisfaction of the goal.

One might suspect in response to this that an agent-neutral goal is just a disguised "me"- or "we"-goal.[12] This is a reasonable suspicion since the conditions of satisfaction remains the same whether they are completely determined by the content of the goal-directed state or whether they are in part also determined by the state's mode. However, this is not merely a terminological move. First, the fact that the conditions of satisfaction includes that condition that the agent herself brings about O, possibly with help of the contributions of others, does not require that the agent has the ability to think of herself as an agent or as acting with others. The role of the self as an agent can merely be implicit in the functional role of the goal-directed state.[13] Secondly, having a goal that allows for the contributions of others is

---

[10] On Searle's view, my condition (1) will never be satisfied by goals that agents have in virtue of what they *intend*. His view implies that there is never a single outcome that satisfies the intention of each. The conditions of satisfaction for our ordinary intentions will never be the same.

[11] Björn Petersson (2015) makes a similar point in the context of a discussion of Bratman's account of shared cooperative activity. Petersson draws on the work of François Recanati (2007).

[12] Thanks to an anonymous reviewer for raising this worry.

[13] See Sober and Wilson (1998, 213–217) for a related discussion about the possibility of "general and impersonal desires" and the evolutionary benefits of self-directed desires.

not equivalent to having the goal that others contribute, that *we* do it. If Ann intends that 10,000 steps is taken, she will adjust her activities in light of what she believes or expects about Bob's activities, but she doesn't *intend* that Bob contributes.

Consider several agents who are out hunting an arboreal prey. Suppose that each agent in the group has the goal "that the prey is captured" or "to capture the prey", where the satisfaction of this goal is compatible with the involvement of others. Furthermore, each performs appropriate actions directed to the outcome that they prey is caught, and they do this in virtue of their beliefs that this single outcome satisfies a goal of each as well as their belief that their goals are dividable. Arguably, their hunting the prey would now be a kind of joint action. If the prey is captured, then they capture it by performing a joint hunt.

What I am suggesting is that perhaps this is the best way of conceptualising chimpanzee group hunting. It would on this view be part of the function of the chimpanzees' goal-directed states to facilitate that each agent contributes to the prey being caught (within the context of the hunting activity of the group). However, each agent need not explicitly intend that *they*, individually or collectively, catch the prey. Of course, this is not to deny that each agent has an interest in getting a share of the prey if one of the agents finally reaches or intercepts the prey. Otherwise they would not have the goal "that the prey is captured". This suggestion is of course hostage to empirical fortune. If Tomasello and colleagues are correct that what determines a chimpanzees' share of the spoils is simply its proximity to the kill, then this would of course be defeasible evidence for the view that each chimpanzee has the goal "that I capture the prey", where this is compatible with the contributions of others. Here, I leave this empirical issue be.

9.5 Empirical Tractability

My aim in this section is to say something about how conditions (1)–(4) are related to empirical studies of performance on tasks that require agents to cooperate and coordinate their actions. Whether or not we should claim that a multi-agent activity is coordinated by the goal-directed states and belief states specified by conditions (1)–(4) will depend on whether those states figure in our best psychological explanation of how the activity came about. This issue is one of inference to the best explanation, so we should not expect or demand a fixed set of behavioural criteria that can be used to decisively determine whether those states are present or not. The best one can hope for is a set of relevant constraints and factors that can be used to judge whether an inference to their presence is justified or not.

One source of evidence is constraints given by the cognitive and conceptual capacities of the agents involved, as suggested in Sect. 9.3. Another source of evidence concerns whether two or more individuals reliably achieve an outcome that is desired by each but which can only be brought about if they coordinate their actions. In such a case, the performance of the subtask of each comes to nothing unless the other also performs his or her subtask. If one can observe that an individual only performs its own subtask (or is more likely to perform it, or performs it at higher rate) when the other individual is likely to carry out its subtask (such as only when the other individual is present for example, or only when the other individual is facing in the relevant direction) then we can infer that this individual has some understanding of the role the other plays for successful task completion. On the basis of this line of reasoning, primatologists, comparative psychologists, and developmental psychologists have tested whether various types of agents are able to coordinate their actions to achieve a common goal.

The following experimental setup from a study by Alicia Melis et al. (2006) is representative. Melis et al. (2006) tested whether chimpanzees were able to appropriately judge when to recruit a collaborator in order to retain food. Each subject was presented with a platform baited with food that was placed behind a railing. A rope was threaded through metal loops on the platform, with both ends of the rope extending through the railing into the test room that the subject was released into. If a subject only pulled one rope end in order to drag the food-baited platform toward the railing, then the rope would unthread through the metal loops and come loose from the platform (and thus make it impossible to retrieve the food). In order to get the platform to get closer to the railing, either both ends had to be pulled at the same time or one end had to be pulled while the other was held steady. In an adjacent locked room another chimpanzee, a potential collaborator, was waiting. By removing a wooden peg, a test subject could release the potential collaborator into the test room. Melis et al. (2006) were interested in how the decision of a subject to release or not to release the other chimpanzee depended on whether it was physically possible for the subject to retrieve the food from the platform or not.

There were two conditions in this experiment: the collaboration condition and the solo condition. In the collaboration condition, the ends of the rope were placed three metres apart so that it was impossible for the subject to hold or pull both ends of the rope at the same time. Hence, to retrieve the food on the platform in the collaboration condition, the subject would have to release the collaborator into the test room. Then each could pull one rope end and together drag the platform toward the railing. In the solo condition, the ends of the rope were placed fifty-five centimetres apart, so that it was possible for the subject to pull both rope ends at the same time. The subject could thus acquire all the food for herself without having to share half of it with the collaborator. The result was that seven of the eight chimpanzee subjects released the collaborator significantly more often in the collaboration condition than in the solo condition. Melis et al. (2006) took this to show that the chimpanzees were sensitive to the fact that both their own contribution and the contribution of the collaborator were required for successful food retrieval.

This experiment gives an either/or measure of subjects' understanding of whether or not a situation calls for cooperation: the chimpanzee can either release the potential partner or choose not to. In other experiments, subjects are repeatedly trying to do something either alone or together with a partner. This allows one to compare the rate at which subjects try to achieve a goal (such as retrieving food) without and with a partner. If a task requires the contribution of a partner and there is a significant difference in the rate of attempts between when the partner is absent and when the partner is present, then this suggests that the individuals are monitoring each other's activity and modulating their activity in light of what the other is doing (see Chalmeau and Gallo 1995; Chalmeau 1994; Visalberghi et al. 2000).

If a subject clearly shows sensitivity to the fact that a task requires another's contribution, then any interpretation that construes the behaviour of these agents as rational will involve attribution of a dividable goal to at least one agent (the subject in Melis et al.'s (2006) task). Purely individualistic goals—goals "that I bring about O on my own"—cannot be satisfied if the agent knows that bringing about O is impossible without the contribution of another agent. It may of course be the case that the goal the agent has in common with another agent, say the goal "to bring the platform to the railing", is one that she only has in virtue of an instrumental desire formed in light of her ultimate desire "to eat all food myself". The higher-level goal that she has in virtue of this ultimate desire will then not be shared with another agent.[14]

Note that the experiments confound the situation where subjects have a common goal but fail to grasp that there is an opportunity to benefit from coordinated action, and the situation where they do have the causal understanding required to benefit from coordinated action but lack a common goal. The capacity to have dividable goals and to recognise that one has a goal in common with another agent can thus be masked by failure to meet other performance requirements of the task. In particular, the task may require a quite sophisticated understanding of causal relationships for an agent to understand that the contribution of another agent can help him or her to achieve the goal. But if a subject only acts in concert with a partner, or only acts when a partner is likely to contribute, then inferring that the subject has the capacity to participate in a form of joint action with another seems to be justified (that is, that the subject is able to appreciate that he and another agent have a common goal and that it is more likely that the goal is achieved if both perform their contributions).

However, note that apparently purposive coordination may be the result of the fact that the agents are embedded in the same environment at the same time, and thus presented with similar constraints and opportunities for action. This may accidentally lead to similar actions being performed roughly at the same time toward the same target object. But we cannot conclude from the fact that coordination is merely accidental that the agents do not have a common goal. Plausibly, some multi-agent activities may be coordinated in virtue of agents typically having agent-neutral goals and high social tolerance, so that they are able to act in parallel in close proximity to each other, and thereby be exposed to the same action constraints and opportunities, potentially leading to them to perform actions directed to a single outcome O (see Petit et al. (1992) on "coproduction"). However, this would not amount their actions constituting a joint action that is caused by the states defined by conditions (1)–(4).

---

[14] Note that I have not considered the issue of how agents share the spoils of their joint action. The account I have given concerning coordination of action toward a common goal, it does not say anything about how agents act once the goal has been achieved.

9.6 Conclusion

The starting point for the account of joint action that I have given in this chapter is an influential definition according to which 'cooperative behaviour' consists of "two or more individuals acting together to achieve a common goal" (Boesch and Boesch 1989, 550). The account specifies an inter-agential pattern of goal-directed states and beliefs that can facilitate the coordination of several agents' actions with respect to a common goal. Here, several agents have a common goal if there is a single outcome that satisfies a goal of each agent (*single outcome condition*) and each agent believes that this is the case (*doxastic single outcome condition*). For their common goal to enable coordinated action directed to the single outcome that satisfies their goals, each must also believe that the goal of each is compatible with the involvement of the others (*doxastic dividable goal condition*). Furthermore, in virtue of these beliefs, each must also believe that each of the others has performed, is performing or will perform actions toward the single outcome in order achieve their goal (*doxastic action condition*).

This account differs from Tomasello and Hamann's account of "true collaboration" and Bratman's account of "shared cooperative activity" in several respects. First, it is relatively cognitively undemanding in comparison to these accounts. In particular, the account doesn't require that participants represent beliefs and understand that others can have a different cognitive perspective on the world. Secondly, the account doesn't require that each agent has the performance of a collective activity as a goal. However, this doesn't mean that each agent has as a self-directed goal that they do something on their own (without the involvement of others). Arguably, agents can also have goals that are "agent-neutral".

The account I have presented is of a kind of joint action that falls somewhere in between robust forms of intentional joint action such as "true collaboration" and "shared cooperative activity" on one side and cases of mere parallel activity such as that exemplified by two strangers walking in parallel on the other side. There are arguably many forms of social coordination and joint action that do not neatly fall into either of these two categories. My hope is that this kind of account can counteract a tendency in both philosophical and scientific discussions about joint action to assume a false dualism between genuinely joint and merely parallel activity. In the beginning of the chapter, I showed how this dualism has structured a debate between Michael Tomasello and Christophe Boesch regarding whether or not chimpanzee group hunting in Taï National Park is a form of joint cooperative hunting. Whether this group hunting behaviour is best explained by the pattern of goals and beliefs specified by the account I have proposed is of course an empirical question that I have left be here.

Acknowledgements Thanks to Catrin Misselhorn and to an anonymous reviewer. This chapter is partly based on research I did during my PhD studies at the University of Edinburgh, made possible by a European PhD Scholarship from Microsoft Research, for which I am grateful. For fruitful discussions during my PhD, I thank Steve Butterfill, Natalie Gold, Suilin Lavelle, Matt Nudds and Till Vierkant. Further work has been made possible by a postdoctoral research grant (DFF—4089-00091) from Danish Council for Independent Research and FP7 Marie Curie Actions COFUND under the 7th EU Framework Programme.

References

Blomberg, Olle. 2015. Shared goals and development. *Philosophical Quarterly* 65(258): 94–101. doi: 10.1093/pq/pqu059

Boesch, Christophe. 1994. Cooperative hunting in wild chimpanzees. *Animal Behaviour* 48: 653–667.

Boesch, Christophe. 2002. Cooperative hunting roles among Taï Chimpanzees. *Human Nature* 13: 27–46.

Boesch, Christophe. 2005. Joint cooperative hunting among wild chimpanzees: Taking natural observations seriously. *Behavioral and Brain Sciences* 28: 692–693.

Boesch, Christophe. 2012. *Wild cultures: A comparison between chimpanzee and human cultures*. Cambridge: Cambridge University Press.

Boesch, Christophe, and Hedwig Boesch. 1989. Hunting behavior of wild chimpanzees in the Taï National Park. *American Journal of Physical Anthropology* 78: 547–573.

Bratman, Michael. 1992. Shared cooperative activity. *The Philosophical Review* 101: 327–341.

Bratman, Michael. 2009. Modest sociality and the distinctiveness of intention. *Philosophical Studies* 144: 149–165.

Brinck, Ingar, and Peter Gärdenfors. 2003. Co-operation and communication in apes and humans. *Mind & Language* 18: 484–501.

Brosnan, Sarah F., Lucie Salwiczek, and Redouan Bshary. 2010. The interplay of cognition and cooperation. *Philosophical Transactions of the Royal Society B: Biological Sciences* 365: 2699–2710. doi:10.1098/rstb.2010.0154.

Brownell, Celia A., Geetha B. Ramani, and Stephanie Zerwas. 2006. Becoming a social partner with peers: Cooperation and social understanding in one- and two-year-olds. *Child Development* 77: 803–821. doi:10.1111/j.1467-8624.2006.00904.x.

Butterfill, Stephen. 2012. Joint action and development. *The Philosophical Quarterly* 62: 23–47.

Butterfill, Stephen A. 2013. Interacting mindreaders. *Philosophical Studies* 165: 841–863.

Butterfill, Stephen A., and Ian A. Apperly. 2013. How to construct a minimal theory of mind. *Mind & Language* 28: 606–637.

Call, Josep, and Michael Tomasello. 2008. Does the chimpanzee have a theory of mind? 30 years later. *Trends in Cognitive Sciences* 12: 187–192. doi:10.1016/j.tics.2008.02.010.

Call, Josep, Brian Hare, Malinda Carpenter, and Michael Tomasello. 2004. 'Unwilling' versus 'unable': Chimpanzees' understanding of human intentional action. *Developmental Science* 7: 488–498.

Chalmeau, Raphaël. 1994. Do chimpanzees cooperate in a learning task? *Primates* 35: 385–392.

Chalmeau, Raphaël, and Alain Gallo. 1995. Cooperation in primates: Critical analysis of behavioural criteria. *Behavioural Processes* 35: 101–111.

Cohen, Philip R., Hector J. Levesque, and Ira Smith. 1997. On team formation. In *Contemporary action theory volume 2: Social action*, ed. Ghita Holmström-Hintikka and Raimo Tuomela, 87–114. Synthese Library, Vol. 267. Dordrecht: Kluwer.

Csibra, Gergely, and György Gergely. 2007. 'Obsessed with Goals': Functions and mechanisms of teleological interpretation of actions in humans. *Acta Psychologica* 124: 60–78.

Gilbert, Margaret. 2009. Shared intention and personal intentions. *Philosophical Studies* 144: 167–287.

Glüer, Kathrin, and Peter Pagin. 2003. Meaning theory and autistic speakers. *Mind and Language* 18: 23–51.

Jacob, Pierre. 2012. Sharing and ascribing goals. *Mind & Language* 27: 200–227.

Kutz, Cristopher. 2000. Acting together. *Philosophy and Phenomenological Research* 61: 1–31.

Melis, Alicia P., Brian Hare, and Michael Tomasello. 2006. Chimpanzees recruit the best collaborators. *Science* 311: 1297–1300. doi:10.1126/science.1123007.

Miller, Seumas. 1986. Conventions, interdependence of action, and collective ends. *Nous* 20: 117–140.

Miller, Seumas. 2001. *Social action: A teleological account*. New York: Cambridge University Press.
Miller, Kaarlo, and Raimo Tuomela. 2013. Collective goals analyzed. In *From individual to collective intentionality: New essays*, ed. Sara Rachel Chant, Frank Hindriks, and Gerhard Preyer, 34–60. New York: Oxford University Press.
Naderi, Sz., Ádam Miklósi, Antal Dóka, and Vilmos Csányi. 2001. Co-operative interactions between blind persons and their dogs. *Applied Animal Behaviour Science* 74: 59–80.
Pacherie, Elisabeth. 2013. Intentional joint agency: Shared intention lite. *Synthese* 190: 1817–1839.
Pacherie, Elisabeth, and Jerome Dokic. 2006. From mirror neurons to joint actions. *Cognitive Systems Research* 7: 101–112. doi:10.1016/j.cogsys.2005.11.012.
Perner, Josef, and Johannes Roessler. 2010. Teleology and causal understanding in childrens' theory of mind. In *Causing human action: New perspectives on the causal theory of action*, ed. Jesús H. Aguilar and Andrei A. Buckareff, 199–228. Cambridge, MA: MIT Press.
Petersson, Björn. 2007. Collectivity and circularity. *Journal of Philosophy* 104: 138–156.
Petersson, Björn. 2015. Bratman, Searle, and simplicity. A comment on Bratman: Shared agency, a planning theory of acting together. *Journal of Social Ontology* 1(1): 27–37.
Petit, Odile, Christine Desportes, and Bernard Thierry. 1992. Differential probability of 'coproduction' in two species of macaque (*Macaca Tonkeana, M. Mulatta*). *Ethology* 90: 107–120.
Recanati, François. 2007. Content, mode, and self-reference. In *John Searle's philosophy of language: Force, meaning, and mind*, ed. Savas L. Tsohatzidis. New York: Cambridge University Press.
Searle, John R. 1983. *Intentionality: An essay in the philosophy of mind*. New York: Cambridge University Press.
Searle, John R. 1990. Collective intentions and actions. In *Intentions in communication*, ed. Philip R. Cohen, Jerry Morgan, and Martha E. Pollack, 401–415. Cambridge, MA: MIT Press.
Seed, Amanda, and Josep Call. 2009. Causal Knowledge for Events and Objects in Animals. In *Rational Animals, Irrational Humans*, ed. Shigeru Watanabe, Aaron P. Blaisdell, Ludwig Huber, and Allan Young, 173–187. Tokyo: Keio University Press.
Sober, Elliott, and David Sloan Wilson. 1998. *Unto others: Evolution and psychology of unselfish behavior*. Cambridge, MA: Harvard University Press.
Sterelny, Kim. 2003. *Thought in a hostile world: The evolution of human cognition*. Malden: Wiley-Blackwell.
Tollefsen, Deborah. 2005. Let's pretend! Children and joint action. *Philosophy of the Social Sciences* 35: 75–97.
Tomasello, Michael. 2014. The ultra-social animal. *European Journal of Social Psychology* 44: 187–194.
Tomasello, Michael, and Katharina Hamann. 2012. Collaboration in young children. *The Quarterly Journal of Experimental Psychology* 65: 1–12.
Tomasello, Michael, Malinda Carpenter, Josep Call, Tanya Behne, and Henrike Moll. 2005. Understanding and sharing intentions: The origins of cultural cognition. *Behavioral and Brain Sciences* 28: 675–691.
Visalberghi, Elisabetta, Benedetta P. Quarantotti, and Flaminia Tranchida. 2000. Solving a cooperation task without taking into account the partner's behavior: The case of capuchin monkeys (*Cebus Apella*). *Journal of Comparative Psychology* 114: 297–301.
Ward, Paul I., and Margit M. Enders. 1985. Conflict and cooperation in the group feeding of the social spider stegodyphus mimosarum. *Behaviour* 94: 167–182.

Chapter 10
Choosing Appropriate Paradigmatic Examples for Understanding Collective Agency

Tom Poljanšek

10.1 Introduction

Over the last three decades, collective agency (CA) has received increased attention from analytic philosophers. However, to date the analytic community has still not agreed on a unified account of CA. Below, I will argue for a pluralistic account of CA, asserting that the current debate focusses too much on developing a unified account that encompasses all relevant types of CA. Although the term 'collective agency' seems to imply that CA is a homogeneous phenomenon, I argue that there are in fact several distinct types of CA which cannot be reduced to a single model. Thus, if our theories of CA are to be empirically adequate, we must distinguish conceptually between these different types of CA.

In the paper's first part, I examine how, within current theories, small-scale cases, involving just a few agents, often serve as paradigmatic examples of CA. My focus is on two of the most prominent accounts of CA in the contemporary debate, namely, those offered by Margaret Gilbert and Michael Bratman. It can be shown that both, to some extent at least, presuppose the generalizability of small-scale cases of CA, while they ignore the possibility of more complex cases. I argue that this methodological approach is problematic, because, in choosing a constricted sample of paradigmatic cases, it neglects several important features of specific cases of CA.

In the second part, I discuss the current 'puzzle form' (Kuhn) of CA theories. By this I intend to show that theories based on taking small-scale cases as paradigmatic for CA fail to account sufficiently for at least three constitutive features of some collective agency types: (1) the *sense of belonging to a specific group*, (2) the *socio-technological environment* (e.g., communication media) which arguably is the constitutive dimension of some types of CA, and (3) specific

---

T. Poljanšek (✉)
Institute of Philosophy, University of Stuttgart, Stuttgart, Germany
e-mail: tom.poljansek@philo.uni-stuttgart.de

features of *cases of trans-temporal CA*. For brevity's sake, I will only focus on the first of these three features.

To begin with, I want to introduce an important distinction, which may contribute to shaping the current debate: the distinction between theories of *collective agency* and theories of *groups* or *plural agents*. While the two types of theories are closely related, I will argue that they are not the same. While the latter focuses on shared action in the narrow sense of the term, the former seems to touch much more closely on questions of social ontology, e.g., "Are there such things as 'groups'?", or "Is a group somehow ontologically independent of its members?".

However, if it can be shown that a group does not need to be constituted by individuals in pursuit of a common goal, the theory of groups would prove to be (at least partially) independent of the theory of CA. Similarly, if it can be shown that not all cases of individuals acting together involve forming a group (in any relevant sense of the term),[1] the theory of CA is independent of the theory of groups. If so, it means that there can be CA without group-formation and group-formation without CA. I present some arguments in support of this assertion in what follows.

10.2 Choosing the Right Paradigmatic Examples

The current debate about CA focuses mainly on small-scale cases of CA. Most of the examples deal with two people doing something together: going for a walk together, preparing a sauce hollandaise together, painting a house together, going to New York together, etc. The choice of these paradigmatic examples reflects the idea that a general conception of CA can be derived from such small-scale cases. Margaret Gilbert explicitly states this idea as follows:

> The idea is that we can discover the nature of social groups in general by investigating such small-scale temporary phenomena as going for a walk together. (Gilbert 1996, 178)

Small-scale cases of CA, Gilbert adds, can thus be "considered a paradigm of social phenomena in general". Michael Bratman's methodological intuitions are quite similar to those of Margaret Gilbert, at least when it comes to the choice of paradigmatic examples. His examples, too, mostly comprise two people doing something together. However, in contrast to Gilbert, Bratman explicitly restricts the scope of his theory to small-scale cases of CA. As he puts it, his theory of CA does not address "larger, institutional forms of shared agency, such as, perhaps, law or democracy." (Bratman 2010, 9) What Bratman wants to explain are cases of "small scale shared intentionality, in the absence of institutional authority relations" (ibid.).

However, Bratman's emphasis on the role of authority relations suggests that by adding such relations more complex cases of CA can be accounted for without changing the basic elements of his theory in a significant way, i.e. introducing new primitive entities (for example Searlian "we-intentions" or Margaret Gilbert's

---

[1] I will try to elucidate what a 'relevant sense' of the term 'group' could be later within this paper.

10 Choosing Appropriate Paradigmatic Examples for Understanding Collective Agency

conception of a "joint commitment", see Bratman 2014, 9). (As we will see below, Scott Shapiro tries to apply Bratman's account to cases of what he calls 'massively shared agency' by adding exactly such authority relations.)

Paradigmatic examples do not merely serve as proofs of concepts and intuition pumps, they can also have a serious theoretical impact. It seems to me that a general problem within current philosophical research is that philosophers do not reflect sufficiently on the theoretical impact of the paradigmatic examples they use (for an exception see Schaub 2010). If the variety of the paradigmatic examples used to develop our theory of a given phenomenon is too narrow, then we may be inclined to overlook aspects of the phenomenon which are marginalized within the chosen examples but may nevertheless be relevant. To illustrate this, let us consider some examples from the 'theory of emotions'. If you take fear of a snake as the paradigmatic example of an emotion (perhaps because it seems to be a simple phenomenon which may be easier to analyze), you will be likely to design a significantly different model of what emotions are (say: that they are in some way similar to perception) than if you had chosen pride of owning a new car, or the way you care for your children as your standard example for an emotion (for an overview see Salmela 2011). Here, as in many other domains, the question how a specific phenomenon is modeled "depend[s] on individual cases" chosen as paradigmatic examples (Salmela 2011, 25). In a famous paragraph of his *Philosophical Investigations*, Ludwig Wittgenstein phrased the methodological trap lying behind such choices as follows:

> A main cause of philosophical disease – a one-sided diet: one nourishes one's thinking with only one kind of example. (Wittgenstein 2009, 164)

The thought that Wittgenstein is expressing here is that within philosophy, we often bias ourselves by only focusing on 'one kind' of example. One might think that this does not pose a huge problem for philosophy, since philosophers are considered to reflect extensively upon their methodology and often make use of counter-examples to show that a given theory is not complex enough to handle specific cases which do in fact fall under its scope. However, the choice of paradigmatic examples not only affects our description of a given phenomenon. It also limits the scope of possible candidates for counter-examples, especially when the phenomenon in question is not considered to form a natural kind (as I claim is the case with CA). What we take to be "intuitively paradigmatic cases" (Gilbert 2006, 166) of a given phenomenon restricts our attention to phenomena which somehow resemble these paradigmatic cases. In the worst-case scenario, we are stuck in a vicious circle of understanding: Our theory merely elaborates on what we presupposed from the beginning.

I suggest that the standard cases we choose to describe a phenomenon function as what Thomas Kuhn originally called a 'paradigm':

> [O]ne of the things a scientific community acquires with a paradigm is a criterion for choosing problems that, while the paradigm is taken for granted, can be assumed to have solutions. To a great extent these are the only problems that the community will admit as scientific or encourage its members to undertake. Other problems, including many that had

previously been standard, are rejected as metaphysical, as the concern of another discipline, or sometimes as just too problematic to be worth the time. A paradigm can, for that matter, even insulate the community from those socially important problems that are not reducible to the puzzle form, because they cannot be stated in terms of the conceptual and instrumental tools the paradigm supplies. (Kuhn 1962, 37)

Kuhn's notion of a 'paradigm' has been criticized for being rather vague (see e.g. Materman 1974). However, whether his description of puzzle-solving within science is generally adequate or not is not relevant here. What I rather want to argue is that we can use Kuhn's notion to elucidate a problematic aspect of philosophical methodology, namely, that the paradigmatic examples we use within philosophical theorizing do in fact play the role of what Kuhn calls a 'paradigm'. The paradigmatic examples we use provide us with a conceptual framework which limits the scope of features and phenomena our theories (seem to) have to address. This is why we will be either inclined to treat certain aspects of a phenomenon as less important if they are not prominent enough in the descriptions of our standard cases, or it may even lead us to completely ignore cases which do not fall under the standard description we derived from these examples. The paradigmatic examples function as attention-markers, which highlight certain aspects of a phenomenon while at the same time eclipsing other potentially important features. Hence, if we restrict our theoretical attention by a limited selection of paradigmatic examples at an early research stage, it will be hard to recognize what we excluded from the beginning. However, it is important to stress that the *number* of examples is not significant in this context. What really matters is the *variety* of the examples we use. We do not only have to use *several* but several different *kinds* of examples.

To summarize what I have argued so far, I see two problems arising from a too constricted selection of paradigmatic examples: First, it is possible that we underestimate the relevance of certain aspects of a phenomenon because we simply don't need to take them into account in order to explain our standard examples. Our selection of paradigmatic examples would then provide us with a certain tendency to model other phenomena in a similar manner for the sake of a unified theory. A second problem arises when what we originally labeled as one unified phenomenon, turns out to comprise several independent phenomena which should be treated independently.

I suggest that we should therefore bear in mind the *hermeneutic circle* that is involved in philosophical inquiry: We chose our paradigmatic examples in order to express a certain preunderstanding of the phenomenon in question, but if we merely stick to the task of making this preunderstanding explicit through our philosophical investigations, we might simply elaborate the prejudices we already have concerning the phenomenon we want to investigate. To overcome this trap, philosophers must put great effort into the first step of their inquiry: choosing a variety of paradigmatic examples. Of course, we are not able to start from scratch when theorizing, but if we direct our attention towards the task of selecting a variety of paradigmatic examples, which, based on our preunderstanding, seem to differ in certain relevant aspects, while falling under the same generic description ("A case of CA/a group."), we might, I suggest, be a little better equipped to avoid the problems described above.

10.3 The Current Puzzle-Form of CA Theories

I will now transfer these general considerations to the question of a unified theory of CA. If the previous considerations are well-founded, then we have to look closely at the paradigmatic examples at hand in order to flesh out the specific 'puzzle-form' – to use Kuhn's notion – of the current theories of CA. I think that at least three aspects are essential here: First, all standard examples of CA deal with just a few agents, who try to espouse a certain goal together: going for a walk, preparing a sauce hollandaise, painting a house, visiting New York, etc. So one major characteristic of the puzzle-form is that it focuses on *shared goal-directed agency*. Furthermore, in most of the cases, the agents acting together stay at the same place while espousing their shared goal together. Thus there seems to be no need for any form of communication beyond speech and face-to-face interaction between them. So the second important characteristic of the puzzle-form is that it deals with cases of *face-to-face interaction*. The third aspect concerns the temporal structure of CA. The standard examples not only focus on agents acting together in shared situations; they also focus on *agents acting together at the same time*. The specifics of trans-temporal cases of CA (scientific inquiry, literary traditions, religion, etc.) thus tend to be neglected as well. I do not claim that these are the only aspects of the specific puzzle-form of the current CA paradigm but rather that they are essential in certain respects that I will now elaborate upon. In the following, I will focus, however, on the first of the three aspects for reasons of space.

The first of the three aspects may appear trivial at first glance: Insofar as we are trying to develop a theory of collective *agency*, a theory of people *doing something together*, it is not surprising that we focus on people who try to realize a shared goal. In the theory of groups, however, there is a long tradition of departing from these kinds of cases as well. As Arthur Bentley, who founded the "theory of groups" within modern political science, remarked: "There is no group without its interest." (Bentley 1908, 211; see also Olson 1968, 7 ff.) However, if we take a closer look at the phenomenon in question, we can see that this focus on shared intentional action already significantly shapes the way in which we capture cases of CA. By starting our investigation with cases of shared intentional agency, we tend to overlook relevant mechanism underlying the constitution of CA which do not necessarily require that individuals (already) have a shared goal.

What are, we could ask, the prior phenomena when it comes to CA? Is it really constitutive for CA that people try to espouse a shared goal together, or might it be the case that there are some human faculties or mechanisms underlying and grounding the constitution of CA such as a 'sense of belonging', identification,[2] a special kind of motor representations,[3] or empathy? It has been shown that humans

---

[2] Sigmund Freud claims that we have to study the processes of interpersonal identification if we want to understand the mechanisms which underlie group formation. See Freud 1949, 70.

[3] See Butterfill (in this volume) for such a proposal. What Butterfill is hinting at in this context seems to me to be similar to a mechanism which the phenomenologist Hermann Schmitz calls 'Einleibung' (encoporation); a kind of bodily empathy or unreflective motor representation of

react significantly different to other humans than to objects (Norris et al 2004). For example, they have a strong sense for social inputs. Consequently, the presence or absence of others has a significant impact on one's emotions and behavior (Cacioppo and Patrick 2008). After all, these kinds of mechanisms, the way people relate to and perceive each other, their sense of being together with others or belonging to certain groups, seem to be much more basic (and perhaps more important) than having and ascribing shared goals when it comes to CA. I do not have the space here to investigate all of these mechanisms; therefore, I will stick to the 'sense of belonging' in the following passage.

Take a group of people who do not have a 'shared goal' in any informative sense of the term, but who do have a 'sense of belonging' to a specific group: a bunch of friends, for example, or a class in school. In his book *The Group Mind* (originally published in 1920), the psychologist William McDougall gives an example of what he calls "fortuitous" or "ephemeral" groups which he distinguishes from "crowds". According to him, the latter "may be either fortuitously gathered or brought together by some common purpose" (McDougall 1927, 88). Although these ephemeral groups are not "brought together by common interest", they would "yet present in simple and rudimentary form some of the features of group life":

> The persons seated in one compartment of a railway train during a long journey may be entirely strangers to one another at the outset; yet, even in the absence of conversation, they in the course of some hours will begin to manifest some of the peculiarities of the psychological group. To some extent they will have come to a mutual understanding and adjustment; and, when a stranger adds himself to their company, his entrance is felt to some extent as an intrusion which at the least demands readjustments; he is regarded with curious and to some extent hostile glances. (McDougall 1927, 88)

In a way, the persons seated in the same compartment developed a sense of belonging together during their journey, although they neither know if they have the same destination, nor if they share any other goal. It is this sense of belonging which makes them react in a common way to the arrival of the stranger. Thus, their sense of belonging has effects on the way they commonly behave, although they do not have a shared goal in the first place. Of course, such groups are *able* to have or develop common plans or goals, but to *define* the 'plural agents' they constitute through such goals would be like putting the cart before the horse. Examples like this one suggest that there are indeed important human faculties and mechanisms underlying group formation which we have to consider when studying CA and before asking how it is possible that people come to have shared intentions. Group formation would thus be independent of having or espousing a shared goal, at least in some cases.

However, the paradigmatic examples we use to theorize about CA already seem to imply a decision concerning this matter: As far as we focus on cases of people trying to realize a shared goal, we are inclined to take the sense of belonging to specific groups – if we even consider it important for a theory of CA – as being in some way derivative of having such a shared goal or plan. Opposing this idea, my

---

others, which Schmitz takes to be constitutive for social interaction in general and CA in particular (Schmitz 1994).

claim is that we have to identify and elaborate the faculties underlying the constitution of relevant cases of CA first if we want to capture and understand the whole range of phenomena which fall under the notion 'collective agency' or 'group action'. Otherwise, we will end up with a reduced theory of CA which cannot account for such cases – and this is, I suggest, exactly what many current theories of CA fall prey to.

In the following, I will therefore argue that we should distinguish between groups which are bound together by a shared goal or plan and groups which are bound together primarily by a sense of belonging. This distinction is inspired by Max Weber who in his famously unfinished work *Economy and Society* distinguishes between two kinds of social relationships, or, as we could say, two types of group formation, namely "communal" and "associative" relationships:

> A social relationship will be called 'communal' (Vergemeinschaftung) if and so far as the orientation of social action [...] based on a subjective feeling of the parties, whether affectual or traditional, that they belong together. A social relationship will be called 'associative' (Vergesellschaftung) if and insofar as the orientation of social action within it rests on a rationally motivated adjustment of interests or a similarly motivated agreement, whether the basis of rational judgment are absolute values or reasons of expediency. It is especially common, though by no means inevitable, for the associative type of relationship to rest on a rational agreement by mutual consent. (Weber 1978, 40)

McDougall distinguishes in a similar manner between "purposive" and "traditional groups" (McDougall 1927, 89). (Current philosophers concerned with CA do not seem to draw very heavily (if at all) on theorists such as Weber and McDougall. Nevertheless, this sociological and psychological tradition is rich in distinctions that could be useful for refining the current debates surrounding CA.)

What I take to be the important aspect of Weber's distinction between communal and associative relationships is the fact that not every group has to be bound together through a shared goal. Yet it is important to hold in mind that these different types of groups do not exclude each other: On the one hand, there can be groups bound together by a shared plan as well as by a sense of belonging. On the other hand, not every case of CA has to involve a feeling of belonging together or belonging to a specific group. There may be cases where people lack exactly this sense of belonging while nevertheless collaborating (maybe, because they came to the conclusion, that it's in their own interest to collaborate) and cases where people even feel 'alienated' from what they are doing and those with whom they are doing it. Nevertheless, we will not be able to distinguish these other types of CA if we only focus on the shared goal-directed agency aspect of CA.

10.3.1 Joint Commitment and Sense of Belonging: Gilbert's Account of CA

McDougall, whose work on *The Group Mind* I discussed above, distinguishes "two essential processes" of the development of what he calls the "group spirit": The "knowledge of the group" (which enables its members to discriminate between members and non-members of the group) and "the formation of some sentiment of attachment to the group as such" (McDougall 1927, 86). The former is important for the group being "apprehended and conceived as such by its members", McDougall claims, but it "is of itself of no effect, if there be not also widely diffused in the members some sentiment of attachment to the group" (ibid.).

I will now elaborate the idea of a 'sentiment of attachment' or what I call a 'sense of belonging' as an important aspect of some cases of CA by turning to Margaret Gilbert's account. I do so for two reason: First, Gilbert explicitly addresses her methodological choice of small-scale cases of CA as paradigmatic examples. Second, she discusses the 'sense of belonging' as an important aspect of CA within her account, but, as I will try to show, fails to deal with it appropriately.

Gilbert tries to develop her account of CA out of the paradigmatic example of two people doing something together; she thus starts with the 'dyad', a concept she borrows from the sociologist Georg Simmel (Gilbert 2006, 100). Nevertheless, dyads cannot simply be considered standard cases for CA, as Gilbert herself admits, insofar as they seem to have special characteristics. One of them is that "each member's presence in the group has the greatest possible significance for its survival as a group." (ibid.) Although Gilbert explicitly addresses this problem, namely, that her paradigmatic case seems to be somehow special, she adds:

> While recognizing that a dyad has special characteristics, I proceed on the assumption that two people can constitute a paradigmatic social group and that two people who are doing something together fall into that category. I argue that exploration of an example of two people doing something together discovers a structure that is constitutive of social groups in general, and political societies in particular. (ibid.)

I suggest however, that there is a problem with the assumption 'that two people can constitute a paradigmatic social group'. The problem is this: It makes a significant difference for someone if she acts as a part of a dyad or as a part of a larger group which is bound together by a 'sense of belonging'. This is not merely due to the fact that in the case of a dyad each of the two participants is necessary for the group to persist. Rather, I suggest that a sense of belonging (to a group) is much less important in the dyad-case than in cases of larger groups: Two people may form a *dyad* (or a *couple*), but they do not form a *group*. What may have sounded like a notional discrimination at first is in fact a basal difference: While the former can involve a strong sense of unity without a sense of belonging to the couple as a 'group' (you normally do not 'take part' in a group venture when you are part of a dyad),[4] a group seems to be much more persistent not only in the sense that it persists if you leave it. Groups (often) seem to us as somehow independent of the people who constitute them. We may even imagine trans-temporal cases of groups that – at least within certain time-spans – do not have any members at all, but we

---

[4] I would claim that this is so because in the case of a couple the shared goal (if there is one) cannot significantly differ from what the persons constituting the couple want to do. This may be a gradual difference, however, I think it is still significant enough to differentiate between couples and groups.

cannot imagine a couple without members.[5] Therefore, groups can be indifferent to exchanges of their members while dyads (normally) are not. This is not necessarily a question of social ontology. I do not have to claim that groups are somehow ontologically independent of their members. What I want to claim is rather that the members of certain groups have the *feeling (or take themselves) to belong to something which exists beyond themselves and the current members who constitute these groups* (and that this feeling significantly shapes the way people behave when acting as parts of groups). If this 'felt existence' is in any way ontologically significant is a question, however, I do not (have to) address in this context.

Dyads doing something together are thus an independent type of CA (and I think we should even consider differentiating between different types of dyads). However, in the current context it is enough to claim that dyads differ from groups insofar as their members (normally) lack the feeling of belonging to something that exists beyond themselves. It follows from this that, at the very least, we must be careful when suggesting that dyads can serve as paradigmatic examples of CA in general.

Leaving behind this methodological issue I will now turn to Gilbert's account of CA. One of the most important features of CA she derives from her standard case is what she calls a 'joint commitment to espouse a goal'. She considers this 'joint commitment' to be a constitutive part of every case of CA: When people constitute what she calls a 'plural agent', they jointly *commit* to espouse certain goals, which is why they face certain rights and obligations towards each other that bear on this joint commitment. Gilbert labels this idea the 'obligation criterion' for joint action (see Gilbert 2006, 106).

Recall that the key question here is if all cases of plural agents really are constituted by such a commitment to espouse a certain goal. In her book on political obligation Gilbert explicitly addresses the objection I am hinting at. "[O]ne might reasonably aver", she says, that

> it is not the case that every social group must have an overarching goal or aim. Consider, for instance, the case of a family. Families may tend to formulate plans and projects and carry them out. However, it is by no means clear that families as such must be characterized by some overarching goal. (Gilbert 2006, 165 f.)

Gilbert's solution to this objection lies in the idea of not restricting the joint commitment to shared actions in the narrow sense of the term:

> People may jointly commit to accepting, as a body, a certain goal. They may jointly commit to intending, as a body, to do such-and-such. They may jointly commit to believing, or accepting, as a body, that such-and-such. (ibid. 136)

However, although she argues that the idea of a 'plural subject' "goes beyond the idea of a plural subject of goal acceptance or, derivatively, of acting together" (ibid. 166), the solution she is offering is not really satisfying. Her suggestion is simply to

---

[5] However, we might well imagine a team (within a company for example, which is defined by a specific goal) composed of two individuals that is in some way indifferent towards the exchange of its members, but I claim that this case is significantly different from Gilbert standard example of two people going for a walk together.

take the notion of a 'goal' in quite a broad or abstract sense: The goal a plural subject is jointly committed to espouse can thus simply be to "believe that...", to "value...", to "feel remorse for...", or to "accept that...". Thus Gilbert phrases her definition of a plural subject as follows:

> A and B (and . . .) (or those with feature F) constitute a plural subject (by definition) if and only if they are jointly committed to doing something as a body—in a broad sense of 'do'. (Gilbert 2006, 144 f.)

I would call such an approach a *grammatical account of plural agency*, insofar as it takes it to be constitutive for a plural subject that it is at least jointly committed *to* espouse *something* in the broadest possible sense of the term: Everything a plural subject can be said to be jointly committed to can then indeed constitute such a plural subject. This said, we can restate the objection Gilbert tries to address as follows: It is not the case that every social group *must have an overarching goal or aim in a broad sense*, or, in other words, it is not the case that *every social group is jointly committed to espouse something* (in a broad sense of 'something').

I will now try to argue that this objection is sound by showing that there are indeed groups which are bound together by nothing more than a sense of belonging. To do so, I have to further elaborate the concept of the 'sense of belonging' that Gilbert indeed takes to be constitutive for CA in general. Gilbert follows Raz basically, when it comes to the definition of the 'sense of belonging'. He defines it as

> a feeling that one belongs, but this feeling is nothing other than a complex attitude comprising emotional, cognitive and normative elements. Feeling a sense of loyalty and a duty of loyalty constitutes, here too, an element of such an attitude. (Raz 1984, 154)

Thus Raz takes it to be a "natural indication of a member's sense of belonging" to have a "belief that one is under an obligation to obey because the law is one's law, and the law of one's country" (ibid.). However, the notion of a 'sense of belonging' I have in mind does not involve any commitment or obligation to obey *per se*.[6] Let us consider another example to make this point a little bit clearer: Martina's sense of belonging to the group of Europeans. Martina may not feel obliged to or be jointly committed in any way to espouse something specific if she has this sense of belonging, but she may nevertheless feel offended if somebody accuses Europeans in general of being "lazy and selfish". Following Raz, Gilbert could try to rephrase this case by saying that the sense of belonging to the group of Europeans comes, among other things, down to the joint commitment (or obligation) of feeling offended when someone attacks Europeans in general (see Raz 1984, 154). However, this description does not seem to capture the phenomenology of the 'sense of belonging' that I have in mind. For example, Martina is not criticizable for not being offended when someone attacks Europeans in general. She has not 'jointly committed' herself to feeling offended if someone does so in any way. On the contrary, it rather seems to be the case that her feeling offended in fact indicates that she has this

---

[6] Raz, like Gilbert, tries to address the question in what way members of a given community are obliged to follow its rules. The point I want to express here, however, is that the "sense of belonging" does not necessarily involve a kind of commitment or obligation.

sense of belonging to the group of Europeans, although she might not have been aware of it before. If this description is correct, it is possible to have a sense of belonging to a specific community or group without being jointly committed to espousing anything. One simply takes oneself to be a part of a larger group; something that seems to exist independently of oneself and the other members who make it up. Such groups do not necessarily have to be defined by something they try to espouse or a collective obligation.

Raz and Gilbert do not take this possibility into account, because they both basically think of CA as shared goal-directed agency. Focusing on shared goal-directed agency implies a theory of CA which must contain a concept of a 'joint commitment' to do something together or at the very least something functionally equivalent to a joint commitment. (I would take Bratman's idea of common knowledge about our mutually intending 'that we J' as such a functional equivalent.) According to such an account, there has to be an element of agreement or commitment (some kind of implicit or explicit 'contract') preceding the formation of a shared action, from which we can then derive norms and standards of criticizability for members of groups who don't act in accordance with the joint commitment. Furthermore, within such a theory the sense of belonging itself is only explicable as derivative of such a commitment or its functional equivalent.

As far as Gilbert tries to elucidate the 'sense of belonging' she has in mind by her concept of a joint commitment to espouse a goal, through which the individuals participating in CA are bound together, the sense of belonging is derivative of this commitment. One reason why she might argue in this way is that in her standard example of two people going for a walk together no sense of belonging of the kind that I described above is involved. Her account is thus somewhat close to a 'contractualist' or 'formal' notion of CA (as indeed I think is the case for many of the current theories of CA): According to Gilbert, a 'rationally motivated adjustment of interests or a similarly motivated agreement' between participants (to use Weber's phrase) is required when a plural agent is constituted. The sense of belonging then somehow seems to evolve directly out of this commitment. I do not question, however, that there might be a joint commitment involved in dyadic- as well as in group-cases. Rather, I claim that the sense of belonging cannot be explained solely in terms of such a commitment. People may well feel alienated from the CA they are involved in meaning that they share a joint commitment without a sense of belonging. At the same time, it seems to be possible for people to feel a sense of belonging to a group (i.e. the group of people who speak French as their mother tongue) without having a joint commitment to espouse a certain goal.

One could nevertheless wonder, whether cases of the latter type are in any way relevant to a general theory of CA. For example, if the sense of belonging isn't in any way linked to a shared project, such cases may not seem to be important when theorizing about CA. I believe, however, that there are good reasons for considering these cases to be relevant to any general theory of CA as a feeling of belonging to certain groups can significantly shape the way people commonly behave without there needing to be a joint commitment to espouse a specific goal (as shown in the example of the train journey).

Having said this, there are parts of Gilbert's account where she seems to at least implicitly address the 'sense of belonging' that I have in mind: She develops the idea that the agents involved in CA try to espouse what they are jointly committed to 'as a body'. CA is thus defined within her account as two or more individuals who "jointly commit themselves to espousing a goal as a body" (Gilbert 2006, 124). However, it is not at all easy to clarify what the qualifier 'as a body' is meant to denote. Gilbert tries to elucidate this phrase by giving the example of a "joint commitment to believe as a body that democracy is the best form of government":

> This can be parsed as follows: the parties are jointly committed together to constitute, as far as is possible, a single body that believes democracy is the best form of government. (Gilbert 2006, 137)

(Interestingly, she ascribes the belief that democracy is the best form of government to the plural agent as a whole, implying that it is possible for none of the group members to share this belief, while the 'single body' they try to constitute does.)

The first problem we encounter when trying to elucidate Gilbert's conception of CA as a property of a "single body" is with the metaphorical use of the term 'body'. Of course, she does not want to say that the people in question really do form a 'single body' in any biologically significant sense. It rather seems that what she is trying to say is that people who form a 'plural agent' act *as if* they would constitute a single organism. But this, again, leaves us with not much more than a metaphor. A further problem arises directly from her description of her example: How can she deal with the difference between alienated and non-alienated participants of CA? Is there a difference, for example, between simply *performing as if* one is part of a single body which espouses a certain goal and actually having *the feeling of being* a part of such a "body"? A difference between *believing* (as part of a body) that democracy is the best form of government and simply *saying* (as part of body) that this is the case (without necessarily believing it)?

I think Gilbert is right in claiming that there is something special about the way people feel and behave when they act as part of groups, but I don't think that her 'as a body' qualifier is able to sufficiently account for this aspect. As already mentioned above, I think the sense of belonging members have towards their groups is described best as the *feeling to belong to something which exists beyond themselves and the current members who constitute these groups*. Belonging to a group is then a specific kind of *experience*, not just something one merely *commits* oneself to.

To sum up what I have argued so far, we could say that although Gilbert explicitly names the 'sense of belonging' as an important feature of some cases of CA, her notion of a 'joint commitment' is not able to account for it appropriately. Thus, if we want to understand CA properly, we not only have to take into account how people can coordinate the things that they are trying to espouse together but also the way in which people can relate to each other and to the groups that they constitute more generally.

I suggest then, that we must therefore distinguish between groups which are indeed (in some way) *constituted by a joint commitment to espouse something* in Gilbert's sense and groups which are *based on a 'sense of belonging'* but which do *not necessarily involve any form of a joint commitment.* To see this more clearly, consider once more the example of a family, a case that Gilbert herself tries to address: It is of course possible that families implicitly or explicitly commit themselves to espouse a certain goal, say, to promote the Christian faith. But this seems to be in no way necessary for being a family, as Gilbert herself concedes. Gilbert could try to describe such cases by simply acknowledging that it is enough for being a family that its members are jointly committed to nothing more than promoting the existence of the plural subject they constitute – so that it does not have to have a specific plan or goal above this commitment. But as far as the notion of a 'plural subject' is nothing more than a "label for those who are jointly committed with one another in some way" (Gilbert 2006, 144), this solution seems to be question begging. What is, we could ask, the object of this joint commitment? It would have to be either nothing ("We are simply jointly committed and as such we constitute a plural subject.") or the joint commitment itself ("We are jointly committed to being jointly committed – as far as a joint commitment is the condition *sine qua non* for the constitution of a plural subject."). So Gilbert's account does not allow for groups which are simply constituted by a sense of belonging without 'being jointly commitment with one another in some way', but this is exactly what we need in order to explain the examples I elaborated above.

10.3.2 Cases of Massively Shared Agency

Let us now turn to the second aspect of the current puzzle-form of CA, namely, that theories of CA are focused on cases of face-to-face interaction. While Margaret Gilbert and Michael Bratman explicitly disagree in their accounts of shared intentional agency when it comes to the question of reductionism, it seems to be the hope of both of them that it is possible to develop an account of large-scale cases of shared agency which is at least based on the theoretical elaboration of small-scale cases. Scott Shapiro, to whom Bratman sometimes refers in regard to the question of large-scale cases of collective agency, tries to apply Bratman's conception of shared intentional agency to the social practice of law as well as to other cases of what he calls 'massively shared agency'. I will argue that even though he explicitly focuses on large-scale cases of collective agency, which go beyond cases of face-to-face interaction, his account is still lacking some important features; features which I take to be important for at least *some* types of CA.

Shapiro sees two reasons why a philosophical theory of action has not yet been able to account for cases of 'massively shared agency'. The first reason is that "action theorists have largely eschewed giving analyses of activities involving authority structures" (Shapiro 2014, 258). The other reason he identifies is that they have "largely concentrated on analyzing shared activities among highly committed participants", who themselves are committed to the success of their shared venture (ibid.). For Shapiro, however, "alienation and massively shared agency usually go hand in hand." (Shapiro 2011, 149)

While the first reason is linked to what I identified as the second aspect of the current puzzle-form of CA (that it is focused on cases of face-to-face interaction), the second reason is concerned with the problem that I discussed in regard to Gilbert's account: that we have to distinguish between alienated and non-alienated participants of CA, respectively, that we have to distinguish between groups with and without a sense of belonging. However, despite Shapiro's explicit consideration of alienation, I do not think that he succeeds any more than Gilbert does in giving a satisfying account of what it is for a participant of CA to be non-alienated, i.e. to have a sense of belonging to a group. Nevertheless, by adding authority relations to the theory of CA, Shapiro is sufficiently addressing a number of cases of CA beyond face-to-face interaction scenarios.

Shapiro's basic idea is that shared intentional agency has to be understood as an activity "guided by a shared plan" (Shapiro 2014, 277), which does not entail the claim that the individuals acting together each know of each other that they mutually intend 'that we J' (to use Bratman's description). He names three criteria for a plan to be shared: The plan has to be "designed with a group in mind", "it is publicly accessible, and it is accepted by most members of the group in question" (Shapiro 2011, 177). Thus, to be guided by a shared plan in the sense Shapiro has in mind here does not imply that every single participant of CA must have the corresponding intention to fulfill the plan as a whole; each does not have to be "committed to the success of the group venture", as he puts it (ibid. 272, Shapiro 2011, 136 f.). The example he uses to illustrate this last point is that of two people painting a house together.

> Suppose Abel wants Baker and Charlie to paint his house. Abel offers one thousand dollars to Baker if Baker does what he tells him to do. Abel offers Charlie the same terms. Baker and Charlie both agree. Abel then tells Baker to scrapes [sic!] off all the old paint and Charlie to paint a new coat on the scraped surface. Charlie waits until Baker scraps the old paint from the front of the house and then proceeds to paint a fresh coat on it. While Charlie paints the front, Baker scrapes the paint off the back of the house. When Baker finishes scraping, Charlie paints the rest of the house. It would seem that both Baker and Charlie have intentionally painted the house together. (Shapiro 2014, 270 f.)

For them both to be guided by a shared plan it is enough that there is at least somebody, an authority for example, who functions as what Shapiro calls a "mesh-creating" mechanism (Abel fulfills this role in the above example); a mechanism that takes care of meshing the different subplans. For Shapiro, the participants involved in CA don't have to know the whole of the plan in which they are fulfilling their roles. However, the shared plan has to be "publicly accessible" to all its participants (Shapiro 2011, 136). In so far as Charlie and Baker are not themselves committed to the success of the whole venture, Shapiro describes them as 'alienated' from the shared action as a whole, although he nevertheless takes them to be painting the house together (Shapiro 2014, 270). One of the main differences between Shapiro's and Bratman's account is, then, that Shapiro wants to account for cases of "shared agency among alienated partners" who are not "highly committed" to the success of their shared plan.

The main problem I see with Shapiro's account is that, even though it also applies to larger groups, it is still strictly focused on shared, goal-directed activities. This seems to be the reason why he doesn't give a satisfying account of how the individuals involved in CA can get a feeling of belonging to the group that they belong to in the first place; an account of what makes them feel 'highly committed' to the group venture. Shapiro does not, as far as I can see, provide an account of how a feeling of belonging together or what he sometimes labels as "enthusiasm towards the group venture" emerges along with its opposite, i.e. alienation. What he proposes instead is that, "[s]hared plans, we might say, bind groups together." (Shapiro 2011, 137) It is this close connection of groups and plans that, again, seems to prevent some further investigation of what it is for a person to have a feeling of being part of a group independently of having a shared plan. So Shapiro does not take into account a sense of belonging to a group which in some way or the other goes beyond mere goal-acceptance.

In response to Shapiro's proposal that it is shared plans that bind groups together, I repeat the argument elucidated above: It is not, or at least not always the case that there has to be a shared plan to bind a group together. Think, again, of the persons seated in the same compartment of a railway train: They do not, by assumption, have a shared plan, but they nevertheless (re)act as a group when a stranger steps in. To elaborate this point a little further, I will now focus on Shapiro's attempt to explain how the lack of a sense of belonging can be handled in cases of 'massively shared agency'. My aim here is to show that because he does not have a rich notion of a sense of belonging, he sees no other means of coordination as authority for cases of massively shared agency.

Let's start with his notion of 'massively shared agency'. Shapiro lists a lot of different examples for the cases of CA that he has in mind here:

> Business corporations, consumer cooperatives, trade unions, research universities, philanthropic organizations, professional associations, standing armies, political parties, organized religions, governments and legal systems, not to mention the collaborative ventures made possible by the digitally networked information and communication technology, such as Wikipedia, massively multiplayer online games (MMOGs), open-source software, and the World Wide Web itself, all harness the agency of multitudes in order to fulfill certain objectives. The modern world, we might say, is one defined by 'massively shared agency'— the agency of crowds. (Shapiro 2011, 149)

Massively shared agency is thus basically the agency of multitudes and crowds. Shapiro's idea is that these crowds cannot be described as cases of CA where the participants are all highly committed to the group ventures they take part in:

> Because the modern world is also characterized by diversity, it is extremely unlikely that large scale ventures can be staffed with individuals who are all committed to the same goals. (Shapiro 2011, 149)

For Shapiro being enthusiastic about the group venture comes down to nothing more than being committed to the group's overall plan. So Shapiro, like Gilbert, does not take into account a feeling of belonging to a group which is not in some way linked to something like a joint commitment. And that is why the only solution

Shapiro suggests when it comes to the question of how to deal with alienated participants of CA is that there have to be some mechanisms which ensure that the alienated participants act *as if* they were committed to the group venture:

> For the task of institutional design in such circumstances is to create a practice that is so thick with plans and adopters, affecters, appliers, and enforcers of plans that alienated participants end up acting in the same way as nonalienated ones. The fact that activities can often be structured so that participants intentionally achieve goals that are not their goals accounts for the pervasiveness of massively shared agency in the world around us. (Shapiro 2011, 150)

Even though I would not deny that this is one of the important tasks of institutional design, we should also consider that institutions, especially companies, do not merely try to create mechanisms (authority relations and surveillance) that make alienated participants act *as if* they were not-alienated. What the person who is alienated (within a company for example) is lacking is not just the commitment to the success of the group venture; after all in such cases companies often try to offer incentives to make employees act as if they were not alienated. What she is lacking is a sense of belonging to the company itself. She does not really experience or take herself to be a real part of it. Thus, what companies do to handle alienation is that they try to evoke what is often labeled as a sense of 'corporate identity' between their employees. Corporate identity seems to be best understood as a feeling of belonging, or a sense of being a real part of something, a company or institution, which exists independently of its members. Such a sense of belonging makes them take part in whatever project or venture the company is about to realize. It is a sense of identification which goes beyond, I would argue, being committed to one or another of the specific aims of the company. The same holds for examples like the hacker-movement, which operates under the name of 'Anonymous'. This movement is not primarily bound together through a shared plan but through a sense of belonging to the movement as a whole – even though nobody of its participants really knows who the other members are (Coleman 2010). So even though Shapiro's account highly depends on the concept of alienation (and accordingly on something like a sense of belonging), he does not really develop a theory of this kind of feeling. It is for this reason, I argue, that his account misses important features of some cases of CA; features which are even present in some cases of massively shared agency.

However, Shapiro does address the second aspect of the current puzzle-form of CA. He explicitly takes into account cases where the participants of CA do not directly interact in face-to-face scenarios. These are cases where we have to consider the 'organizational structure' of a group as a mediating mechanism of group formation (for more on this important aspect, see in particular List and Pettit 2011, 60 ff.). Although Shapiro only mentions authority relations as examples for what he calls 'mesh-creating' mechanisms, I think we could use and refine this conception in order to describe and differentiate such types of CA. Here it would be of particular importance to take a closer look at the socio-technological structures and environments, which have a great influence on how different types of CA are structured.[7]

---

[7] I try to describe the way CA is shaped within social networks like Facebook and Twitter in Poljanšek (2014b).

Nowadays, new types of CA seem to evolve, which are based on specific techniques and interfaces[8] enabling new types of communication (i.e. many-to-many vs. one-to-many communication). While I cannot elaborate upon these ideas any further here, I hope it is nevertheless clear that there is still some theoretical work to be done to account for these specific cases of CA. We not only have to take into account authority relations when we want to describe these types of CA but also the specific types of communication media (e.g. social networks) which enable them.

Focusing on such mechanism which enable specific types of CA could then also proof useful to handle the third aspect of the current puzzle-form of CA, namely, that the agents involved in our standard cases of CA act together at the same time. One problem linked with this third aspect is that, similar to what I just described concerning the focus on face-to-face interaction, it tends to make us overlook the way in which specific types of CA are coordinated through time and space. It would be important here as well to take a closer look at the way trans-temporal coordination of groups is realized within specific socio-technological environments and through specific technical interfaces.[9] However, for reasons of space, I cannot elaborate this point here any further.

10.4 Concluding Remarks and Future Perspectives

I have argued that the tendency to use small-scale cases of CA as paradigmatic examples restricts the scope of our theories of CA. The discussion of the accounts of CA that Gilbert and Shapiro have recently put forward led to the conclusion that it seems especially important to distinguish between groups which are in some way *constituted by a shared goal or plan* and groups which are *based on a 'sense of belonging'* and therefore do not necessarily involve a shared goal or plan. Current theories of CA, I have suggested, focus mainly on CA of the former kind; they focus on cases of *shared goal-directed activity* and, for the most part, ignore the possibilities of *groups* which are merely constituted by a sense of belonging. In this context, I argued that the sense of belonging is best to be understood as a *feeling to belong to something which exists independent of its members*, whether this felt existence is ontologically significant or not. My main conclusion is therefor that we must distinguish between different types of CA, each of which should be modeled separately.

---

[8] For further considerations concerning the importance of interfaces and the way they shape our everyday lives and social interactions see Poljanšek 2014a.

[9] Barry Smith recently argued that we have to take into account the role documents (like, for example, a musical score of an orchestra) play as mesh-creating mechanisms in cases of massively shared agency (Smith 2013). I think this role of documents as 'mesh-creating' mechanisms is even more significant in cases of long-term CA. Think just of the ongoing progress within the scientific community: Without documents laying the ground for ongoing research by comprising and stabilizing data it would be hard to imagine how it could take place at all. Here it would be especially important to take into account different kinds of 'mesh-creating' mechanism that enable cases of massively shared agency and long-term cases of CA. Again, I'm not able to elaborate these thoughts within this paper.

In this regard, the distinction between groups with and without a sense of belonging on the one hand and groups with or without a shared goal on the other hand is to be understood as a first typological distinction between such different types of CA. There will of course be other types of CA, which are in need of closer elaboration. So in order to flesh out this line of thought, it would be especially important to come up with a variety of paradigmatic examples for CA, which, based on our pretheoretical intuitions, seem to differ in relevant aspects. Thus, we have to give up the idea that there is one simple phenomenon from which we could derive all the important features that we need in order to describe CA as well as the idea that we can develop *one* theory of CA which fits all different types of CA.

References

Bentley, Arthur. 1908. *The process of government. A study of social pressures*. Chicago: The University of Chicago Press.

Bratman, Michael. 2010. Agency, time and sociality. *Proceedings and Addresses of the APA* 84(2): 7–26.

Bratman, Michael. 2014. *Shared agency. A planning theory of acting together*. Oxford: Oxford University Press.

Cacioppo, John T., and William Patrick. 2008. *Loneliness. Human nature and the need for social connection*. New York/London: W. W. Norton & Company.

Coleman, Gabriella. 2010. Anonymous – From the Lulz to collective action. *New Significance*. From http://www.thenewsignificance.com/2011/05/09/gabriella-coleman-anonymous-from-thelulz-to-collective-action/ Accessed 12 May 2014.

Freud, Sigmund. 1949. *Group psychology and the analysis of the ego*. Authorized Translation by James Strachey. London: The Hogarth Press Ltd.

Gilbert, Margaret. 1996. Walking together: A paradigmatic social situation. In *Living together. Rationality, sociality and obligation*, ed. Gilbert Margaret, 177–194. Lanham: The Rowman and Littlefield.

Gilbert, Margaret. 2006. *A theory of political obligation. Membership, commitment and the bonds of society*. Oxford: Clarendon.

Kuhn, Thomas. 1962. *The structure of scientific revolutions*. Chicago: The University of Chicago Press.

List, Christian, and Philip Pettit. 2011. *Group agency. The possibility, design, and status of corporate agents*. Oxford: Oxford University Press.

Materman, Margaret. 1974. Die Natur eines Paradigmas. In *Kritik und Erkenntnisfortschritt*, ed. Imre Lakatos and Alan Musgrave, 59–88. Braunschweig: Vieweg.

McDougall, William. 1927. *The group mind. A sketch of the principles of collective psychology with some attempt to apply them to the interpretation of national life and character*. Cambridge: Cambridge University Press.

Norris, Catherine J., E. Elinor Chen, et al. 2004. The interaction of social and emotional processes in the brain. *Journal of Cognitive Neuroscience* 16: 1818–1829.

Olson, Mancur. 1968. *Die Logik des kollektiven Handelns. Kollektivgüter und die Theorie der Gruppe*. Tübingen: Mohr (Siebeck).

Poljanšek, Tom. 2014a. Benutzeroberflächen: Techniken der Verhüllung des Technischen. In *Haut und Hülle. Umschlag und Verpackung. Techniken des Umschließens und Verkleidens*, ed. Ute Seiderer and Michael Fisch, 102–117. Berlin: Rotbuch.

Poljanšek, Tom. 2014b. Kritisieren heißt Filtrieren. Eine Kritik kritischer Praxis in sozialen Netzwerken. *Juridikum. Zeitschrift für Kritik – Recht – Gesellschaft. Schwerpunkt "Internet und Freiheit"* 4/2014.

Raz, Joseph. 1984. The obligation to obey: Revision and tradition. *Journal of Law, Ethics and Public Policy* 1: 139–155.

Salmela, Mikko. 2011. Can emotions be modeled on perception? *Dialectica* 65(1): 1–29. doi:10.1111/j.1746-8361.2011.01259.x.

Schaub, Miriam. 2010. *Das Singuläre und das Exemplarische. Zu Logik und Praxis der Beispiele in Philosophie und Ästhetik*. Zürich: Diaphanes.

Schmitz, Hermann. 1994. Der gespürte Leib und der vorgestellte Körper. In *Wege zu einer volleren Realität. Neue Phänomenologie in der Diskussion*, ed. Großheim Michael, 75–92. Berlin: Akademie Verlag.

Shapiro, Scott. 2011. *Legality*. Cambridge/London: The Belknap Press of Harvard University Press.

Shapiro, Scott. 2014. Massively shared agency. In *Rational and social agency. Essays on the philosophy of Michael Bratman*, ed. Manuel Vargas and Gideon Yaffe, 257–293. New York: Oxford University Press.

Smith, Barry. 2013. Diagrams, documents, and the meshing of plans. In *How to do things with pictures: Skill, practice, performance*, ed. András Benedek and Kristóf Nyíri, 165–180. Frankfurt: Peter Lang.

Weber, Max. 1978. Communal and associative relationships. In *Economy and society. An outline of interpretive sociology*, ed. Guenther Roth and Claus Wittich. Berkeley: University of California Press.

Wittgenstein, Ludwig. 2009. *Philosophical investigations* (trans: Anscombe, G.E.M., P.M.S. Hacker, and J. Schulte). Malden: Blackwell Publishing Ltd.

Chapter 11
Can Artificial Systems Be Part of a Collective Action?

Anna Strasser

11.1 Introduction

To answer the question of whether artificial systems may count as agents in a collective action,[1] I will argue that a collective action is a special kind of an action and show that the sufficient conditions for playing an active part in a collective action differ from those required for being an individual intentional agent. Firstly, I will clarify the needs of a theory of collective actions and give a description of the phenomenon of collective action referring to the contemporary philosophical debate about shared intentionality. In this section I will describe different kinds of collective actions including the actions of homogenous and mixed groups of participants (1).

In the second section I will suggest that it could be desirable in general to have a broader notion of action not excluding non-living-beings, infants and animals from the outset as the most common philosophical notion of action does[2] and therefore being able to embrace different kinds of actions. In opposition to the common notion of action I will argue for a more fine-grained differentiation of classes of events to develop a notion of action enabling us to differentiate between simple and complex action. In principle a collective action with diverse participants could be described as a combination of simple and complex actions (2).

Coming back to the question of what makes a collective action a special case of action, I focus on the conditions agents have to fulfill to have an active part in a

---

[1] Unfortunately one cannot observe a consistent usage of the concept 'collective action'; in some debates (cp. Bratman 2014, 10) notions like joint action are used to distinguish from collective agency. I will use the notion 'collective action' and 'joint action' interchangeably.

[2] Many philosophers claim that actions are always 'intentional under some description' (cp. Davidson 1980). There is wide agreement that this is true for individual human actions.

A. Strasser (✉)
Berlin School of Mind and Brain, Humboldt-Universität zu Berlin, Berlin, Germany
e-mail: anna.strasser@hu-berlin.de

collective action. This will show that playing an active role in a collective action does not necessarily presuppose being able to be an agent of an individual action in the standard philosophical view explained above – in short maybe one does not have to require that each participant must fulfill the established demanding conditions of intentionality. If this is true the chances for artificial systems to play an active part in a collective action get much better (3).

Another very important difference between collective and individual actions becomes obvious when analyzing what kind of social abilities are required for collective action. I claim that the coordination between the participants of a collective action presupposes something that we could label as 'social cognition'. In this section I will outline paradigmatically specific features of coordination, which involve social cognition abilities occurring in collective actions. And it will be questioned how demanding the requested abilities have to be (4).

Having relieved artificial agents from the burden of deep intentionality in Sect. 11.2, I will discuss in what sense they might succeed in having social cognition abilities to coordinate their (simple) actions with human agents (5).

11.2 The Phenomenon of Collective Actions

All actions are events; collective actions are events in which two or more agents do something together non-accidentally. A theory of collective action should fulfill at least three requirements to enable us

1. to distinguish such cases from those that look similar but are in fact only accidentally parallel actions rather than collective ones.
2. to distinguish collective actions from cases we should better describe as tool-use or simple stimulus–response behavior.
3. to clarify how demanding presupposed abilities of participants should be.

Prototypical examples often refer to actions an agent could not have carried out on her own, such as carrying something heavy, painting a house, dancing a tango or playing tennis. In the philosophical discussion we mainly find examples of collective action (often termed as 'joint action') only involving small groups. I will follow this line even though mass phenomena[3] might shed another light on the question of what the sufficient and necessary conditions of a single participant in collective actions are. But to generalize from one instantiation of a collective action of small groups of human adults might be a risky undertaking. In this paper I want to cover at least different kinds of collective action with small groups as agents. I will take into account homogenous and mixed groups acting collectively not excluding infants, animals and artificial systems from the start.

1. Non-accidentally doing something together emphasizes that the behaviors of all participants of a collective action are closely related to one another. How social

---

[3] Mass phenomena should be an object of future research.

cognition enables the required relationship of coordination will be focused on in Sect. 11.4. The ability to coordinate will ensure that one is able to distinguish collective actions from accidentally in parallel individual actions.[4]
2. The possibility of a broader notion of action being able to draw fine-grained distinctions will be discussed in the next section.
3. The main claim of this paper is that those conditions are not identical with the conditions we require for individual agency. And this question should not be confused with the epistemological question: How can we be sure that we are justified in ascribing such abilities to an observed agent?[5]

To meet such requirements we consequently need to have a definition of the notion of a collective action. According to Bratman (1992, 1997, 2014) a collective action (he uses the notion 'joint action') of human adults must fulfill several conditions.

> ... sufficient conditions for our shared intention to J:
> *Intention condition*: We each have intentions that we J;...
> *Belief condition*: We each believe that if the intentions of each in favor of our J-ing persist, we will J by way of those intentions and relevant mutual responsiveness in sub-plan and action;...
> *Interdependence condition*: There is interdependence in persistence oft he intentions of each in favor of our J-ing.
> *Common knowledge condition*: It is common knowledge that A. – D.
> *Mutual responsiveness condition*: our shared intention to J leads to our J-ing by way of public mutual responsiveness in sub-intention and action that tracks the end intended by each of the joint activity by way of the intentions of each in favor of that activity." (Bratman 2014, 152)

Roughly summarized one could claim that

- the agents must have a common goal – 'an intention to J';
- their sub-plans must be coordinated – and there should be a mutual willingness to support one another; and
- each agent must possess mutual knowledge about the other concerning the action.[6]

From those claims it follows that every possible agent has to be able to

1. have and share goals with the other agent (according to Bratman this includes both having intentions of one's own, and recognizing the intentions of others);
2. plan and coordinate planning with the other; and
3. possess knowledge about the knowledge state of the other and additionally be able to entertain higher-order representations of the kind 'knowing that the other one knows that he knows'.

---

[4] Compare Butterfill's example of two strangers painting a large bridge red in his contribution to this volume.

[5] The second question will not be addressed in detail in this paper.

[6] This is not meant to be a detailed exegesis of Bratman's theory – I just want to highlight some essential requests.

Interestingly, the question of whether less demanding conditions might be sufficient at least for some participants of a collective action has barely been discussed so far.[7] Bratman (2014) discusses those concerns shortly in his fourth chapter, mainly making the point that the postulated conditions (cp. footnote 6) might also be fulfilled in an implicit way. The answer of the fulfillment of the requested conditions then becomes a more or less empirical question.

As mentioned above, the paradigmatic cases that serve as a basis of those debates focus on collective actions we can observe between human adults using cognitive abilities. This might be one reason why up to now it seems to be a necessary presupposition that each participant is required to fulfill the same demanding conditions.

The aim of this paper is to show that there are cases where we do not have to require the same conditions of all participants in a collective action.

This idea is based on three observations:

- If one takes a developmental perspective on the phenomenon of collective actions one will be confronted with cases in which at least one agent does not fulfill the same demanding conditions as the other. I will call such cases *asymmetric cases* because the developed abilities of the participants differ. Imagine a mother playing together with her little child – intuitively nobody would claim that this child is not an agent in this collective action and instead describe this case as a case of tool-use. By the notion of tool-use I refer to simple stimulus-responsive behavior claiming a rigid coupling in between input and output. Children are not tools, they are socially interacting beings. Nevertheless it is questionable whether a child fulfills the required conditions as an adult does. Children never or rarely have sophisticated knowledge about the knowledge of others concerning action and probably are not able to entertain complex higher-order representations. Due to their more limited reserves of working memory and executive functions they may be altogether unable or significantly less likely than adults to deploy such knowledge in deciding what to do. And there are philosophical positions even denying that children have intentionality in a deep sense. Still children are able to play an active part in collective actions and they do coordinate with the action of the others. To explain this ability we have to take into consideration that there are varieties of conditions, which enable an agent to participate in a collective action. The same might be true for certain interactions between humans and animals, again to describe those interactions as merely tool-use would at least be very counter-intuitive.[8]
- In other cases the unequal distribution of abilities even motivates the collective action because a division of labor can make outcomes possible that a group consisting of individuals all with identical abilities could never reach. For example

---

[7] One exception is Butterfill (2012). This paper objects that Bratman's view is unable to explain simple cooperative interactions between very young children, who do not yet have a full understanding of other minds. Further contributions on these questions can be found in Tollefsen (2005), Pacherie (2013). Cp. Bratman (2014, 104ff).

[8] A possible objection could suggest claiming a gradual development of the ability to act. But then the discrimination in-between behavior and action will become blurred.

in many situations in which we can identify a learning goal, a teacher will take on a leadership role in that interaction while other participants nonetheless share her goal and coordinate their actions with the teacher's. In such cases it is nonetheless appropriate to speak of a collective action. Nonetheless, the pupil may not fulfill the demanding conditions for agency in such cases.
- In cases of actions that mainly involve motor coordination it is particularly questionable how demanding the cognitive abilities required in such cases must be. For example, if two people shake hands the coordination process obviously is not a cognitive one. Rather, these processes can be localized at a lower level.

Obviously, collective actions performed by small homogenous groups of human adults are only one possible category of collective actions. Depending on the composition of the group involved in a collective action, the distribution of abilities might differ. Bratman's description of a collective (joint) action covers only some cases with human adults as agents. But mixed groups involving adults and infants or animals or even artificial systems cannot easily be captured by his description.

It might seem quite challenging to jump from mixed groups consisting of adults and infants or animals to mixed groups consisting of humans and artificial systems. But to argue for the claim that not all participants have to fulfill the demanding requirements the contemporary notion of action poses it might be more convincing if one occupies oneself with systems to which we do not tend to ascribe all sorts of implicit abilities. A collective action involving artificial systems as agents can serve as a showcase for very minimal conditions.

We can sum up that so far existing positions about collective actions do not consider mixed groups of participants in particular. Including above described asymmetric cases in a theory of collective action one has to make sure to find the right constraints to ensure that two actions are not merely accidentally in parallel. To exclude contingency we need reasons to claim that both agents coordinate their actions with the aim of successfully completing a collective action. Consequently, our constraints have to guarantee that the actions of each participant are coordinated with the others.

In some cases this coordination might be realized through strong conditions like mutual knowledge about the intentions of the others. In other cases – like when we dance the tango or play table tennis – it can be fulfilled by successful social interaction, which can be realized with the help of motor representations (Butterfill and Sinigaglia 2014). And there might be many in-between cases as well.

Coming back to the question whether artificial systems might be able to be agents in a collective action it is obvious that a clear-cut boundary between tool-use events and cooperative events, which should rather count as collective actions, is difficult to draw – at least with existing terminology.

This should lead us to consider whether there are reasons to broaden the notion of action. A major problem for the subsumption of artificial systems under philosophical theories of action consists in the very demanding understanding of intentionality. Non-living beings, and maybe also infants and animals, seem to be doomed to remain mere tools (capable of producing only behavior – not action), because only intentional living beings with high cognitive abilities are considered to be able to act.

In the next section I will sketch ways in which one could broaden the notion of an action. In opposition to common notion of action I argue for a more fine-grained differentiation of classes of events to develop a notion of action – namely a simple action – that is not reserved for living beings from the outset. Whether contemporary artificial systems are actually able to fulfill the notion of a simple action has then to be discussed on a case-by-case basis (Cp. Strasser 2004).

11.3 Kinds of Events: Kinds of Actions: A Broader Notion of Action

In the following I suggest a broader notion of action. Every action is an event, but not all events are actions: some are just cases of behavior or of other events like a thunderstorm. Every action is an instance of a behavior. What I am claiming is that there are events that are more complex than simple stimulus–response behavior events but not complex enough to fulfill contemporary notions of action. To be able to describe such events I suggest differentiating between simple and complex behavior and between simple and complex action (Fig. 11.1).

For the purpose of this paper I will not go into details concerning complex behavior (Cp. Strasser 2004) – the goal here is to sketch only a notion of simple action.

The main reason that the philosophical notions of action exclude artificial systems as potential agents and others lies in the understanding of intentionality. To ascribe intentionality several implicit conditions must first be fulfilled: First of all an ability to generate goals, second the resulting behavior must be described as goal-directed, third possession of some kind of a free will. Last but not least consciousness and self-consciousness are postulated implicitly in many cases.

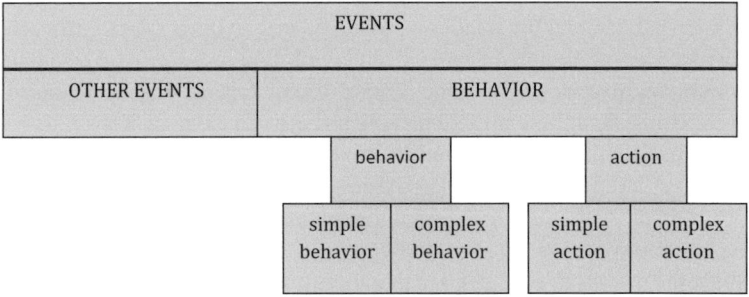

Fig. 11.1 Categories of events (There are two categories of behavior: Firstly, behavior (differentiated into simple and complex forms) and secondly, actions (again differentiated into simple and complex forms). Seen from left to right each event gets more demanding)

In the following I want to show how one could interpret conditions for actions and agency in a weaker sense to construct a broader notion of an action, namely the notion of a simple action in contrast to the notion of a complex action.[9]

I The Ability to Generate Goals The generation of a goal does not necessarily have to occur in the system that performs the act. A goal could also be generated in one system and then transferred into another. Of course we need goals if we want to have actions but acting systems only have to be able to receive goals *as* goals and not to generate goals. I do not see why this should be a necessary condition. Assumed we live in a world with at least some goal generators performing complex actions, the necessity of goal generation does not hold for simple actions.

II Goal Directedness For something to be described as goal-directed it is not obvious that it is necessary to have intentionality. To sustain this condition it might be sufficient to claim that the on-going information process is goal-directed. For example if an acting system is able to plan and anticipate then it is goal-directed.[10]

III Free Will As I claimed above, goal generation does not necessarily have to take place within the acting system. If this is true the same can hold for free will.

IV and V Consciousness and Self-Consciousness The fourth and fifth conditions mainly characterize properties of living beings and are not necessary for simple action. Self-consciousness gives us the impression of being informed about the information processing processes as well as the goal generation process – if there is one. But consciousness of a process is neither a precondition for that process being a process nor a necessary condition of information processing processes in general.

I now specify the necessary conditions for simple action. That requires an explanation of which conditions relating to the abilities to perceive and process information have to be fulfilled to legitimately say that an individual is the agent of a simple action.

Perceiving Information Any theory of action has to claim that the acting system is able to perceive information. Again in the case of perceiving information, having an input does not depend on consciousness. Additionally the acting system has to be able to perceive something *as something*, e.g. it must be able to perceive a goal state *as* a goal state. Looking at this ability we see that perception already involves information processing and cannot be considered separately. What kind of perception is needed depends on the type of action that is performed. We will see that there are cases of collective actions in which the involved agents have to be able to read social hints. In a minimal case the acting system should at least be able to perceive the state of the relevant environment at the beginning of an action – the start state – as

---

[9] For details see Strasser 2004.
[10] There might be opponents claiming that planning requires intentionality. I will use a broader notion of planning claiming that if a system is able to find a way from a starting state to a goal state it is legitimate to call this information processing planning.

well as the effects of its own effectors – so to say the outcome of an action. For example, a singer should be able to hear.

Information Processing Second, I claim that the condition of information processing should involve the ability to plan, the ability to have representations, the ability to anticipate and the ability to learn.

No system is able to find a way to reach a given goal state without an ability to plan, unless it is manipulated by another acting and planning system. This would be a remote-controlled tool use, not an action. The other alternative to planning would be the case if the steps towards the fulfillment of the goal state were already given as a rigid coupling between input and output. But in this case we would no longer talk about an action. As actions involve cognition and cognitive information processing includes the idea of a flexible coupling between input and output. Enabling the system to behave sensitive to changes to its environment and in the case of collective actions to coordinate its action with the actions of another agent. To fulfill the condition of cognition in our complex environment it is necessary to require the ability to learn as well. The notion of learning unfortunately has the disadvantage that it is not really well defined. What kinds of learning would come into play will not be discussed in this paper. Concerning the ability of artificial systems to play an active part in certain collective actions I assume that the necessary lessons have been learned already.

Effectors to Perform Third, to be the cause of an action the acting system needs effectors to perform the action, as well as the ability to recognize goal states *as* goal states and knowledge about how to reach the goal. Systems without effectors are not able to act. Again what kind of effectors a system needs depends on the type of action the system has to perform.

11.4 Towards a Definition of Collective Actions

If one is willing to broaden the notion of action in the manner described above a first step can be taken. One could then claim that a collective action might consist of simple and complex actions, which have to be coordinated. But even if one wants to hold on to the demanding notion of an individual action it can be argued that collective actions are a different kind of action.

Regardless of the result of the discussion of a broader notion of action, I argue that collective actions are different in that we do not need to demand similar abilities of all participating agents. Additionally, I will show how social cognition enables the necessary coordination of collective actions.

First, I will transfer the conditions for being an agent of an individual action to the conditions we should claim for being an agent in a collective action. Second, I will examine further additional conditions we have to claim so that we can ensure that there is coordination, which prevents the collective action from being merely accidental.

11 Can Artificial Systems Be Part of a Collective Action?

In talking about action theories, one can refer to the well-known belief-desire-intention model. This model postulates that every agent is able to have the following mental states:

1. BELIEF: With the right perceiving and information processing abilities the acting system can plan how to reach the goal. Being able to plan results in a **belief** (knowledge) about what has to be done to reach the goal.
2. DESIRE: Assuming that we have a starter (a goal generator) and that the goal is transferred to our system, all this system has to be able to do is to recognize this goal *as* a goal and start to plan the action. In the BDI model this is what the **desire** represents. Having a certain goal does not necessarily imply that it was generated by the system itself. And at least in the artificial intelligence debate it does not require the possession of phenomenological states, emotions, etc.
3. INTENTION: The aim of reaching that goal, which can be shown by being goal-directed.

Setting aside for now the additional conditions to be claimed for collective actions one could transfer above conditions to collective actions:

1. BELIEF (knowledge of each participant about what has to be done to reach the goal)
2. DESIRE (recognizing the goal *as* a goal – having a shared goal)
3. GROUP-INTENTION/PARTICIPATORY/WE-INTENTION (shared aim to reach the goal or aim to reach that goal together)

Concerning the above conditions it is not quite clear to whom we should ascribe them:

- Should knowledge about how to reach a goal be ascribed to the group or to each individual? Or should it be postulated that each part of the group has only knowledge about what itself as an individual must do?
- Does each individual have a goal, or is it the group as an entity that has a shared goal?

Obviously, the question of how to understand this is especially pressing if one is interested in the ascription of intentionality.

Concerning the ascription of intentions there are several positions in the philosophical debate (Cp. Searle 1990; Bratman 2014; Butterfill 2012; Schmid 2009):

On one side we find the position claiming that shared intentions should be ascribed to the group as a whole (Schmid 2009). On the other side there is also the position that collective actions should be described as actions in which each individual has her intention and additionally has a participatory intention regarding the collective action, or the participants are entertaining so-called we-intentions (Searle 1990).

I do not claim a preference for ascribing intentions to groups, because the introduction of new entities to which we can ascribe mental states poses more new problems than it solves old ones. However, for the purpose of this paper this question does not have to be decided – and so I will simply remain neutral on the question of whether groups have intentions of their own. What I will show could convince advocates of all positions.

What is at stake is the question of whether all participants in a collective action necessarily have to fulfill the demanding conditions of a philosophical notion of action apart from the question of whether the group as an entity might have further mental states.

But what makes a mere transfer of conditions insufficient in the case of collective actions is not only the open question of whether all participants have to fulfill symmetric conditions but the question of whether there are any additional conditions.

I claim that what makes collective actions special is the fact that they demand successful social interaction. Only if the agents are able to coordinate we can be sure that we are not simply observing accidentally parallel actions. What it means to be able to coordinate cannot be explained by only referring to self-knowledge and pure observational knowledge about the world. This is the point at which social cognition comes into play. To coordinate we need knowledge about the behavior of the other agents, we should have an understanding of their behavior and their abilities to plan to be able to anticipate their behavior. This 'understanding' has been termed as social cognition.

To coordinate oneself with other agents in a collective action one must possess the ability to anticipate the actions of the other participants. One way to succeed in this task is to ascribe mental states to the others. As I will be arguing later, that does not necessarily mean that the other actually has to have a mental state. Another possibility consists in coordination processes on a bodily level. Such cases will not be discussed in this paper (Cp. Knoblich et al. 2010).

11.5 Coordination Through Social Cognition

As we have seen in the above sections one essential aspect of a collective actions lies in the question of how the actions of the participating agents are coordinated.

Following the hypothesis that social cognition can enable participants to coordinate their actions one has to clarify what kind of abilities have to be there so that we are legitimate in ascribing social cognition. This is relevant for my claim of us not needing the demanding conditions of individual agency because conditions, which are discussed in relation to social cognition, tend to be quite demanding as well.

Unfortunately the notion of social cognition is not defined in a strict way, many phenomena are described as instances of social cognition. The question of how to define social cognition and what instances of social cognition should be differentiated cannot be addressed in this paper in full length. I will limit myself to the specific needs of social cognition in some specific cases of collective actions.

Generally speaking we can talk about social cognition if we have good reasons to ascribe firstly the ability to encode, store, retrieve, and process information relating to conspecifics and perhaps also to virtual agents. This ability might be realized through conscious or unconscious processes. Additionally, the system should be able to express information, which can be understood by the others.

Concerning the coordination of collective actions, I would sketch several social cognition abilities, which play an important role. This is not meant to be a complete list. I will use paradigmatic examples from different categories to approximate a clearer picture. By doing that I will take into account whether required conditions necessarily have to be very demanding.

First of all, in the category of being able to perceive and process information *joint attention* for example can deliver important information about the goal-directedness of the participating agents. For many collective actions we should 'understand' whether we are attending to similar objects. Consequently one should claim that a special kind of information perception consists in the ability to recognize what the other is attending to.

Furthermore I claim that in the category of ongoing information processing the ability to read *social hints* plays an essential part. For examples, in collective actions of communication gestures or prosody could be instances of such hints. To understand that you are expected to answer a question you must be able to recognize a phrase as question. You should also be able to monitor whether the other is listening and understanding.

Last but not least the category of *anticipation* plays an important role. We would not talk about a collective action if one agent were continually surprised by the behavior of the other participating agents. In gaining mutual knowledge about the other, using theory of mind abilities and theory of emotion abilities can play a functional role. Participants in a collective action need anticipation abilities by which they can predict, 'understand', and explain the behavior of others to some extent. Ascribing mental states to the other agents is one way to avoid surprise. But as we have seen in analyzing the example of a mother playing together with her child there might be different solutions to coordinate a collective action.

The importance of forms of social cognition might revive the discussion about how demanding such conditions have to be. Consequently to my claim that you do not need individual agency in collective actions I would argue that no full-blown ascription of propositional attitudes is required.

I claim that coordination is an essential criterion by which we can judge whether two agents are involved in a collective action. The ability to coordinate your actions presupposes the ability to 'understand' the behavior of others. What this 'understanding' requests from the potential agents varies depending on the kind of collective action being at stake. In the case of bodily collective actions this understanding will surely be realized in a different way than for high cognitive collective actions.

In the following I will concentrate on communication.

11.6 Cases of Collective Actions

Theoretically, each participant in a collective action must be able to *perceive information* about themselves and the world. The information about the world has to imply the ability to perceive and conclude what the other agents perceive as well as the so-called social hints they send.

The *information processing* condition can be specified such that each agent needs the ability to process perceived information to enable herself to anticipate the behavior of the other agent. For example the social input might lead to a Theory of Mind to predict behavior by attributing beliefs, desires and intentions. In specific cases of collective actions it might be necessary to be able to move from the social input to a Theory of Emotional States.

To play a role in a collective action one must have *effectors* by which one is able to express social hints that are readable for the other agents as well. Most important in the case of artificial system is the point that one does not to have to claim that an agent needs to have mental or even emotional states. To coordinate actions it is sufficient to send the right information by expressing 'social hints'. Being able to express adequate social reactions is something that makes collective actions much more successful even if other levels of coordination might be used in collective actions. Looking at examples like dancing motor representations including mirroring the motor representations of the other might do the job of coordination.

As I am concerned with artificial systems in this paper I will not discuss cases of embodied coordination. The prototypical examples of a collective action I want to discuss are cases of communication. I claimed above each speech act is an action and a dialog is consequently a collective action. There is a long history of research developing so-called chat-bots, conversational machines, avatars that are able to serve as a communicative partner. Besides the interesting debates about the Turing test[11] asking the question of whether we can be tricked into falsely believing that we are communicating with humans while we are talking to an artificial system, there is a field of research developing artificial systems which do not try to trick us but try to satisfy the needs we have when having a conversation.

The example I will rely on is the avatar 'Max'. Max was developed by the research group of Ipke Wachsmuth, at University of Bielefeld.[12] Max is able to perceive several kinds of information. He processes acoustic and visual input, and he is able to send acoustic as well as visual information. This is more information than we have in any chat conversation on Facebook.

Through the visual information output Max can show gestures, mimic, including mimicking eye-movement behavior. He is able to express states of being attentive, bored, happy and so on. The acoustic channel enables him to produce speech acts. But the most important question is whether his information processing abilities lead to the ability to coordinate his speech acts with ours in such a way that we would judge this to be a communicative situation we want to call a collective action.

As claimed above, the ability to anticipate the behavior of the other is an essential condition for coordinating behavior. The most basic condition for coordinated communication might consist in the general rule that if one is speaking the other knows that he is expected to listen and the other way around. Additionally the mimic should somehow show that you are listening or even that you do understand what

---

[11] For more see Saygin et al. 2000.

[12] See the contribution of Wachsmuth to this volume. More references can be found: http://pub.uni-bielefeld.de/person/73476

the other says. Furthermore, the hearer must recognize what kind of speech act is in play. That means one should be able to recognize the intention (the goal) a speech act has. This can be very sophisticated if ironic or sarcastic features play a role. But in general cases the differentiation between questions, statements and so on might be sufficient to begin with. The ability of intention-reading is an essential part of coordinating speech acts.

Naturally, virtual conversational machines are limited in their successful communication behavior but there is already a lot realized which might lead to the assessment that something different from mere tool-use is happening if one engages in a dialog with one of those artificial agents.

Using the terms of computer science the behavior can be described as follows

Virtual agents are able to produce several kinds of outputs, which can be differentiated in several speech functions (speech acts) like an offer, a statement, or a question. On the perception side they are able e.g. with the help of a so-called BDI system (Cp. Braubach et al. 2005) to choose what might be the appropriate answering behavior. That means they recognize a question *as* a question and so on. Furthermore they have a variety of other speech functions by which for example they can monitor whether the other agent agrees. Another feature enables conversational machines to elaborate on the content by posing continuative questions using memory and knowledge base. To show that they are attentive they do express social hints like the utterance 'Mmm'(Mattar and Wachsmuth 2012). They can express agreement and disagreement and other speech functions of reciprocation. A description of the technical details and the consequently generated abilities virtual agents could in principle show that they are able to coordinate their speech acts according to the speech acts of their dialog partners. And furthermore, that they expect a certain communicative behavior from them in return. This seems to be categorically different to tool-use as tools do not tend to behave context-sensitive and do not express emotional or intentional states.

11.7 Conclusion

Summarizing my argument regarding whether artificial agents may play an active part in a collective action I argued for two claims.

First, I showed that there are cases of collective actions in which the distribution of abilities among the participating agents is not equal. I concluded from this that not all participants must necessarily have intentionality in the deep sense that the traditional notion of an action claims. Additionally, I suggested broadening the notion of action to be able to distinguish between simple and complex actions. So that one could in principle describe a collective action of mixed groups as a combination of simple and complex actions.

Second, I examined how we can be sure that we do not judge accidentally parallel actions as collective actions. Here I argued that if we are able to show that actions of a collective action are coordinated then we are legitimate in talking of a collective action. Social cognition abilities have been suggested as a way of enabling an agent to coordinate its actions. Focusing on conversational machines I demonstrated how coordination could be realized by recognition of speech functions, using social hints, expressing emotional states and producing context-adapted answers.

References

Bratman, Michael E. 1992. Shared cooperative activity. *The Philosophical Review* 101(2): 327–341.
Bratman, Michael E. 1997. I intend that we J. In *Contemporary action theory*, Social action, vol. 2, ed. R. Tuomela and G. Holmstrom-Hintikka, 49–63. Dordrecht: Kluwer.
Bratman, Michael E. 2014. *Shared agency: A planning theory of acting together*. Oxford: Oxford University Press.
Braubach, Lars, Alexander Pokahr, Daniel Moldt, and Winfried Lamersdorf. 2005. Goal representation for BDI agent systems. In *PROMAS 2004, LNAI 3346*, ed. R.H. Bordini et al., 44–65. Heidelberg: Springer.
Butterfill, Stephen. 2012. Joint action and development. *The Philosophical Quarterly* 62(246): 23–47.
Butterfill, Stephen, and Corrado Sinigaglia. 2014. Intentions and motor representation in purposive action. *Philosophy and Phenomenological Research* 88(1): 119–145.
Davidson, Donald. 1980. *Essays on actions and events*. Oxford: Oxford University Press.
Knoblich, Günther, Stephen Butterfill, and Natalie Sebanz. 2010. Psychological research on joint action: Theory and data. In *Psychology of learning and motivation*, vol. 51, ed. B. Ross, 59–101. Burlington: Academic.
Mattar, Nikita, and Ipke Wachsmuth. 2012. Small talk is more than chit-chat: Exploiting structures of casual conversations for a virtual agent. In *KI 2012: Advances in artificial intelligence*, Lecture notes in computer science, vol. 7526, ed. Birte Glimm and Antonio Krüger, 119–130. Berlin: Springer.
Pacherie, Elisabeth. 2013. Intentional joint agency: Shared intention lite. *Synthese* 190(10): 1817–1839.
Saygin, Ayse P., Ilyas Cicekli, and Varol Akman. 2000. Turing test: 50 years later. *Minds and Machines* 10(4): 463–518.
Schmid, Hans Bernhard. 2009. *Plural action: Essays in philosophy and social science*. Dordrecht: Springer.
Searle, John. 1990. Collective intentions and actions. In *Intentions in communication*, ed. P.R. Cohen et al., 401–415. Cambridge, MA: MIT Press.
Strasser, Anna. 2004. *Kognition künstlicher Systeme*. Frankfurt: Ontos-Verlag.
Tollefsen, Deborah. 2005. Let's pretend: Children and joint action. *Philosophy of the Social Sciences* 35(75): 74–97.

Chapter 12
Is Collective Agency a Coherent Idea? Considerations from the Enactive Theory of Agency

Mog Stapleton and Tom Froese

12.1 Introduction: The Enactive Approach

The enactive approach[1] to cognitive science is characterized by being grounded in biology, phenomenology and principles like self-organization and autonomy that were developed in the second-order cybernetics movement. Whereas cybernetics focused on the observation and study of feedback systems, second-order cybernetics tried to also account for the possibility of the observer herself (Froese 2010). Enactivism is an inherently interdisciplinary approach to cognition rather than a

---

[1] Note that the term "enactivism" has recently come to be used in several ways. Here we use it to refer to the paradigm heavily influenced by Maturana and Varela (1987) and formally instigated with the introduction of the term in Varela et al. (1991). This has been described as "autopoietic enactivism" by Hutto in order to distinguish it from his theory which he calls "radical enactivism" (Hutto and Myin 2013) and from sensorimotor enactivism (Noë 2004). While it is useful to distinguish these streams of research, the term "autopoietic enactivism" is somewhat misleading as although the theory of autopoiesis has been a strong inspiration for researchers in this paradigm, not all accept that autopoiesis is necessary and/or sufficient for cognition (for this debate see Froese and Di Paolo 2011; and the discussions in Thompson 2011; Wheeler 2011). It is therefore perhaps better to refer to it as "biological enactivism" in order to distinguish it from the other streams. For the purpose of this paper we do not draw on these other streams and will use the term "enactivism" as it was originally introduced and as it continues to be used by the main propagators of this approach (Varela et al. 1991; Thompson 2007; Di Paolo 2005; 2009a; Di Paolo and Thompson 2014).

M. Stapleton (✉)
Institute of Philosophy, University of Stuttgart, Seidenstraße 36, Stuttgart 70174, Germany
e-mail: stapleton@philo.uni-stuttgart.de

T. Froese
Instituto de Investigaciones en Matemáticas Aplicadas y en Sistemas/Centro de Ciencias de la Complejidad, Universidad Nacional Autónoma de México, Mexico City, Mexico

mere bringing together of several disciplines. This is reflected in the variety of disciplines in which the enactivist paradigm is used to guide research, such as artificial life and robotics (Di Paolo 2003; 2005; Barandiaran et al. 2009; Morse et al. 2011), social and developmental psychology (Reddy 2008; De Jaegher et al. 2010; Froese and Fuchs 2012) psychiatry (de Haan et al. 2013), sociology (Protevi 2009), and philosophy of mind (Thompson 2007; Gallagher and Zahavi 2008).

The enactivist paradigm differs from the received view in these disciplines because it does not condone starting our investigation into the mind by abstracting away from our biological body, as has been the standard (cognitivist and functionalist) approach in the cognitive sciences broadly construed. Instead, our biology is taken seriously, as the basis from which to start an investigation into the nature of mind. Cognition is thus, for enactivists, a fundamentally embodied phenomenon in the strongest sense of the term. As we will see, agency is likewise grounded in this "deep embodiment".[2] And, herein lies the crux: if cognition and agency are fundamentally embodied phenomena how could there be such a thing as "collective agency"?

While enactivism has its roots in the theoretical framework of second-order cybernetics (see Froese 2010), the starting point for enactivism proper was the research into the autopoietic organization of living cells (e.g., Maturana and Varela 1987). It was argued that the cell is the minimal living (and cognitive) system because it metabolically produces itself as an individual in its own right along within that individual's domain of possible interactions. What this means is that the processes within the cell create the very boundary which enables these processes to continue to produce both themselves and the boundary, while also maintaining viable interactions with the environment. It is this self-organising and self-producing organization of matter that is defined as "autopoiesis." Abstracting from the material instantiation of this organization – but maintaining the property of operational closure, i.e., conditions whereby processes depend on each other for their continuation – yields a form of organization which is defined as "autonomous." This organization was first derived by Varela from a study of cellular metabolism and other biological networks, but it does not need to exclusively be instantiated in natural biological systems (for an introductory overview of the concepts of autopoiesis and autonomy see Di Paolo and Thompson 2014). The upshot of this is that despite the enactive approach being grounded in the theory of autopoiesis and thus being a fundamentally biological, and embodied, approach, it remains an open question as to whether artificial systems (or indeed a collection of biological systems) could instantiate an autonomous or autopoietic organization (Froese and Ziemke 2009).

---

[2] The term "deep embodiment" is taken from Ezequiel Di Paolo's (2009) ShanghAI Lecture available at http://shanghailectures.org/guest-lectures/43-presenter/177-ezequiel-di-paolo. It refers to the fact that embodiment is taken as ontologically essential for mind, rather than as just a contingent functional extension of mind that could be separated from it, like a tool.

12.2 The Enactive Theory of Agency

What is the relevance of this self-organising and self-producing organization for agency? To see this, consider our pre-theoretical notion of agency. At the very least the term 'agent' implies (1) an individual, and (2) a capacity for action. This minimal notion is the one often used in informatics and robotics where 'agent' is standardly used to refer to any robot (or indeed software) that has a particular function and 'acts' so as to achieve this. There are two things to note here in regard to the use of 'agent' to refer to these systems. Firstly, the demarcation of the system in question is heteronomous, because its boundaries are only defined externally. That is to say, what counts as the individual agent depends upon an outside perspective and the interests of the observer. Secondly, such "agents" have of course been designed and programmed by humans to act in whatever way they do, be it by explicit design or by artificial evolution. They do not act in order to satisfy intrinsic needs and so their existence as agents does not directly depend on what they do; whether their movements happen to satisfy the conditions of successful behavior is decided by an external observer. Thus, even though nobody may be touching or controlling them remotely as they 'act', these 'actions' can still be seen as something external to that system (see Froese et al. 2007, for a related discussion of how the concept of 'autonomy' is used in robotics).

Enactivism offers a principled way of grounding individuality and action in a system. As outlined above, an autopoietic organization can be defined as a system's capacity to produce its own boundary. This constitutes a system as an individual in its own right with its own domain of interactions. Take the paradigmatic case of the bacterium. Even though, depending on what our explanatory project is, we can zoom in to look at parts within its body, or zoom out to include parts of the environment such as the whole Petri dish in which it is swimming, it is nevertheless the case that the bacterium constitutes itself as an individual in a way that the systems of the other two perspectives do not. There is nothing that indicates those other systems as such to be anything but dependent on our distinctions. Only when we distinguish the bacterium as being delimited by its self-produced network of metabolism and boundary does the bacterium appear to us as self-distinguishing; it is ontologically individuated from the environment. Ontological individuation in this respect refers to more than an appropriate posit relative to an explanatory project; it is a strong claim about the fundamental status of minimal living systems. On this view, ontological individuation is necessarily based on self-individuation, of which autopoiesis is one fundamental example.

In theory at least, it is possible that the material constituents of an autopoietic system could come together with the right kind of organization, and if all its metabolic needs were provided for in its direct environment then that organization could be maintained without the need for it to move. Such a system however needs an extremely stable environment to survive (an example of this in practice are some endosymbiotic bacteria, see Barandiaran and Moreno 2008). As soon as the external environment is anything short of benevolent the system will disintegrate and no longer exist. This sharp binary between life and death in autopoiesis has been addressed by Di Paolo (2005) who introduced the notion of adaptivity to enactive theory.

Following Di Paolo (2005), if we understand the viability set of a system to be the set of changes that can happen in the environment and within the system itself without the system's organization breaking down, then adaptivity is the property that a system has of being both sensitive to and able to regulate these changes such that if they are leading towards dissipation the system can change itself (adapt) in order that it can evade dissolution. This could happen in two ways. The system could adapt itself internally such that it, for example, improves its chemotactic ability by incorporating a new compound into its metabolism (Egbert et al. 2012). Or, and this is the property of interest to us here, it could regulate its interactions by moving itself away from a noxious environment and/or towards a beneficial environment (Barandiaran and Moreno 2008). This is not a mere passive movement of a system that is subject only to external modulations but the action of a system moving itself according to its intrinsic needs. This is the foundation of the enactive account of agency.

Barandiaran et al. (2009) develop this idea of intrinsic agency in adaptive systems and propose an operational definition of agency on which there are three necessary and (jointly) sufficient conditions: individuality, interactional asymmetry, and normativity. The requirement of individuality falls out of what we discussed in the preceding paragraphs; the claim is that to be an agent a system must distinguish itself as an individual rather than merely having individuality thrust upon it from our perspective. The enactive theory of agency argues that systems with autopoietic organization are a paradigm case of genuine individuality because they generate their own boundaries which allow the very processes which generate these boundaries to continue, in a circularly causal spiral. The second condition, interactional asymmetry, is the requirement that the system must be able to change its relationship to the environment, and that changes be actively generated more from within rather than being a passive result of external forces. Finally, the third condition, normativity, is the requirement that action is guided by internal norms. These internal norms are the goals that arise as a result of the intrinsic needs of the system. A bacterium, for example, may be guided to move up a sugar gradient because of its metabolic needs; to continue existing and not succumb to dissolution the bacterium must metabolise the sugar. The normativity lies in the intrinsic value that these goals (in this case metabolizing sugar) have for the continued existence of the system.

In summary, an enactive theory of agency (following Barandiaran, Di Paolo & Rohde) proposes that an agentive system is one which creates itself as an individual, adapts to changes in the environment by manipulating itself in that environment more than the environment manipulates it, and is moved to do so in order to satisfy needs that arise internally, needs that result from the system undergoing self-generation under far-from-equilibrium conditions, i.e., conditions which continually threaten its existence.

12.3 Multi-agent Systems, Multi-system Agents, and Multi-agent Agents

The enactive theory of agency provides us with conditions for a system being an agent. It might seem that a minimal agentive system, according to these criteria, must be autopoietic and adaptive; in short: a biological organism (even if a mere single-celled one). While this may turn out to be the case, it does not directly follow from the Barandiaran et al. proposal, indeed they specifically state that "agency does not have to be subordinated to biological/metabolic organization but can appear at different scales responding to a variety of autonomous processes…" (pp. 8–9). While all autopoietic systems instantiate an autonomous organization there might also be systems which instantiate this autonomy without necessarily being grounded in autopoiesis in the chemical domain. We can therefore see that even though the enactive conception of agency is grounded in biological embodiment it nevertheless allows for the possibility of 'collective agents'. That is to say, on the face of it there does not seem to be a contradiction between the conception of collectivity or distributedness and the enactive conception of agency: as long as the collective system fulfills the conditions of individuality, interactional asymmetry, and normativity it may be considered to be a genuine agent.

12.3.1 Multi-agent Systems vs. Multi-system Agents

Recall that we are here using the term collective agency in the sense of a system made up of agents, which itself – as a system – is agentive. We must therefore distinguish collective agency from two other similar concepts that are easy to conflate: multi-agent systems and multi-system agents. Multi-agent systems are systems which are composed of multiple independent software 'agents', each of which has its own goal or function to fulfill. These 'agents' interact – or at least communicate – with each other in order to negotiate their activity and make a contribution to (what at least on the surface might seem to be) the 'goal' of the larger system. In informatics, because the term 'agent' is used so broadly, there is no problem in identifying such a system as a multi-agent agent – or 'collective agent'.

But, if we insist upon a deep notion of agency, such as the enactive one proposed by Barandiaran et al., then these systems will fail to be agents at both levels: neither the subsystems nor the group 'agent' appropriately distinguish themselves ontologically from the environment so as to fulfill the criteria of autonomous (rather than heteronomous) individuality, nor are their interactions and goals endogenously created, serving their persistence as a system. This is not to say that it is in principle impossible for us to design the conditions of a multi-'agent' system whose emergent properties satisfy the enactive conditions of agency, but in practice this has not been achieved so far (see Froese and Di Paolo 2008 for an initial attempt). The problem is that we are faced with the task of engineering second-order emergence (Froese and Ziemke 2009): the emergent behavior of the interacting 'agents' has to provide the conditions for the emergence of a new agent. And even if we managed to overcome this problem in practice, we would still only end up with a multi-system agent, i.e., a genuine agent at the emergent group level but composed of systems which do not fulfill the criteria for agency. Such a system would fall short of genuine collective agency.

12.3.2 Swarms as Multi-agent Systems

Consider however a case where the individual agents of a multi-agent system really are agents in the full (enactive) sense of the term, as for example in the case of swarms of insects, birds, or fish. Seemingly very complex behavior can emerge at the group level as a result of local interactions between these individual agents. Do we have evidence however, to think that the swarm as a whole is a well-defined entity, that it regulates its domain of interactions, and that it generates its own norms, according to which it's activity is guided? Although it may be possible to view swarms as generating themselves (it is after all the own dynamics of, say a tornado that keeps itself going) is it right to think of them as generating themselves as individuals?

In the case of simulated flock behavior (Reynolds 1987), which presumably generalizes to biological systems, it has been shown that swarm behavior can be obtained by combining three rules: collision avoidance, velocity matching, and flock centering. If moving agents follow these three rules, then (presumably with the addition from time-to-time of some deviance from within or outside the system) this seems to be sufficient for generating the patterns in flocking behaviour that strike us as seeming so complex as to be somehow a movement of a whole rather than of an aggregation of individuals (think for example of the murmuration of starlings). Of course such groups do not generate the individuals that compose them (in the timescale of the swarm), but that would only be a prerequisite for *autopoietic* organization and not for *autonomous* organisation which is what underpins the criterion of individuality. Likewise, there may not be a physical boundary of the system in place, but it does seem right to think of such swarms, or flocks, as nevertheless forming themselves as a system such that their systemhood is not heteronomous; it is not just thrust upon them by an observer but rather is a "real pattern" (see Dennett 1991). It is an open question as to whether such real patterns fulfill the criterion of individuality, after all if we are to allow for the possibility of artificial agents then we cannot stipulate, from the outset, that all individuality must result from a system creating its own boundary in autopoietic fashion.

Similarly, it does not seem entirely unintuitive to think that a flock's interaction with the environment might, for the most part, be generated more from within the flock itself. Whether or not it is right to think of a swarm as having any goal of its own or not, it's movement is not solely due to environmental perturbations, though these may play an important role in stimulating, modulating, or even dissipating, the group behavior. Even such minimal rule following as is suggested by Reynolds (1987) appears to be sufficient for producing a collective system that does not merely hang together as it is passively shunted around the environment, but rather moves itself.

So, we cannot rule out that a swarm may satisfy the criterion of individuality, and it seems likely that it at least satisfies the criterion of interactional asymmetry. What then of normativity? Recall that the criterion of normativity stipulated that for a system to have genuine agency it is not enough that it has goals and acts so as to achieve those goals. The goals of a genuine agent arise from the needs of the system itself. This is where the normativity comes from; the system *should* act in such a way that brings it closer to achieving these goals, not because any external force or agency wills it or designs it to be so, but in order for it to continue its existence as that system. Can we see this kind of normativity instantiated in simple swarms or flocks?

It is not clear how it could be. The constituent interactions of such a system are such that it rather seems that either they are all instantiated by the components of the system (whether these are genuine or 'as if' agents) in which case they give rise to swarm behavior, or they are not, in which case swarm behavior does not emerge. This swarm binary does not leave any room for adaptivity, which – as we saw in the previous section – is what grounds normativity. If there are no grades of survival then the system has only one need; exactly the right internal and environmental conditions to survive. If it were going to regulate these conditions either internally or externally, as would be adaptive, then it would need to somehow alter its internal dynamics. But altering the internal dynamics – when the only dynamics are the three rules that together give rise to swarminess, would result in altering those rules, and therefore no swarminess arising. In other words, in this kind of biological multi-agent system there are emergent dynamics at the group level, which may even fulfill the enactive definition of an autonomous system (Thompson 2007; Froese and Di Paolo 2011), but not the stricter requirements of agency.

12.3.3 Multi-agent Agents: Towards a Genuine Collective Agency

There are however, multi-agent systems that present more plausible candidates for instantiating agency at both the individual and group levels. Consider the eusocial insects, such as some kinds of ants, bees and termites, whose interactions give rise to colonies and hives with highly complex emergent properties. In fact, these groups of agents are so impressively complex and coordinated that they can appear to us as strongly animalistic; so much so that they are often referred to as "superorganisms" (Sterelny and Griffiths 1999, Chap. 8) In contrast to simple swarms, in eusocial systems there is a clear division of labor. The individuals are so specialized in their morphology and behavior that they cannot survive in isolation; they depend on each other for their existence. In addition, there is a clear colony boundary in place. Not every individual of the same species can join any colony, since colonies are individuated by means of chemical markers. Indeed, in contrast to simple swarms, individuals act in the interest of the whole even to their own detriment, rather than just for the sake of their own lives.

If we are to grant that swarms satisfy the conditions of individuality and interactional asymmetry then we should be willing to grant that superorganisms also do so. They are not only self-organising systems (and thus generate themselves in the manner of keeping themselves going as a result of the dynamics that the system itself generates) but they also literally generate new components to keep the system going; they have offspring which are incorporated in to the system to fulfill specific functions. And, while it may seem that swarms interact with the environment more than the environment acts on the swarm, in the case of superorganisms this clearly seems to be the case as they manipulate the environment around them to provide suitable living and breeding space for the group. Could these groups of organisms then also generate their own internal normativity, and thus satisfy the final enactive criterion of agency?

The colony as a whole does indeed have a variety of irreducible properties, such as levels of food supplies, external threat and internal temperature, which emerge from the interactions of the individuals and at the same time modulate the behavior of those individuals so as to keep the global properties within the colony's range of viability. For example, there is a clear inside-outside division of a colony, which is normatively enforced by specialized individuals at dedicated boundary points. If the colony is under attack by intruders, the individuals will sacrifice themselves in order to neutralize the threat to the colony, similar to the function of some white blood cells in our bodies.

It could therefore be argued that such a group of social insects does manage to satisfy the enactive criteria of agency, and that it qualifies as a collective agent in a strong sense of the term. Their colonies are composed of individual organisms whose normativity as individuals is subsumed under the normativity of the colony as a whole. Yet it must also be noted that the higher-level of agency is only behaviorally integrated and individuated, in contrast to forming a single material structure. This is especially evident in terms of the colony's movement. If an ant colony shifts to a new location this is achieved by means of all ants individually moving to that new location. In other words, movement is not something specifically realized at the level of the colony as a whole. We therefore have an intermediate example of integration and individuation between (unintegrated) simple swarms and materially integrated systems such as multicellular organisms.

Multicellular organisms instantiate a stronger form of higher-level individuation. Typically multicellular organisms might be thought to be multi-system agents; built as they are out of cells, and modularized components such as organs. However, on the enactive view of agency it is not obvious that multi-cellular organisms are indeed mere multi-system agents rather than multi-*agent* agents. After all, as outlined above, a solitary bacterium satisfies Barandiaran et al.'s conditions for minimal agency: identity, interactional asymmetry, and normativity. Might it therefore be the case that some of the single cells in our body also satisfy these conditions?

The cells in our bodies constitute themselves through self-producing (autopoietic) means, just as was described earlier in the case of bacteria, such that they form themselves as an identity independently of an observer's perspective. It may not be so obvious however that they easily satisfy the asymmetry and normativity conditions for agency. Recall that the asymmetry condition is that adaptive regulation of the system must be powered – in general – more from intrinsic processes than environmental ones. In the case of a bodily cell it seems that much of this regulation is constrained by the bodily environment. In extreme cases the body can even cause some of its cells to commit 'suicide' (a process known as apoptosis) – an action which goes against the most basic of biological values, namely self-preservation.

To some extent such subordination must of course be the case – if the cells were not constrained sufficiently, no physically integrated whole would be possible in the first place. It is also of course partially in their "interest" to remain so constrained. If their replication, development and behavior are not subordinate to the needs of the whole system then they themselves will not survive for as long as may otherwise have been possible. For example, cell replication run amok is not conducive to maintaining bodily homeostatic balance, as is illustrated by aggressive forms of cancer. Do such internal environmental constraints, however mean that the relation between the cell and the body is asymmetric, with the locus of agency predominantly on the side of the overarching system (i.e., the body rather than the cell) such that the body effectively confiscates the subordinate cell's agency? And does this confiscation also mean that the activity of the cell is subordinate to the normativity of the body as a system, i.e. to the goals of the whole system rather than acting according to its own intrinsic norms?

It is difficult to provide definite answers to the questions raised in this discussion, but some general trends can nevertheless be identified. While a swarm may be too little individuated at the group level, and not generate enough endogenous normativity, to count as a multi-agent *agent*, a multicellular organism may be too tightly integrated at the group level, and its endogenous normativity overly constraining on the cells that make it up, to count as a multi-*agent* agent. A colony of eusocial insects however, seems to be situated somewhere along the middle of this spectrum and may therefore provide us with the clearest example of a genuine kind of collective agency. Each of these forms of interaction has advantages and disadvantages regarding the relative capacities of the parts and the wholes. Future work could apply the enactive concept of agency to the social world of humans, where cultural principles of integration are the predominant factor. For example, Steiner and Stewart (2009) have highlighted that from the autonomous perspective of human individuals, social norms appear as heteronomous, that is, as externally determined by cultural traditions. Interestingly, human societies have addressed the potential instabilities of a purely behaviorally integrated group level of agency by forming social institutions that are independent of the people passing through them, and which in the modern world even have legal representation as individuals in their own right. The foundation of a country creates a new individual entity that goes beyond its founding members, that has its own domain of interactions (for example with other countries), and that has its own normativity. However, this still falls short of the traditional metaphor of society being an organism in its own right, and that is fortunate for us. As Di Paolo (2009b) has pointed out, we should be wary of trying to push the ideal of living within a super-organism too far. Moving human society along the collective agency spectrum by increasing the powers of the institutional agent as a whole, i.e. by shifting it from a multi-agent system toward a multi-agent agent, implies a significant reduction of our personal liberties.

12.4 Collective Agency Is Not an All-or-Nothing Concept

The temptation is to see agency in black and white terms. Either a system fulfills the conditions for minimal agency or it does not. Indeed by talking of necessary and sufficient conditions for agency it might seem that we are implicitly propagating this kind of black-and-white thinking. While it may be right to think that a system that never satisfies these conditions is not agentive, a system may not have to always satisfy these conditions in order to be an agent. It seems right to think of ourselves as agents (in this minimal sense) even when we allow ourselves to go with the environmental or psychological "flow", that is, we do not lose our agentive nature just because – for a period of time – our interaction with the environment is either non-asymmetric or asymmetrically powered by the environment. Consider even the example of minimal agency being instantiated in a bacterium: it does not unceasingly act one-sidedly in its environment but rather oscillates between actively moving "in search" of higher sugar gradients (or away from noxious substances) and being passively modulated by the environment in which it currently finds itself (see also Di Paolo and Iizuka 2008). Agents can also *actively* become what according to Barandiaran et al.'s definition would seem to be non-agents for a period of time and temporarily let the environment control them, as a lizard does when it plays dead until the cat gets bored of playing with it. Similarly, agents can allow external social norms to take precedence over their intrinsic norms, and then later return to following their own norms.

Our inclination towards a binary concept of agency may arise from the tendency to view systems at just one time-slice or in abstraction. When viewed statically either the system satisfies the conditions or it does not. But this does not give us an accurate description of the system. Agentive systems are dynamic; they exist through time not only in the trivial sense that all things that exist do so in time, but rather it is part of the very concept of 'action' that it takes place over a period of time. If we define agency as the regulation of interaction according to norms then we are dealing with an extended process by definition. This of course raises the question of what time period is appropriate to take into account if we want to assess whether a system is agentive. From our everyday point of view, those systems which unfold in the timescales that we are used to observing actions in will be the most likely to be attributed agenthood by us. That is, actions that unfold in seconds, minutes, days, or perhaps months. Actions that unfold in very tiny time scales, or over many years, decades or centuries do not intuitively seem agentive to us. We do not naturally think of plants and trees for example as 'acting'. And yet, when we view films of plants growing in fast-forward this intuition begins to be a little undermined and they seem quite animalistic; they can move, they can climb, they can strangle other plants, they can shoot seeds, etc. To be sure, these actions may be more limited than those of animals, but the notion of interactional asymmetry does not have to be an all-or-nothing concept, either. It seems plausible that there are different grades of asymmetry. We must therefore be careful to not pre-judge agency on the basis of what seems to us to be the relevant time-scale. This is once more just a call for looking for an autonomous perspective rather than attributing agency on the basis of an observer's external considerations, except this time in reverse: rather than a false tendency to attribute agency to systems *that do not in fact have* intrinsic agency, our intuitions falsely guide us away from attributing agency to systems *that in fact do have* intrinsic agency.

Relatedly, the appearance of a conflict between sub-agents and the superordinate agent may arise as a result of viewing both at the same 'level' rather than acknowledging each in terms of their own ecological niche. The body taken as a whole must have interactional asymmetry with its *Umwelt*, and perhaps in so far as one of its cells ever becomes that *Umwelt* – in cases of cancer perhaps – the body must interact asymmetrically with that cell. But in normal functioning when we take the perspective of the cell, it is not interacting with 'the body' at all. Rather its *Umwelt* happens to be *in* the body. This body of course presents constraints on the cell's possibilities for action but surely this is not a case of the environment directly and irrevocably dictating the behavior of the cell, but rather just the presentation of a more constrained environment for that cell. From the cell's point of view, i.e. once we zoom in to consider the cell's activity in its own tiny *Umwelt*, we may no longer be inclined to think that it does not interact asymmetrically with its environment and therefore it is no longer unintuitive to think that it might be a genuine agent that partially constitutes a super-ordinate "collective" agent.

This is an important point and bears elaborating upon a little more. Our considerations of agency in this context reveal that agency in each system may be more, or less, visible at different levels of analysis. This means that it may not be the case that a genuine collective agency strikes us – from a single level of analysis, and at a single time period – as agentive both at the level of the group/collective and at the level of the components that make up that group. Add to this that, as we have argued, agency is a spectrum varying along dimensions rather than an all-or-nothing concept. We can therefore see that even if we define collective agency as a multi-agent agent in contrast to either multi-agent systems or a multi-system agents, we nevertheless are faced with a variety of kinds of agency in collective agents: agency varies in each case along the dimensions of individuality, interactional asymmetry and normativity both at the level of the individuals composing the group, and of the group itself. Depending on what time-slice you use to analyse the system, agency may be more or less visible. But nevertheless if a system is to be a genuinely collective system, according to our definition and as can be seen by the examples that we have presented, the variations along the dimensions of agency at each level mutually enable and constrain those at the other. Although what we consider to be the most compelling examples of what might be a genuine collective agency are to be found in nature since the enactive concept of agency is not fully realized by current robotic 'agents', artificial examples are nevertheless not excluded by definition. Furthermore, cultural institutions that act as individuals in the social domain, and may even interact with their component individuals in, for example a court of law, may potentially fit the criteria.

12.5 Collective Agency Versus Collective Subjectivity

If we are to consider the possibility of genuine collective agency, rather than using 'agency' as a mere metaphor when it comes to the collective level, then we must also open ourselves up to considering whether such a thing as collective subjectivity may also exist. Even if we were to approach the topic from a non-enactive viewpoint this would seem warranted because in our paradigm case of agents – humans – the concept of agency is tightly interwoven with the interrelated concepts of mindedness, subjectivity, intentionality, and consciousness. The burden of proof should perhaps then be on showing that agency, or the processes upon which agency depends, does *not* entail subjectivity rather than the reverse. But of course the idea of collective subjectivity is a much harder pill to swallow than even that of collective agency. Let us then separate the question of subjectivity from that of consciousness, and address the question of whether a genuine collective agency on the enactive account might imply collective subjectivity even if not a collective consciousness. According to the enactive approach an organism's subjectivity is it's lived perspective, or in other words, it's meaningful point of view on the world (e.g., Weber and Varela 2002; Thompson 2007). Is there something it is like to be an ant colony or even a country like Germany? Do they have a subjective perspective and *Umwelt* of their own? Common intuition would deny this, but on what basis?

Let us consider the case of dyadic human interaction. Each of the individual participants is certainly a subject with a lived perspective, but what about the dyad as such? After many years of methodological individualism in cognitive science, which holds that the proper level of analysis is the individual (or even just their brain), there are now many paradigms emphasizing a deeper dynamic interconnectivity between people (Kyselo and Tschacher 2014; De Jaegher and Di Paolo 2007; Oullier and Kelso 2009; Riley et al. 2011). To give an example from our own work, Froese et al. (2013) used the minimal cognition approach developed by Beer (1996) and others to show that interacting robots become different kinds of systems while interacting. In this case each robot was only equipped with one artificial 'neuron', but via the interaction process they managed to extend each other's capacities such that the neurons exhibited oscillations and chaos – properties that are in principle impossible for a 1D continuous system, as realized by each robot's single neuron. For oscillations a minimum of two dimensions are necessary, while chaos requires a minimum of three. The fact that these properties were observed in each robot's neural activity therefore implies that the dimensionality of their artificial 'brains' became mutually extended via their embodied social interaction. The two brains and their bodies and their interaction via the environment formed one system, thereby making it impossible to reduce a robot's neural activity to its neural system. What this minimal robotic model shows is that, in principle at least, nothing stands in the way of an extended body realizing a socially extended mind (Froese and Fuchs 2012).

One might think that this kind of socially extended mind suggests that it is at least a legitimate question to ask whether it is possible to also share each other's subjective perspectives, i.e. to give rise to a genuinely second-person perspective with its own social actions. It is important not to conflate these second-person interactions with joint action as conceived under the title of the "we-mode" (Galotti and Frith 2013). The idea behind Galotti and Frith's we-mode is that when we engage in joint actions we do not need to represent our individual goal, and the goals of the others with whom we act (indeed this may in some circumstances be counter-conducive). Rather we enter a particular "mode" in which we represent the task as one that we – as a couple or a group – are doing together so that the representation that guides each of us is a "we-representation" rather than an "I-representation". But this is not to say that the representation is "shared" between individuals in any interesting way. The we-representation is still a representation inside the individual and although each member of the group that is acting jointly must have a we-representation to be cooperating successfully in a joint action, and presumably these we-representations must have broadly similar contents, nevertheless it is certainly not "the same" representation that is shared by the actors. Galotti and Frith's "we-mode" therefore does not yield any interesting notion of shared or collective agency that is relevant to the question of whether the individuals do truly come together to form some kind of supra-individual agentive system with its own unique subjective perspective.

A possibly genuine form of shared intentionality and lived perspective is suggested by a psychological study conducted by Froese et al. (2014). Making use of a minimalistic virtual reality setup first proposed by Auvray and colleagues (2009), which has come to be known as the 'perceptual crossing' paradigm (Auvray and Rohde 2012), they explored the conditions under which people come to be aware of the presence of another person on the basis of differences in interaction dynamics alone. The interaction between a pair of spatially separated participants was mediated by a human-computer interface that reduced the scope of their interaction to moving an avatar in an invisible linear virtual space and receiving tactile feedback while overlapping with a virtual object. Participants were instructed to try to help each other to locate each other in the virtual space while avoiding distractor objects. One of these objects was static while the other was an exact, but unresponsive, instantaneous playback of the partner's avatar movements. They had 15 trials in which they were asked to click when they felt that they had succeeded in finding the other, and after each trial they were asked to rate the clarity of the experience of their partner if they had clicked. No feedback about the correctness of a click was provided during the experiment. Despite the reduced scope for social interaction and the presence of distractor objects, most pairs of participants were successful at solving the task in a majority of trials.

Two aspects of this outcome are of particular relevance to our discussion here. Firstly, it seems that success was largely a cooperative achievement. It was more common that both participants clicked correctly on the other than for one participant to succeed alone, and the delay between such jointly successful clicks was often less than a few seconds. In other words, although participants could not be directly aware of each other's clicks, the recognition of the other participant was highly synchronized between the participants, to the extent that we could interpret this as evidence of genuinely shared intentionality, i.e., mutual recognition of each other. Secondly, there was an experiential difference between two kinds of correct clicks, namely clicks occurring in jointly successful trials and clicks occurring in individually successful trials. Participants were much more likely to rate their experience of the other as being most clear after jointly successful clicks. The clarity of social awareness therefore had less to do with the objective presence of the other, and more with whether that awareness was mutually shared by both participants. In other words, it is suggested that the co-regulation of mutual embodied interaction also gave rise to a shared lived perspective, i.e., a second-person perspective.

This kind of collective, *second*-person subjectivity already goes beyond the traditional constraints of internalist-individualist cognitive science (Froese and Fuchs 2012). But does this study enable us to talk about the interactive constitution of a new subject with its own *first*-person perspective? This does not seem to be the case because, although there is a deep integration of minds, the participants do not lose their individual point of view on their shared situation during this process. We are left with the idea of a genuine second-person perspective, but not a collectively constituted first-person perspective. It is doubtful that it is any different for social interaction processes involving more participants, such as a football team or a nation state: under certain conditions of mutual interaction a multi-person perspective may be formed (the famous 'mob mentality'), but this is not sufficient for attributing a collective first-person perspective. The concepts of collective agency and collective subjectivity, where the latter refers to a collective agent with its own unique, unified and meaningful point of view on the world, may therefore not always coincide. The former seems necessary but not sufficient for the latter.

But if we allow the possibility of collective agency, why are we more skeptical about the possibility of collective subjectivity? Partly this has to do with the fact that, according to the enactive approach, subjectivity has more requirements than just agency. We have discussed the relationship between agency and subjectivity at length elsewhere (Stapleton and Froese ms.), so we will be brief here. Essentially, subjectivity depends on a specialized subsystem for the monitoring and regulating of internal processes, which we interoceptively experience as valence or emotion. In animals this system is realized as a special neural system that is spread throughout the entire body. It would be difficult to realize an operationally equivalent system on the basis of social interaction without some kind of physical integration that allows for the structuring of the necessary processes of monitoring and regulating in a stable manner. In other words, one important reason why we, as individual human beings, are collective subjects in addition to being collective agents is that our collective agency is realized as one physical living body. Nevertheless, on closer inspection it may well turn out that we are dealing with another spectrum of possibilities, and that we should avoid becoming trapped in a way of thinking that assigns an absolute categorical difference between first- and second-person perspectives. One or the other type of perspective may become more or less prevalent depending on conditions. For an extreme example, we can consider that reports of how others are experienced by people with schizophrenia demonstrate how that phenomenology can vary from complete autistic isolation (Stanghellini and Ballerini 2004) to normal co-presence (during periods of relative well-being) and to complete fusion with others and self-dissolution (Lysaker et al. 2005).

12.6 Conclusion

Whether collective agency is a coherent concept depends on the theory of agency that we choose to adopt. We have argued that the enactive theory of agency developed by Barandiaran et al. (2009) provides a principled way of grounding agency in systems to which we already attribute it: biological organisms. The instantiation of the necessary and jointly sufficient conditions of individuality, interactional asymmetry, and normativity give rise to a system that is ontologically demarcated, endogenously active, and generates its own needs; its agency does not depend on the viewpoint of an observer nor does it exist only relative to a particular explanatory project. Enactivism, and therefore also the enactive theory of agency, is however grounded in biological embodiment which might lead one to be skeptical as to whether artificial systems or collectives of individuals could instantiate genuine agency. To explore this issue we contrasted the concept of collective agency with the ideas of multi-agent systems and multi-system agents, and argued that a genuine collective agency would instantiate agency at both the collective level and at the level of the component parts. We argued that although swarms present impressively complex behavior at the level of the collective system, this collective nevertheless fails to instantiate genuine agency because it does not generate its own normativity. We then considered the case of eusocial insect colonies, sometimes termed 'superorganisms' and multicellular systems like ourselves. Eusocial colonies, unlike swarms, may be seen to not only instantiate individuality and interactional asymmetry but also to generate endogenous normativity. While this would bring us to what might be considered to be a paradigm case of collective agency, we questioned whether the behavioural (rather than material) integration and individuation was quite strong enough to instantiate the individuation required for genuine agenthood at the collective level. In contrast, we questioned whether the material integration of cells in multicellular organisms such as ourselves might actually be *too tight*, and the normativity generated by the collective *too strong* to allow for genuine agency of the components of the collective. We proposed that agency cannot be judged at a single time-slice and that we should therefore understand agency as a spectrum that varies along dimensions of individuality, interactional asymmetry, and normativity rather than as an all-or-nothing concept in which necessary and sufficient conditions either are – or are not – instantiated. On such an understanding agency is not necessarily lost when, for example, interactional asymmetry is temporarily reversed or absent. Furthermore, it can help explain how agency may be being instantiated even if it may not be clearly visible at both the level of the collective and the component at the same time as both the agency of the components, and the agency of the collective are spectra, and both will differ individually not only along the dimensions of individuality, interactional asymmetry, and normativity but also through time.

We highlighted that our own collective agency, based on our existence as multicellular organisms, coincides with collective subjectivity, that is, we have a lived perspective of concern, as do the individual organisms that make up our bodies (albeit in a much more minimal form than ourselves). But subjectivity (even when understood minimally as having a meaningful point of view on the world) coinciding with agency as it does in multicellular organisms such as ourselves, seems to be the exception rather than the rule. We argued that while it may be possible to genuinely share one's individual lived perspectives with other agents by forming a second- or multi-person perspective, to fully satisfy the additional operational conditions of first-person subjectivity at the level of the collective system as a whole, collective agents may have to be more than merely collective; they must be materially integrated into one living body.

References

Auvray, Malika, and Marieke Rohde. 2012. Perceptual crossing: The simplest online paradigm. *Frontiers in Human Neuroscience* 6(181). doi:10.3389/fnhum.2012.00181.
Auvray, Malika, Charles Lenay, and John Stewart. 2009. Perceptual interactions in a minimalist virtual environment. *New Ideas in Psychology* 27: 32–47. doi:10.1016/j.newideapsych.2007.12.002.
Barandiaran, Xabier, and Alvaro Moreno. 2008. Adaptivity: From metabolism to behavior. *Adaptive Behavior* 16: 325–344. doi:10.1177/1059712308093868.
Barandiaran, Xabier, Ezequiel Di Paolo, and Marieke Rohde. 2009. Defining agency: Individuality, normativity, asymmetry and spatio-temporality in action. *Adaptive Behaviour* 17(5): 367–386.
Beer, R.D. 1996. Toward the evolution of dynamical neural networks for minimally cognitive behavior. In *From animals to animats 4: Proceedings of the fourth international conference on simulation of adaptive behavior*, ed. P. Maes, M. Mataric, J. Meyer, J. Pollack, and S. Wilson, 421–429. Cambridge, MA: MIT Press.
de Haan, Sanneke, Erik Rietveld, Martin Stokhof, and Damiaan Denys. 2013. The phenomenology of deep brain stimulation-induced changes in OCD: An enactive affordance-based model. *Frontiers in Human Neuroscience* 7(653). doi:10.3389/fnhum.2013.00653.
De Jaegher, Hanne, and Ezequiel Di Paolo. 2007. Participatory sense-making: An enactive approach to social cognition. *Phenomenology and the Cognitive Sciences* 6: 485–507.
De Jaegher, Hanne, Ezequiel Di Paolo, and Shaun Gallagher. 2010. Can social interaction constitute social cognition? *Trends in Cognitive Sciences* 14: 441–447. doi:10.1016/j.tics.2010.06.009.
Dennett, Daniel C. 1991. Real patterns. *The Journal of Philosophy* 88: 27–51.
Di Paolo, Ezequiel. 2003. Organismically-inspired robotics: Homeostatic adaptation and natural teleology beyond the closed sensorimotor loop. In *Dynamical systems approach to embodiment and sociality*, ed. K. Murase and T. Asakura, 19–42. Adelaide: Advanced Knowledge International.

Di Paolo, Ezequiel. 2005. Autopoiesis, adaptivity, teleology, agency. *Phenomenology and the Cognitive Sciences* 4: 429–452. doi:10.1007/s11097-005-9002-y.
Di Paolo, Ezequiel. 2009a. Extended life. *Topoi* 28: 9–21. doi:10.1007/s11245-008-9042-3.
Di Paolo, Ezequiel A. 2009b. Chapter 3 Overcoming autopoiesis: An enactive detour on the way from life to society. In *Advanced series in management*, vol. 6, ed. Rodrigo Magalhães and Ron Sanchez, 43–68. Bingley: Emerald.
Di Paolo, Ezequiel A., and Iizuka Hiroyuki. 2008. How (not) to model autonomous behaviour. *Biosystems* 91: 409–423. doi:10.1016/j.biosystems.2007.05.016.
Di Paolo, Ezequiel, and Evan Thompson. 2014. The enactive approach. In *The routledge handbook of embodied cognition*, ed. Shapiro Lawrence. New York: Routledge Press.
Egbert, Matthew D., Xabier E. Barandiaran, and Ezequiel A. Di Paolo. 2012. Behavioral metabolution: The adaptive and evolutionary potential of metabolism-based chemotaxis. *Artificial Life* 18(1): 1–25.
Froese, Tom. 2010. From cybernetics to second-order cybernetics: A comparative analysis of their central ideas. *Constructivist Foundations* 5: 75–85.
Froese, Tom, and Ezequiel Di Paolo. 2008. Can evolutionary robotics generate simulation models of autopoiesis? In *Cognitive science research paper*, vol. 598. Brighton, University of Sussex.
Froese, Tom, and Ezequiel Di Paolo. 2011. The enactive approach: Theoretical sketches from cell to society. *Pragmatics and Cognition* 19: 1–36.
Froese, Tom, and Thomas Fuchs. 2012. The extended body: A case study in the neurophenomenology of social interaction. *Phenomenology and the Cognitive Sciences* 11: 205–235. doi:10.1007/s11097-012-9254-2.
Froese, Tom, and Tom Ziemke. 2009. Enactive artificial intelligence: Investigating the systemic organization of life and mind. *Artificial Intelligence* 173: 466–500. doi:10.1016/j.artint.2008.12.001.
Froese, Tom, Nathaniel Virgo, and Eduardo Izquierdo. 2007. Autonomy: A review and a reappraisal. In F. Almeida e Costa, L. M. Rocha, E. Costa, I. Harvey & A. Coutinho (Eds.), *Advances in Artificial Life: 9th European Conference, ECAL 2007* (pp. 455–464).
Froese, Tom, Carlos Gershenson, and David A. Rosenblueth. 2013. The dynamically extended mind – A minimal modeling case study. In *2013 IEEE Congress on Evolutionary Computation* (pp. 1419–1426), IEEE Press.
Froese, Tom, Iizuka Hiroyuki, and Takashi Ikegami. 2014. Embodied social interaction constitutes social cognition in pairs of humans: A minimalist virtual reality experiment. *Scientific Reports* 4(3672). doi:10.1038/srep03672.
Gallagher, Shaun, and Dan Zahavi. 2008. *The phenomenological mind: An introduction to philosophy of mind and cognitive science*. New York: Routledge.
Gallotti, Mattia, and Chris D. Frith. 2013. Social cognition in the we-mode. *Trends in Cognitive Sciences* 17: 160–165. doi:10.1016/j.tics.2013.02.002.
Hutto, Daniel D., and Erik Myin. 2013. *Radicalizing enactivism: Basic minds without content*. Cambridge, MA: MIT Press.
Kyselo, Miriam, and Wolfgang Tschacher. 2014. An enactive and dynamical systems theory account of dyadic relationships. *Frontiers in Psychology* 5(452). doi:10.3389/fpsyg.2014.00452.
Lysaker, Paul Henry, Jason K. Johannesen, and John Timothy Lysaker. 2005. Schizophrenia and the experience of intersubjectivity as threat. *Phenomenology and the Cognitive Sciences* 4: 335–352.
Maturana, Humberto R., and Francisco J. Varela. 1987. *The tree of knowledge: The biological roots of human understanding*. Boston: New Science Library/Shambhala Publications.
Morse, Anthony F., Carlos Herrera, Robert Clowes, Alberto Montebelli, and Tom Ziemke. 2011. The role of robotic modelling in cognitive science. *New Ideas in Psychology* 29(3): 312–324. doi:10.1016/j.newideapsych.2011.02.001.
Noë, Alva. 2004. *Action in perception*. Cambridge, MA: MIT Press.
Oullier, Olivier, and J.A. Scott Kelso. 2009. Social coordination from the perspective of coordination dynamics. In *Encyclopedia of complexity and systems sciences*, ed. Robert A. Meyers, 8198–8212. Berlin: Springer.

Protevi, John. 2009. *Political affect*. Minneapolis: University of Minnesota Press.
Reddy, Vasudevi. 2008. *How infants know minds*, Cambridge, MA: Harvard University Press.
Reynolds, Craig W. 1987. Flocks, herds and schools: A distributed behavioral model. In *Proceedings of the 14th annual conference on computer graphics and interactive techniques*, 25–34. New York: ACM. doi: 10.1145/37401.37406.
Riley, Michael A., Michael Richardson, Kevin Shockley, and Verónica C. Ramenzoni. 2011. Interpersonal synergies. *Movement Science and Sport Psychology* 2: 38. doi:10.3389/fpsyg.2011.00038.
Stanghellini, Giovanni, and Massimo Ballerini. 2004. Autism: Disembodied existence. *Philosophy, Psychiatry, & Psychology* 11: 259–268. doi:10.1353/ppp.2004.0069.
Stapleton, Mog, and Tom Froese. ms. The enactive philosophy of embodiment: From biological foundations of agency to the phenomenology of subjectivity. Manuscript submitted for publication.
Steiner, Pierre, and John Stewart. 2009. From autonomy to heteronomy (and back): The enaction of social life. *Phenomenology and the Cognitive Sciences* 8: 527–550. doi:10.1007/s11097-009-9139-1.
Sterelny, Kim, and Paul Griffiths. 1999. *Sex and death: An introduction to philosophy of biology*. Chicago: University of Chicago Press.
Thompson, Evan. 2007. *Mind in life: Biology, phenomenology, and the sciences of mind*. Cambridge, MA: Harvard University Press.
Thompson, Evan. 2011. Reply to commentaries. *Journal of Consciousness Studies* 18: 5–6.
Varela, Francisco J., Evan Thompson, and Eleanor Rosch. 1991. *The embodied mind: Cognitive science and human experience*. Cambridge, MA: MIT Press.
Weber, Andreas, and Francisco J. Varela. 2002. Life after Kant: Natural purposes and the autopoietic foundations of biological individuality. *Phenomenology and the Cognitive Sciences* 1(2): 97–125.
Wheeler, Michael. 2011. Mind in life or life in mind? Making sense of deep continuity. *Journal of Consciousness Studies* 18: 148–168.

Part IV
Simulating Collective Agency and Cooperation

Chapter 13
Simulation as Research Method: Modeling Social Interactions in Management Science

Roland Maximilian Happach and Meike Tilebein

13.1 Introduction

Traditional research frameworks within organizational studies and management science originate from a closed system view (Porter 2006). Taking the perspective of the closed system view, scholars analyzed firms and organizations as separate rationally acting institutions that operate independently of their environments (ibid.). Organizations and firms were perceived as isomorphic to their environments that were represented as exogenous variables, and research was based on stable idealized predictable events (Baum and Singh 1994; Porter 2006). This assumption of stable and predictable events provided researchers with a broad foundation for theory development and led to the emergence of a variety of different theories and paradigms.

The emergence of different yet conflicting paradigms led to a paradigm war in organizational studies and management science (McKelvey 1997). Tackling the complex and idiosyncratic nature of firms and organizations, scholars provided an integrated approach in which organizations and their environments are intertwined and adapt to each other. This interplay is also referred to as coevolution (Baum and Singh 1994; McKelvey 1997; Lewin et al. 1999). Coevolution means that organizations evolve as environments evolve around them while sociological, political and economic factors drive changes in certain directions (Simon et al. 2008). Usually firms and organizations not only adapt to those changes but also initiate some of them, which results in a dynamic coevolution.

In order to cope with this complexity and to provide a new foundation for theory development, scholars suggested using idealized models (McKelvey 1997) or idealized reasoning (Ketokivi and Mantere 2010). Idealization creates necessary

---

R.M. Happach (✉) • M. Tilebein
Institute for Diversity Studies in Engineering, University of Stuttgart, Stuttgart, Germany
e-mail: maximilian.happach@ids.uni-stuttgart.de; meike.tilebein@ids.uni-stuttgart.de

simplification which enables empirical evaluation (Lomi and Larsen 2001). Empirical evaluation is especially difficult to conduct in highly changing and dynamic environments (Davis et al. 2007). For this reason, the acceptance for modeling and simulation methods as a research approach has been rising recently among management scientists. They use simulation methods in order to address the methodological question of how to better cope with complex problems on the one hand and how to use idealization, on the other hand.

In this paper, we present simulation as an appropriate research method to handle dynamics and complexity within the field of management science. In particular, we focus on social interactions. Firms and organizations are driven by social interactions such as decision processes, strategy making, negotiations, and operations. Those interactions are composed of multiple simultaneous dynamically evolving processes with several different agents. The complexity and interplay of these processes and agents highly affect the outcome of social interactions and thereby the performance or any kind of output of the organization. Yet, performance is not only influenced by social interactions but, in turn, performance also influences these processes and hence evokes interdependency and coevolution. In consequence, this interdependency results in a dynamic coevolution of social interactions and organizational performance.

This article is organized in five sections. Following the introduction, Sect. 13.2 presents a broad overview of social interactions in management science. In Sect. 13.3, we present simulation as research method. We describe a roadmap by Davies et al. (2007) on how to conduct research using simulation. Further, we present prerequisites, advantages and challenges of simulations and show that simulation is an appropriate and additional tool for conducting research. In Sect. 13.4, we present two major simulation methods and their use for investigating social interaction. We then conclude in the last section that simulation can act as an additional tool aside of other research methods. Regarding the advantages of simulation, we promote simulation as a bridging tool for theoretical and empirical research.

13.2 Social Interactions in Management Science

The purpose of this section is to describe the nature and variety of social interactions in management science. Although the perspective of social interactions is not in the focus of traditional management science, often scientists transfer concepts and paradigms from related fields of science. In the case of management science, economics and psychology play an important role in the perspectives of how humans interact. Further, these perspectives evolved over the last decades. However, this section does not aim at presenting the historical development of the respective concepts and paradigms. This section rather gives an overview of the different perspectives. In short, the traditional ubiquitous paradigm of management science describes the human as a rational omniscient agent, i.e. *homo oeconomicus*. However, when management science proceeded and the shortcomings of this paradigm set limits to further scientific progress, a new paradigm emerged that assumed bounded rationality on the side of the agents.

Studying economic behavior requires knowledge about how humans make choices. John Stuart Mill (1848) delivers a model of the human agent assuming rational behavior. This model has been the basis of economic theories for decades. It is known as *homo oeconomicus* or *economic man*.[1] Mill defines an economic man as money accumulating, desiring luxury and leisure as well as procreation (Persky 1995). These four distinct characteristics provide "just enough psychological complexity to make him interesting" (ibid., p. 224). The concept of economic man lays the basis for profit maximization, cost minimization and the rational choice of the best alternative (Simon 1955).

Profit maximization and cost minimization play an important role for microeconomics to analyze a firm's behavior. These concepts are used to determine optimal market prices or optimal production quantities (cf., e.g. Varian 2010). It is assumed that firms have complete information about costs and demand. Thus, firms have an ultimate cost and demand function with which they could calculate the appropriate quantity of production in order to maximize their profit or minimize their cost, respectively (Simon 1955). The interaction between these market agents lies therefore in the fact that they share the same cost and demand functions. Further, David Ricardo (1821) analyzes the trade between nations using the concept of comparative advantage to explain the advantage of trade. This concept shows that trade between two agents is advantageous if one agent can produce goods with lower marginal or opportunity cost than the other agent.

A more comprehensive concept than just maximizing profit is maximizing utility. Instead of maximizing the monetary value, economists substituted money by utility (cf., e.g. Samuelson 1937). In that way, the preference of economic man lays not only in accumulating money but also in increasing leisure time or other preferred aspects of utility. For these kinds of analysis, economists assume stable and well-defined preferences and a given utility function of economic man (Thaler 1988). For instance, Becker (1974) analyzes the utility function of households and incorporates effects of personal characteristics, e.g. race, religion, etc., on the utility function. Schelling (1969, 1971) investigated segregation in cities and the discrimination by households. Schelling finds that the desire of households of having at least one neighbor of the same race/gender/religion may lead to extreme segregation.

The introduction of advanced mathematics into the field of economics then enabled researchers to analyze the interaction between market agents. Von Neumann and Morgenstern (1944) introduced game theory which allowed analyzing the noncooperative interaction between two chess players and their strategies. Von Neumann and Morgenstern argue that complete information and enough computational power would turn chess into a trivial game (von Neumann and Morgenstern 1944). Complete information results in a complete decision tree of possible answers in

---

[1] Persky (1995, p. 222) argues that the Latin term *homo oeconomicus* stems from Vilfredo Pareto (1906). However, he also argues that Schumpeter (1954, p. 156) highlights Frigerio's (1629) book on the *economo prudente* as an ancestor of the economic man.

which players could always choose the preferred responding strategy to the preceding move (Simon 1972).

The emergence of game theory lead to a new perspective: Economists treated market interactions as games (Manski 2000). This perspective opened a field of empirical studies of market behavior. A variety of different games and empirical studies were designed to explore this aspect of social interactions. Some of them did not correspond to the paradigm of economic man. The most famous among these games is the ultimatum game (Güth et al. 1982). The ultimatum game is a two-player game. Player 1 is given a certain amount of money and needs to divide this amount between herself and the other player (whose identity is unknown to player 1). Thus, player 1 makes an offer to player 2 and player 2 decides whether to accept the offer or to reject it. In case of acceptance, player 2 receives whatever player 1 offers and player 1 gets the rest of the amount. In case of rejection, both players will receive nothing. The paradigm of homo oeconomicus predicts that player 2 accepts any amount higher than zero, following the rule of maximizing profit. However, as Güth et al. (1982) show, outcomes of the ultimatum game reveal that decision makers in this case behave "non-rational" as they consider fairness as an important aspect beyond maximizing profit. Only if the persons playing the role of player 2 consider the offers of player 1 as fair or just, they accept it. In case the amount is too small they sacrifice this amount to punish player 1. The ultimatum game is only one of many games that reveal the limits of the paradigm of homo oeconomicus.

The idealistic assumption of rationality and complete information as in the paradigm of homo oeconomicus was heavily criticized by Simon (1955, 1972). He argues that individuals do not have enough knowledge to choose the best alternative or do not possess enough cognitive capabilities to process all available information and therefore do not maximize. Individuals rather choose to act on what they think is best. Therefore, they choose an option out of the limited knowledge they have. This may or may not lead to an optimal choice. The corresponding paradigm proposed by Simon (1955) is called bounded rationality. Consequently, Simon (1957) suggests that organizations aim at processing complex tasks and therefore provide structures in which individuals can cooperatively process more information than individually. In such a way, individuals overcome their cognitive limitations and accomplish complex tasks.

The awareness of bounded rationality led to further analysis of decision making. Kahneman and Tversky (1979) suggest a theory of human decision making they coined prospect theory. They incorporated the specifics of human behavior, e.g. regarding risk aversion or enforcement of social norms shown in experiments, e.g. the ultimatum game described above. Tversky and Kahneman (1981) further designed experiments to show that the framing of choices affects the final choice or that humans make different choices when a sequence of games is isolated and thus their relation is disregarded (Tversky 1972).

Research on the drivers of human decision making and the role of social interactions in management science as shown above have initiated new streams of research in the broad field of behavioral economics. Research within these new streams of research include the effects of groupthink (Janis 1972), the spreading of information through different social groups and networks (Granovetter 1973), the emergence of cooperation between individuals and organizations (Axelrod 1984), the effect of procedural justice on decision-making teams (Korsgaard et al. 1995), the contribution of diversity to innovation processes (Tilebein and Stolarski 2009), or the effect of diversity on the innovativeness of new product development teams (Kreidler and Tilebein 2013). This research focuses on the outcome of simultaneous, interrelated, dynamical social interactions. Since these kinds of social interactions are difficult to measure and capture with real experiments, alternative research methods are needed.

13.3 Using Simulation as Research Method

Simulation is a widely used term and refers to role gaming as well as computer supported calculations to project past and future outcomes. In the following we focus entirely on computer based simulations. In line with previous research, we define simulation as using computer software for building and coding a quantifiable and operating model of real world processes, systems, facilities, or events (Law and Kelton 2000; Davis et al. 2007), whose variables can be manipulated (Berends and Romme 1999; Law and Kelton 2000) and the resulting output can be observed immediately (Bratley et al. 1987; Axelrod 1997a; Davis et al. 2007). Simulation models are consequently computational representations of an underlying theoretical logic that links constructs together (Davis et al. 2007). Computer software for building simulation models is based on mathematics. Therefore, simulation models are mathematical models (Law and Kelton 2000). However, representations of real world systems as mathematical models often become so complex that an analytical solution is impossible (Berends and Romme 1999). For instance, Forrester (1961) argues that typically analytical solutions of mathematical models with more than five interrelated variables become impossible. Computer simulations possess higher computational capacities than humans. This is why they are usually applied in situations of high mathematical complexity that have no analytical solutions. This is the reason why simulations are frequently used in engineering and natural sciences. To open up this approach to other fields of science, simulation software can translate mathematical equations into less abstract symbols and provide a more intuitive and less mathematical representation that is yet rigorously based on mathematical modeling (Berends and Romme 1999).[2]

According to Malcolm (1960) simulation in management science traces back to at least 1950. The first articles on the use of simulation examined material flows like

---

[2] Therefore, pure spreadsheet software does not represent computer software for simulation. Even though some of the software packages provide users with a programmable interface which enables simulation, the major use of spreadsheets lies in the support of calculations without taking into account feedbacks, delay or nonlinearities. Spreadsheet software consequently supports analytical solutions. For that reason, we stress out that only under highly specific circumstances research conducted with spreadsheet software can be called simulation research.

inventory oscillations or production bottlenecks (cf. e.g., Morehouse et al. 1950; Cobham 1954; Jackson 1957). These analyses focused on tangible units that could easily be compared to real material flows and validated against real world data. Almost in parallel, scholars started to focus on simulation of rather intangible units. For instance, Luce et al. (1953) analyze the flow of information in groups, Maffei (1958) uses simulation to reflect management decision rules on bounded rationality, Forrester (1958) simulates the flows of material and information in an industrial organization and Newell and Simon (1961) simulate human thinking. Hence, almost with the emergence of the first programmable computer in the 1940s, scholars used simulation to analyze a variety of aspects in management science, i.e. tangible and intangible factors. It is therefore surprising that such a long history of using simulation in management science did not lead to a more widespread use of this research method. This fact is supported by Harrison et al. (2007) who show that only a few articles using simulation-based research were published in top ranked social science journals. The authors account less than 8 % of the articles published in the period of 1994–2003 to use simulation and suspect that a reason may be that simulation is not well understood as a research method in management science.

13.3.1 *Roadmap for Using Simulation as Research Method*

With the purpose of clarifying when and how to use simulation as a research method, Davis et al. (2007) develop a roadmap for using simulation in management science. Their roadmap consists of seven steps which guide simulation users through a theory development process. Though it is suggested to apply the roadmap chronically, it may be required to iterate to steps carried out before. The seven steps of the roadmap are summarized in Table 13.1.

As can be inferred from Table 13.1, the first step addresses the research question. The research question should be embedded in existing literature and represent a theoretically intriguing research question. Thus, the definition of the research question shows deep understanding of existent theories, studies and analysis (Davis et al. 2007). In the second step of the process, Davis et al. (2007) refer to the identification of simple theory. They define simple theory as "undeveloped theory that involves a few constructs and related propositions with some empirical or analytical grounding but that is limited by weak conceptualization, few propositions, and/or rough underlying theoretical logic" (ibid., p. 484). The purpose of including simple theory into studies that use simulation as research method is therefore to strengthen the significance of the study. Existing theories, studies and analyses support the modeling process because they provide assumptions and logic and show that the assumptions of the resulting simulation model are not just based on the modeler's knowledge but are more widely accepted.

The third step concerns the selection of an appropriate simulation approach. For a proper use of simulation, it is crucial to understand that the research question defines which simulation approach to choose. Simulation approaches usually

Table 13.1 Roadmap for developing theory using simulation methods (Davis et al. 2007)

Step	Rationale
Begin with a research question	Focuses efforts on a theoretically relevant issue for which simulation is especially effective
Identify simple theory	Forms basis of computational representation by giving shape to theoretical logic, propositions, constructs, and assumptions. Focuses efforts on theoretical development for which simulation is especially effective
Choose a simulation approach	Ensures that the research uses an appropriate simulation approach given the research at hand
Create computational representation	Embodies theory in software, provides construct validity, improves internal validity by requiring precise constructs, logic, and assumptions and sets the stage for theoretical contributions
Verify computational representation	Confirms accuracy and robustness of computational representation and confirms internal validity of the theory
Experiment to build novel theory	Focuses experimentation on theory development and builds new theory through exploration, elaboration, and extension of simple theory
Validate with empirical data	Strengthens external validity of the theory

include theoretical logic and assumptions and therefore give a framework and direction of how to conduct research using a certain simulation approach (ibid.). The authors of the roadmap present five distinct simulation approaches and show differences in the focus, in key assumptions and theoretical logic (cf. Davis et al. 2007). In this article, we will present two major simulation approaches (system dynamics and agent-based modeling). However, we want to stress out that other simulation approaches do exist and that the simulation approach should be carefully chosen according to the problem under investigation when conducting simulation based research.

Closely related with the choice of the simulation approach is the creation of a computational representation, which is step 4 according to Table 13.1. The computational representation involves operationalization, building algorithms and specifying the assumptions (ibid.), i.e. step 4 stands for the creation of the simulation model. The operationalization includes a precise definition of the computational measures. The precise definition avoids errors or noisy measurement. Further, building algorithms concerns the translation of theoretical logic into the coded language of the simulation approach. Last, the specification of assumptions relates to boundary and scope condition definitions (ibid.). This step also links to the identification of simple theory (step 2) since the theoretical constructs should be coherent with existing literature and extant theories.

Step 5 alludes to the topic of verification of the computational representation. Within this verification process, researchers compare simulation outcomes with the theories underlying the computational representation. The simulation results should confirm the propositions of the theories and show an inherent logic when applying extreme values, sensitivity analysis and robustness checks. In case of mismatch, the simulation model may be erroneous or, in some cases, may reveal shortcomings of the underlying theories (ibid.).

The next step addresses experimentation. Experimentation includes varying values of the constructs, varying assumptions and boundaries, breaking down the simulation model into components or more detailed constructs and adding new structures or components (ibid.). By experimenting with the model, researchers may identify new theoretical relationships and novel theoretical logic.

Validation of the findings is the final step. The findings of step 6 should be compared with empirical data. Davis et al. (2007) argue that the extent of this validation process depends on the source of theories used within the research process. If the extant literature is based mainly on empirical data, then the validation process can be considered less important than if the literature used is primarily theoretically derived.

13.3.2 The Prerequisites, Advantages and Challenges of Simulations

The roadmap above provides guidelines for conducting research using simulation in management science. For a better understanding, however, when to use simulation and what results to expect, it may be important to describe the prerequisites, advantages and disadvantages of simulation. In this section, we present the three main prerequisites of using simulation. We then present the most important advantages of simulation and then discuss the challenges.

13.3.2.1 Prerequisites

Before using simulation as research method, the prerequisites should be met. Simulation requires domain-specific knowledge, as well as methodological knowledge. As mentioned in step 1 and 2 of the roadmap, researchers should acquire thorough understanding of the subject under investigation, i.e. domain-specific knowledge, in order to define the aim of the study and thereby the aim of the simulation model. Simulation models require a clear formulation and definition of measurements and relationships (Cohen 1960; Davis et al. 2007). The required precision leads to an exact formulation of the problem under investigation, the relationship between variables and the problem of how to measure either variables or indicators of them. Most measures and formulations derive from existing literature. Therefore, domain-specific knowledge is a very important prerequisite for following the process described in the roadmap. This could even include nontangible variables influencing the system (Shubik 1960). Furthermore, the precise formulation of the problem and the system lead to the exclusion of variables when boundaries are defined (Davis et al. 2007) and therefore clarifies which variables are included in and which variables are excluded from the simulation model. A change of these defined boundaries may cause differences in the outcome of the simulation (ibid.). Therefore, a thorough analysis of existing literature and theories plays a key role in the beginning of the research process.

By precise formulation of the relationships within the structure of the simulation model and by the definition of measurements, the simulation model provides a transparent representation of the underlying real world structure (Sterman 2000; Davis et al. 2007). This clear logic and transparency is also called white box modeling (Sterman 2000) since simulation models focus on direct and indirect relationships rather than correlational relations (Davis et al. 2007). The simulation model consequently provides users with the underlying theories used in the model and enhances therewith the general understanding of the problem simulated (Schultz 1974). It requires, however, that modelers and users know the programming language as well as have some level of knowledge about the theories included in the model (Davis et al. 2007).

Furthermore, domain-specific knowledge includes the assessment of the level of aggregation for the problem under investigation. The level of aggregation plays an important role because it predefines the granularity or level of detail included in the model. Usually, a more aggregated analysis contains fewer details than a very explicit analysis. Furthermore, agents can be allocated on different levels: it is possible to formulate aggregated models to analyze whole economies or rather to model the interaction of firms within a country, as well as the interaction of individual humans in a team. As soon as an initial simulation model exists, it is possible to alter the level of aggregation (Cohen 1960; Shubik 1960; Cohen and Cyert 1961). In addition, simulation models provide the possibility to analyze the response of the whole system (Davis et al. 2007), as well as the change of different parts of the simulation model. However, before experimenting (step 6), researchers need to conceptualize and start with an initial simulation model, which includes the precise formulation of the variables and relationships. Only in subsequent steps, alteration is possible.

Domain-specific knowledge also includes data collection and parameter estimation (Cohen 1960). Simulation models are mathematical models and therefore data is needed in order to run the simulation. Furthermore, real world data can then be used for validation (ibid.). There is, however, the claim of some researchers using simulation that exact data is not needed since the underlying relationships define the behavior of the simulation model (Sterman 2000). This claim is justified within the field of some simulation methods but does not abandon the need of some data as reference or estimations of values for the simulation.

Additionally, a simulation model requires assumptions about variables, structure and boundaries (Davis et al. 2007). Thus, domain-specific knowledge will provide facts to derive assumptions. Empirical and theoretical studies often reveal correlations and significant effects of certain variables, possible structures and relationships.

The second prerequisite of simulations aside of the domain-specific knowledge explained above is methodological knowledge. Methodological knowledge consists of knowledge of (1) how to formulate a simulation model, (2) how to experiment with a simulation model, and (3) how to interpret the outcomes of a simulation model. As can be inferred, methodological knowledge is especially important for steps 3–7 of the roadmap described in Table 13.1. Methodological knowledge differs from domain-specific knowledge because it concerns the operationalization but not the numerical and theoretical content of the simulation model. It is, however, closely connected to domain-specific knowledge.

First, the question of how to formulate a simulation model refers to the mathematical and symbolic representation of the content of a simulation model. Simulation methods differ in the use of mathematics: some mainly use integral equations, while others are based on differential equations. Furthermore, the mathematical basis also defines whether nonlinear relationships can be formulated or whether only linear relationships can be included. Different methods also use different symbols for different mathematical constructs. In addition, the method also defines whether the simulation model investigates correlations or causalities. Thus, after the formulation of an appropriate research question, it is important to know which simulation method to use in order to achieve the aim of the study.

Second, the question of how to experiment with a simulation model addresses method-specific experimentation techniques. Davis et al. (2007) claim that experimentation is at the heart of simulation. However, they do not explicitly state the importance of the methodological knowledge needed for experimenting. Since different simulation methods offer different mathematical representations, it is not given that every defined variable can be altered in value or in the underlying equation. Thus, the choice of simulation methods may limit the scope of experimentation. Moreover, researchers need to understand which experimentation is possible. Several experimentation techniques can be identified. They include extreme values, Monte Carlo Simulations, sensitivity analysis, structural changes, time series comparisons, etc. The different experimentation techniques reveal different aspects of the simulation. Furthermore, experimentation also requires the transfer to reality: How can changes in a value or in the structure be operationalized in real world?

Third, the question of how to interpret the outcomes of a simulation models refers to the ability to draw conclusions and to reflect the simulation model regarding its technical boundaries. Since simulation methods offer a variety of mathematical representations, they also include technical boundaries or assumptions. These boundaries and assumptions limit therefore the outcomes of the simulation model and need to be taken into account when drawing conclusions. Aside of reflection on theoretical or numerical assumptions, which is based on domain-specific knowledge, the reflection on technical boundaries reveals characteristics of the simulation context and the design of the experiments and scenarios used in the simulation model.

In summary, the prerequisites of simulation are similar to prerequisites of all scientific research methods: domain-specific and methodological knowledge is needed and mutually obligatory. Domain-specific knowledge is needed to define an appropriate research question, to justify assumptions and hypothesis. Methodological knowledge is needed to properly apply a method and to know if a method can be used to answer the research question as well as if a method is consistent with the assumptions and hypothesis of the study.

13.3.2.2 Advantages

Many researchers pointed out the advantages and risks of computer simulation (Forrester 1958; Cohen 1960; Shubik 1960; Cohen and Cyert 1961; Schultz 1974; Sterman 2000; Davis et al. 2007). The following summary of the advantages shows that simulations (1) can cope with complexity, (2) enable flexibility, (3) can be used for testing, (4) compress time, (5) possess locational independence, (6) support communication, and (7) lead to structured data.

Simulation models can cope with complexity. The computational power of simulation models exceeds human computational capacity and provides therefore a support for formulation of models that are more realistic than static or equilibrium models. Economic models are often criticized of not being realistic since they build on linear assumptions (Schultz 1974). Simulation models allow researchers to include more complex mathematical relationships. Those relationships may include nonlinearities, time delays and feedback (Sterman 2000; Davis et al. 2007). The inclusion of those relationships, on the one hand, makes analytical solutions impossible, but on the other hand, it makes simulation models more realistic.

Simulation models provide a flexible approach, especially in situations in which empirical data is hard to acquire. A fully formulated simulation model can easily be altered in structure (Cohen 1960; Davis et al. 2007). Moreover, assumed values for parameters can be manipulated in different directions (Shubik 1960) and differences in the outcomes of the simulation can then be analyzed. By changing either structure, i.e. the relationships within the model, or by changing the assumed values, "What if"-questions can be addressed, and alternative strategies and policies can be formulated according to the outcomes of the model (Schultz 1974). Thus, simulation models represent computational laboratories (Davis et al. 2007) that support the experimentation with different values.

The advantage of experimentation is not only used to explore "What if"-questions, but also to test real world data. This is a distinct fact because it does not concern the exploration of new alternatives but tests existing ones (Schultz 1974). Beliefs about theories and future development of the system modeled can easily be reproduced by the simulation model and conclusions regarding the likelihood of occurrence can be evaluated.

Both previously mentioned advantages – flexibility and real world data testing – base on the advantage of time compression (Cohen 1960; Shubik 1960). Simulation models provide the researcher with the ability to compress time in the experimental setting. In contrast to empirical studies, simulation models can generate data for longitudinal studies (Davis et al. 2007) within seconds. In addition, data patterns and dynamics are made visible through variation in time (Cohen and Cyert 1961). Especially, the change of intangible variables is made transparent (Cohen 1960). By compressing time, simulation models offer a cost efficient alternative to empirical studies.

Furthermore, simulations are locational independent. The precise formulation of mathematical relationships enables a simulation set-up in a computer and thus specific habits, environmental features or other influences are defined within the simulation model. The simulation can consequently be carried out anywhere on any computer and does not depend on the location of the real system it models (Davis et al. 2007).

Simulation models are also tools of communication. The graphical user interfaces of advanced simulation software packages provide researchers with a non-mathematical language. This offers to simulation users a transparency of the underlying theories used (Sterman 2000). Further, simulation output is often presented in graphs or other visual modes of display other than only numbers. This may be of advantage for collaboration between researchers with different methodological backgrounds, i.e. quantitative and qualitative methods or researchers with strong mathematical backgrounds and those who do not need such a high skill level in mathematics (Cohen 1960). Moreover, simulation software offers an additional research language. Cohen and Cyert (1961) distinguish between verbal descriptions, pictorial geometry and formal mathematics as languages of economic research. They argue that simulation models provide an additional one by introducing a mixture of verbal descriptions, symbols and mathematics.

Shubik (1960) also mentions the advantage of structuring data. In empirical studies it is often the case that some data cannot be measured and that the amount of data collected needs a considerable amount of time for structuring and ordering to process the data as needed. Simulation models will output data in a given structure and therefore avoid long processing times for data representation.

13.3.2.3 Challenges

However, simulation does not only bear advantages. Researchers mainly mention four different weaknesses of simulation: The primary challenge is validation (Davis et al. 2007). Closely related to the issue of validation are the potential weaknesses of parameter estimation (Cohen 1960), and precision in formulation (Cohen and Cyert 1961) as well as the need for assumptions (Davis et al. 2007).

Firstly, the process of validation needs close attention. Thereby both internal and external validity have to be met. Internal validity requires simulation models to consist of a coherent logic and a structure that is either based on existing theories or at least supported by empirical studies. Given that the model's structure is precisely formulated and the model's boundaries are well defined, internal validity is usually given (Davis et al. 2007). The internal validity is therefore especially important when combining simulation with empirical research, since internal validity is a major concern in empirical research (ibid.). External validity, on the contrary, refers to the comparison of the simulation output with empirical data, i.e. the goodness of fit of historical data and the expected future development. Thus, real world data is needed to validate the simulation model's output. In some cases, the collection of this data may be a major problem (Cohen 1960). As mentioned above, precise formulation of the simulation model may require the formulation of intangible variables. Often however, intangible variables cannot be measured directly in the real system but are measured by proxies instead. Consequently, the first issue about external validity is that appropriate data is needed.

Further, external validity is a primary source of weakness since a high goodness of fit is by definition difficult to achieve. Simulation models shall represent reality but nevertheless are still models, i.e. simplified representations of reality that reduce complexity. The reduction of complexity in turn calls for parameter estimation, precision in formulation, and the need for assumptions, which are further challenges mentioned above. The precise formulation of the simulation model requires the definition of boundaries and the focus of the simulation model on the problem and key aspects of the analysis. Drawing boundaries may therefore lead to a behavior that is different from real world data, which in turn lowers the goodness of fit (Davis et al. 2007). Furthermore, the precise definition and formulation is a problem by itself since the relationships between variables need to be specified (Cohen and Cyert 1961). This specification requires knowledge about the physics of reality or assumptions about the relationships are needed (Davis et al. 2007).

The challenge of external validity seems therefore a major threat for simulation-based research. Models may easily fail to give precise forecasts. However, most economic models are also only successful in making propositions about directional change rather than numerical magnitudes (Cohen and Cyert 1961). Therefore, other research methods experience the same disadvantage. In contrast, the precise formulation of a simulation model gives internal validity which is an additional challenge in other research approaches, especially empirical methods (Davis et al. 2007).

To conclude, the application of simulation methods requires the same scope of knowledge as other research methods: domain-specific knowledge as well as methodological knowledge. Simulation methods also experience similar challenges as other research methods: external validity. In contrast to other methods, using simulation as research method offers internal validity, the capability to cope with complexity, flexibility, testing, compressed time, locational independence, communication, and structured data.

However, these advantages do not mean that simulation methods are always superior to other research methods. Simulation has its limits, too. Especially in purely theoretical research including only intangible non-quantifiable variables, simulation cannot be easily applied. Further, empirical data can reveal more data than simulation, since simulation is precisely defined. Empirical data represents real world data and therefore includes fewer assumptions and limits. In addition, simulation methods also need time to be applied. In some case, mathematical equations or pure verbal reasoning may lead to the same insights in less time.

Therefore, in line with other researchers, we advocate simulation methods as an additional tool to conduct research (Cohen 1960). Simulation can bridge empirical and theoretical research because it builds on both of them to create internal validity and it complements empirical work (Cohen and Cyert 1961).

13.4 Social Interactions in Simulations

In the following, we present how simulation is used to investigate outcomes of social interactions. As stated above, social interactions are multifold and subject to different research approaches. Step 3 of the roadmap in Table 13.1 highlights the

Table 13.2 Summary of studies using simulation methods as research method

Study	Object of investigation	Research question	Use of simulation for	Simulation method
Forrester (1958)	The interaction of retailer, distributor and factory by orders, inventory level and production	What causes the oscillation in production, inventory and orders within a supply chain?	Explaining an empirical phenomenon	System dynamics
Meadows et al. (1972)	Economic growth and human welfare	Will economic growth persist under the constraint of limited natural resources?	Scenario analysis for complex longitudinal global data	System dynamics
Kreidler and Tilebein (2013)	Innovativeness of new product development teams	What is the effect of diversity on the performance of new product development teams?	Testing two theories of intangible variables	System dynamics
Schelling (1971)	The racial segregation of cities and the evolution of areas of racial majorities	How does racial and segregation in cities emerge?	Explaining an empirical phenomenon	Agent-based modeling
(Axelrod 1997b)	The emergence of cooperation strategies of organizations and social norms	Can cooperation emerge even though there is an incentive for non-cooperative behavior?	Explaining an empirical phenomenon	Agent-based modeling
Tilebein and Stolarski (2009)	Information processing in diverse top management teams	What is the effect of time on the relationship of information processing and diversity of top management teams?	Analyzing an effect of intangible variables	Agent-based modeling

importance of choosing an appropriate simulation method. Previous research showed the differences in simulation methods (Berends and Romme 1999; Davis et al. 2007; Harrison et al. 2007). We focus on two major simulation methods (system dynamics and agent-based modeling) and their application to social interactions. We especially distinguish two levels of social interaction: Interaction between individuals and a more aggregated perspective on interactions of several agents. In the following, we present system dynamics and agent-based modeling and a number of studies that used the respective simulation method. We intend to show why simulation was especially useful in these studies. Table 13.2 presents a summary of the studies described.

13.4.1 System Dynamics

System dynamics is a simulation method that focuses on causal relationships (Forrester 1958; Sterman 2000). The aim of system dynamics models is to analyze how these causal relationships, i.e. the structure of the system, determine behavior. Hence, system dynamics focuses on the behavior of the system rather than on the behavior of the single agents or parts of the system (Harrison et al. 2007). System dynamics therefore represents a more aggregated perspective. Usually, system dynamics models consist of circular relationships, so called feedback loops, which basically means that a variable A influences a second variable B and at the same time, variable B influences variable A (Davis et al. 2007). While it is possible to understand single feedback mechanisms, it is impossible to infer the dynamic behavior caused by the interrelationship of several feedback loops (Forrester 1958; Sterman 2000; Davis et al. 2007).

System dynamics was developed by Jay Wright Forrester (1958) and is one of the oldest approaches to analyze non-linear systems in management science (Simon et al. 2008). The approach is heavily influenced by systems theory and cybernetics (Harrison et al. 2007; Simon et al. 2008). The coding of the system dynamics method consists of different variables. There are state variables, called "stocks", which accumulate over time (Sterman 2000). They therefore serve as buffers, delays, and summarize values of the time steps before (Sterman 2000; Davis et al. 2007). Stocks are influenced by rate variables, called "flows". Flows link consequential stocks and consist of functional equations (Simon et al. 2008). Stock and flow constructions represent flows of material or information which is physically conserved. That means that within the constructs of stocks and flows, units are usually kept constant. Another kind of variables is called auxiliary. Auxiliaries are linked to the rest of the system by arrows. The arrows represent information links. Material and information is usually transformed by auxiliaries. Extant studies analyzing social interactions with system dynamics include the interaction by orders of different agents within a supply chain (Forrester 1958), interaction of the world population with limited natural resources and pollution (Meadows et al. 1972) and the innovativeness of heterogeneous new product development teams (Kreidler and Tilebein 2013).

Forrester (1958) examines the fluctuations of production and ordering within a supply chain. He focuses on customer demand, retailer, distributor and factory. At the time of the study and still today, matching demand and production is a costly issue because holding inventory and lacking demand fulfillment produce cost. The study is based on the empirical phenomenon of oscillating numbers of orders, production and inventory. Real world data existed for some companies but a whole data set of a whole supply chain is rather hard to collect. Within the study, Forrester focuses on three agents (retailer, distributor and factory). He shows that the orders evoke production and unfulfilled orders evoke an even higher production which than leads to a high inventory and lower production. The interaction between the agents is shown in the orders given by one agent in the supply chain to its supplier. Thus, changes in the customer demand lead to changes in the orders of the retailer to the distributor, who in turn orders to the factory. The delay between orders and production and order fulfillment build up oscillations through the supply chain. Forrester (1958) suggests that faster order handling, better sales data tracking and a gradual inventory adjustment will stabilize the performance of the supply chain. Note, that the aim of the study is to understand the fluctuations of a whole system of a supply chain and to analyze the effect of changes within the system (e.g. by adding a more efficient sales data tracking).

Meadows et al. (1972) show in their famous study "Limits to Growth" how people exploit natural resources and project the depletion of natural resources on economic growth and human welfare. They use a highly aggregated view and treat the population as one stock, as well as the economic growth, which is shown as capital investments, and pollution. They do not specify their variables under investigation in detail: population (regions of the world), capital investments (machinery, digital performance, education) or pollution (soil, water, earth). However, using scenario analysis, the highly aggregated view on the interrelationships of the exploitation and welfare shows that the constraint of limited resources will eventually cause a decline in welfare. The authors argue that humans will have to invest a considerable amount for battling the constraints. That will lead to less leisure time, to fewer capital held by the population, and to a lower population growth or even a decline in population. Note, that the social interaction in this study is highly aggregated and not the main focus of the study. However, the study is a cornerstone for research on sustainability and social policy analysis (cf. e.g. Hirsch 1995). Meadows et al. (1972) use simulation to show possible future outcomes with computer generated global data which, at the time of publication, was not existent and contains intangible data. Nevertheless, the study provided the basis for a variety of studies about the outcomes of the simulation model.

Kreidler and Tilebein (2013) analyze the contradicting literature on the effect of diversity on new product development team performance. In a thorough literature analysis they show that new product development teams, on the one hand, should be interdisciplinary and cross-functional. The diversity of the team provides more perspectives and hence a wider variety of potential solutions. On the other hand, the authors also identify literature that argues that diversity is a risk because diversity can evoke barriers of communication and diminishes cohesiveness. With the intention of evaluating both theories, Kreidler and Tilebein (2013) created a simple system dynamics simulation model. This simulation models show the amount of problem solutions created and accepted by the new product development team as distinct stock variables. Furthermore, they define "diversity of mental models" as another stock which influences the creation of problem solutions and the acceptance rate of the solutions. The study shows that both contradicting theories may be true since there are two different causal relations. A higher diversity of mental models increases the creation of problem solutions but hinders the acceptance of the created solutions to be accepted within the team. However, the more solutions are accepted, the lower the diversity. In summary, Kreidler and Tilebein (2013) created a simulation model in order to test existing theories and to reveal causal relations of the effect of diverse new product development teams. Moreover, they used the simulation model to create data about the two existent theories.

13.4.2 Agent-Based Modeling

Agent-based modeling originates from research on artificial intelligence (Simon et al. 2008). It is a bottom-up simulation method that focuses on behavior of internally autonomous elements, so called "agents" (Simon et al. 2008; Tilebein and Stolarski 2009). Agents influence one another by interactions which are based on some predefined simple rules and preferences (Harrison et al. 2007; Tilebein and Stolarski 2009). Thus, by these interactions, the simulation method focuses on the individual parts of a system and derives the overall system's state from the sum of all interactions as an emergent phenomenon. Examples include Schelling's segregation model (Schelling 1971), the evolution of strategies for organizations and the evolution of social norms (Axelrod 1984, 1997b), as well as the effect of diversity on information processing (Tilebein and Stolarski 2009).

Schelling (1969, 1971) analyzes the empirical phenomenon of racial and financial segregation in cities in the 1960s. The author defined agents to have a threshold preference, i.e. a threshold value that evokes action. The agents hold the predefined preference that a certain amount of neighbors that share the same characteristic, i.e. the same race, need to live in the spatial area directly surrounding the agent. In case this amount of neighbors is not met, the agent becomes unhappy and will move on to an area where the preference is met. Thus, in case the threshold is met, the agent stays, if not, then the agent moves to another area. Note, that Schelling initially uses penny coins and not computer simulation to analyze the problem under investigation. The coins and space used by Schelling, however, represent the heart of an agent-based model and later studies based on Schelling's paper mainly use agent-based modeling to extend the model (cf. e.g. Macy and Willer 2002). Simulation is used in this study to collect data derived from a simple behavior in a short time period. Schelling finds that clear segregation takes place even if the preference for same race neighbors amounts to only 35 %. Consequently, segregation is a natural emergent outcome of the simple wish that a certain amount of neighbors should share some characteristics with one another.

Axelrod (1997b) investigates the emergence of different cooperative strategies among agents. The study is based on his famous analysis "The evolution of cooperation" (Axelrod 1984) which is based on game theory. The initial study shows that cooperation can evolve even in games like the prisoner's dilemma. The prisoner's dilemma is a two-player game in which each player has the incentive for non-cooperative behavior because it would maximize the agent's utility. However, if both agents do not cooperate, their pay-off will be lower than if they had cooperated. This state represents an equilibrium of the game (Nash equilibrium). Axelrod shows that the repetition of the game and the possibility of a Tit-for-Tat strategy of the other agent may evoke cooperation of both agents. Simply put, the threat of one agent to never cooperate again serves as an incentive for the second agent to cooperate. Axelrod (1997b) transfers the findings to a multitude of agents. He defines agents as either companies or people with certain cultural similarities. The simulation method proves useful since the aim of the study is to understand how cooperative behavior of individuals emerges over time. The author shows that social norms, strategic alliances between companies and even the cooperation of countries in World War II can be explained by this approach.

Tilebein and Stolarski (2009) examine the effect of time on processing information and diversity with agent-based modeling. In contrast to the study above (Kreidler and Tilebein 2013), this study focuses on agreement on strategies taken by top management teams. The authors show that different believes and preferences hinder the advancement within the decision process but that time affects communication and thus adaptive behavior. Within this study the teams are modeled as a set of different agents. A goal is set and the agents aim to reach the goal. However, there are several decision steps to be taken before reaching the goal. Each decision step is therefore a layer and is represented as spatial distance. The decision step consists of different decision alternatives. Hence, each agent faces different possible decisions in order to reach the next decision step. Within the model, the agents move spatially forward. However, in order to take a decision, the agents need to agree to choose a decision. Since all agents have different preferences about which decision to take, communication and adaption is needed in order to derive a decision. This study uses simulation in order to compress time of a longitudinal study for team performance. Further, simulation supports the authors to focus on different preferences, which are usually intangible variables. The simulation focuses on individual adaptive behavior.

13.5 Conclusion

In this paper, we promote simulation as an additional research method aside of theoretical reasoning and empirical studies. We first give an overview about social interactions in management science and show that social interactions are complex, non-linear, simultaneous and multifold. Addressing the methodological question of how to better cope with the complexity of social interactions, their simultaneous sequence and non-linear effects, we present simulation as an appropriate tool. The prerequisites of simulation are domain-specific knowledge as well as methodological knowledge, which correspond to the prerequisites of other research methods. Advantages of the use of simulation for conducting research comprise the ability to (1) cope with complexity, (2) enable flexibility, (3) test theory, (4) compress time, (5) be locationally independent, (6) support communication, and (7) structure data. The major concern of simulation methods is external validity. This major concern corresponds, however, to other research methods as well. Thus, simulation methods represent an additional tool for conducting research. Researchers in management science should turn to simulation methods to solve problems that are based on social interactions and are of complex and dynamic nature.

References

Axelrod, Robert M. 1984. *The evolution of cooperation*. New York: Basic Books.
Axelrod, Robert M. 1997a. Advancing the art of simulation in the social sciences. *Complexity* 3(2): 16–22.
Axelrod, Robert M. 1997b. *The complexity of cooperation: Agent-based models of competition and collaboration*, Princeton studies in complexity. Princeton: Princeton University Press.
Baum, Joel A.C., and Jitendra V. Singh. 1994. Organization-environment coevolution. In *Evolutionary dynamics of organizations*, ed. J.A.C. Baum and J.V. Singh, 379–402. New York: Oxford University Press.
Becker, Gary S. 1974. A theory of social interaction. *Journal of Political Economy* 82(6): 1063–1093.
Berends, Peter, and Georges Romme. 1999. Simulation as a research tool in management studies. *European Management Journal* 17(6): 576–583.
Bratley, Paul, Bennett L. Fox, and Linus E. Schrage. 1987. *A guide to simulation*. New York: Springer.
Cobham, Alan. 1954. Priority assignment in waiting line problems. *Journal of the Operations Research Society of America* 2(1): 70–76.
Cohen, Kalman J. 1960. Simulation of the firm. *The American Economic Review* 50(2): 534–540.
Cohen, Kalman J., and Richard M. Cyert. 1961. Computer models in dynamic economics. *The Quarterly Journal of Economics* 75(1): 112–127.
Davies, Jason P., Kathleen M. Eisenhardt, and Christopher B. Bingham. 2007. Developing theory through simulation methods. *The Academy of Management Review* 32(2): 480–499.
Forrester, Jay W. 1958. Industrial dynamics: A major breakthrough for decision makers. *Harvard Business Review* 36(4): 37–66.
Forrester, Jay W. 1961. *Industrial dynamics*. Cambridge, MA: MIT Press.
Frigerio, Bartolomeo. 1629. *L'economo prudente*. Roma: Appresso Lodovico Grignani.
Granovetter, Mark S. 1973. The strength of weak ties. *American Journal of Sociology* 78(6): 1360–1380.
Güth, Werner, Rolf Schmittberger, and Bernd Schwarze. 1982. An experimental analysis of ultimatum bargaining. *Journal of Economic Behavior and Organization* 3(4): 367–388.
Harrison, J. Richard, Zhiang Lin, Glenn R. Carroll, and Kathleen M. Carley. 2007. Simulation modeling in organizational and management research. *The Academy of Management Review* 32(4): 1229–1245.
Hirsch, Fred. 1995. *Social limits to growth*. London: Routledge.
Jackson, James R. 1957. Simulation research on job shop production. *Naval Research Logistics Quarterly* 4(4): 287–295.
Janis, Irving L. 1972. *Victims of groupthink: A psychological study of foreign-policy decisions and fiascoes*. Oxford: Houghton Mifflin.
Kahneman, Daniel, and Amos Tversky. 1979. Prospect theory: An analysis of decision under risk. *Econometrica* 47(2): 263–292.
Ketokivi, Mikko, and Saku Mantere. 2010. Two strategies for inductive reasoning in organizational research. *Academy of Management Review* 35(2): 315–333.
Korsgaard, M. Audrey, David M. Schweiger, and Harry J. Sapienza. 1995. Building commitment, attachment, and trust in strategic decision-making teams: The role of procedural justice. *The Academy of Management Journal* 38(1): 60–84.
Kreidler, Anja, and Meike Tilebein. 2013. *Diversity and innovativeness in new product development teams: Addressing dynamic aspects with simulation*. 31st international conference of the System Dynamics Society, Cambridge, MA, 21 July 2013.
Law, Averill M., and W. David Kelton. 2000. *Simulation modeling and analysis*, McGraw-Hill series in industrial engineering and management science, 3rd ed. Boston: McGraw-Hill.
Lewin, Arie Y., Chris P. Long, and Timothy N. Carroll. 1999. The coevolution of new organizational forms. *Organization Science* 10(5): 535–550.

Lomi, Alessandro, and Erik R. Larsen. 2001. *Dynamics of organizations. Computational modeling and organizational theories.* Menlo Park: American Association for Artificial Intelligence.

Luce, R. Duncan, Josiah Macy Jr., Lee S. Christie, and D. Harvie Hay. 1953. *Information flow in task-oriented groups.* Research laboratory of electronics, Massachusetts Institute of Technology technical report No. 264.

Macy, Michael W., and Robert Willer. 2002. From factors to actors: Computational sociology and agent-based modeling. *Annual Review of Sociology* 28: 143–166.

Maffei, Richard B. 1958. Simulation, sensitivity, and management decision rules. *The Journal of Business* 31(3): 177–186.

Malcolm, Donald G. 1960. Bibliography on the use of simulation in management analysis. *Operations Research* 8(2): 169–177.

Manski, Charles F. 2000. Economic analysis of social interactions. *Journal of Economic Perspectives* 14(3): 115–136.

McKelvey, Bill. 1997. Quasi-natural organization science. *Organization Science* 8(4): 352–380.

Meadows, Dennis H., Donella L. Meadows, Jorgen Randers, and William W. Behrens III. 1972. *The limits to growth: A report for the club of Rome's project on the predicament of mankind*, Potomac associates books. New York: Universe Books.

Mill, John S. 1848. *Principles of political economy with some of their applications to social philosophy*, 1909 edn.

Morehouse, N. Frank, Robert H. Strotz, and S.J. Horwitz. 1950. An electro-analog method for investigating problems in economic dynamics: Inventory oscillations. *Econometrica* 18(4): 313–328.

Newell, Allen, and Herbert A. Simon. 1961. Computer simulation of human thinking. *Science* 134(3495): 2011–2017.

Pareto, Vilfredo. 1906. *Manual of political economy.* Oxford: 2014 Reprint. Montesano, Aldo, Alberto Zanni, Luigino Bruni, John S. Chipman, Michael McLure. Oxford University Press.

Persky, Joseph. 1995. Retrospectives: The ethology of homo economicus. *Journal of Economic Perspectives* 9(2): 221–231.

Porter, Terry B. 2006. Coevolution as a research framework for organizations and the natural environment. *Organization & Environment* 19(4): 479–504.

Ricardo, David. 1821. *The principles of political economy and taxation.*

Samuelson, Paul A. 1937. A note on measurement of utility. *The Review of Economic Studies* 1(2): 155–161.

Schelling, Thomas C. 1969. Models of segregation. *The American Economic Review* 59(2): 488–493.

Schelling, Thomas C. 1971. Dynamics models of segregation. *Journal of Mathematical Sociology* 1(2): 143–186.

Schultz, Randall L. 1974. The use of simulation for decision making. *Behavioral Science* 19(5): 344–350.

Schumpeter, Joseph A. 1954. *History of economic analysis.* London: Allen and Unwin.

Shubik, Martin. 1960. Simulation of the industry and the firm. *The American Economic Review* 50(5): 908–919.

Simon, Herbert A. 1955. A behavioral model of rational choice. *The Quarterly Journal of Economics* 69(1): 99–118.

Simon, Herbert A. 1957. *Administrative behavior: A study of decision-making processes in administrative organizations.* New York: The Free Press.

Simon, Herbert A. 1972. Theories of bounded rationality. In *Decision and organization*, ed. C.B. McGuire and R. Radner, 161–176. Amsterdam: North-Holland Publishing Company.

Simon, Henrik, Sven Meyer, and Meike Tilebein. 2008. *Bounded rationality in management research: Computational approaches to model the coevolution of organizations and their environments.* International Federation of Scholarly Associations of Management (IFSAM) 9th world congress, Shanghai, 26 July 2008.

Sterman, John D. 2000. *Business dynamics: Systems thinking and modeling for a complex world.* Boston: Irwin/McGraw-Hill.

Thaler, Richard H. 1988. Anomalies: The ultimatum game. *The Journal of Economic Perspectives* 2(4): 195–206.

Tilebein, Meike, and Vera Stolarski. 2009. *The contribution of diversity to successful R&D processes*, Wien, 21 June 2009.

Tversky, Amos. 1972. Elimination by aspects: A theory of choice. *Psychological Review* 79(4): 281–299.

Tversky, Amos, and Daniel Kahneman. 1981. The framing of decisions and the psychology of choice. *Science* 211(4481): 453–458.

Varian, Hal R. 2010. *Intermediate microeconomics. A modern approach*, 8th ed. New York: W.W. Norton & Co.

von Neumann, John, and Oskar Morgenstern. 1944. *Theory of games and economic behavior*. Princeton: Princeton University Press.

Chapter 14
How Models Fail

A Critical Look at the History of Computer Simulations of the Evolution of Cooperation

Eckhart Arnold

14.1 Introduction

Simulation models of the Reiterated Prisoner's Dilemma (in the following: RPD-models) are since 30 years considered as one of the standard tools to study the evolution of cooperation (Rangoni 2013) (Hoffmann 2000). A considerable number of such simulation models has been produced by scientists. Unfortunately, though, none of these models has empirically been verified and there exists no example of empirical research where any of the RPD-models has successfully been employed to a particular instance of cooperation. Surprisingly, this has not kept scientists from continuing to produce simulation models in the same tradition and from writing their own history as a history of success. In a recent simulation study – which does not make use of the RPD but otherwise follows the same pattern of research – Robert Axelrod's (Axelrod 1984) original role model for this kind of simulation studies is praised as "an extremely effective means for investigating the evolution of cooperation" and considered as "widely credited with invigorating that field" (Rendell et al 2010b, 208–209).

According to a very widespread philosophy of science that is usually associated with the name of Karl Popper (1971) science is distinguished from non-science by its empirical testability and right theories from wrong theories by their actual empirical success. Probably, most scientists in the field of social simulations would even agree to this philosophy of science at least in its general outlines.[1] However, RPD

---

[1] A referee pointed out to me that there is a tension in my paper between the reliance on a Popperian falsificationism and the implicit use of Kuhn's paradigm concept. However, both can be reconciled if the former is understood in a normative and the latter in a descriptive sense. Popper's falsificationism requires, though, that paradigms are not completely incommensurable. But then, there are

E. Arnold (✉)
Digital Humanities Department, Bavarian Academy of Sciences, Munich, Germany
e-mail: arnold@badw.de

models of the evolution of cooperation have not been empirically successful. So, how come that they are still considered as valuable?

In this paper I am going to examine the question, why the continuous lack of empirical success did not lead the scientists working with these simulation models to reconsider their approach. In the first part I explain what RPD-models of the evolution of cooperation are about. I show that these models failed to produce empirically applicable and tenable results. This will be done by referring to research reports and meta-studies, none of which comes up with an example of successful empirical application.

In the second part of the paper, I highlight a few example cases that show why these models fail. In this context I examine the framing narratives with which scientists justify their method. Such framing narratives form an integral part of any scientific enterprise. My point is not to criticize simulation scientists for employing narratives to justify their method, but I believe that the typical framing narratives that RPD modelers in the tradition of Axelrod employ of are badly founded and I show that in each case there are good arguments against accepting the narrative.

In the third part of this paper I take this analysis one step further by discussing typical arguments with which scientists justify the production and use of unvalidated "theoretical" simulations. Most of the arguments discussed here do usually not form the central topic of scientific papers. Rather, they appear in the less formal communication of scientists, in oral discussions, in small talk, eventually in keynote addresses (Epstein 2008). One may object that if these arguments are never explicitly spelled out, they may not be worth discussing. After all they have never been cast into their strongest imaginable form. Why discuss dinner table talk, anyway? But then, it is often this kind of communication where the deeper convictions of a community are expressed. And it is by no means true that these convictions are without effect on the scientific judgments of the community members. Quite to the contrary, general agreement with the underlying convictions is silently presupposed by one's scientific peers and adherence to them is usually taken for granted by supervisors from their PhD students and often expected by referees from the authors of the papers they review. Therefore, the informal side-talk of science should not at all be exempt from rational criticism.

In the last part of the paper, I relate my criticism to similar discussions in a neighboring (if not overlapping) science, namely political science. It seems that there exist structural similarities in the way scientific schools or research traditions deal with failures of their paradigm. Rather than admitting such a fundamental failure (which, as it touches one's own scientific world view, is obviously much harder than admitting the failure of a particular research enterprise within a paradigm) they retreat by adjusting their goals. In the worst case they become so modest in their achievements (which they, by an equal adjustment of their self-perception, continue to celebrate as successes) that they reach the verge of irrelevance. Green and Shapiro

---

many good reasons that speak against a strong reading of the incommensurability-thesis, anyway. (See the very enlightening remarks about Kuhn and Duhem-Quine in the case study by Zacharias (2013, 11ff., 305ff.).)

(1994, 44f.) have described this process of clandestine retreat for the case of rational choice theory in political science.

14.2 The Empirical Failure of Simulations of the Evolution of Cooperation

14.2.1 Axelrod's Evolution of Cooperation

One of the most important initiators of the research on the RPD-model was Robert Axelrod. The publication of his book "The Evolution of Cooperation" popularized the simulation approach to studying the evolution of cooperation. At the core of Axelrod's simulation lies the two person's Prisoner's Dilemma game. The two person's Prisoner's Dilemma is a game, where two players are asked to contribute to the production of a public good. Each player can choose to either contribute, that is, to cooperate, or not to contribute, that is to defect. If both players cooperate they both receive a reasonably high payoff. If neither player cooperates, they both receive a low payoff. If one player tries to cooperate while the other player defects, the player who tried to cooperate receives zero payoff, while the successful cheater receives the highest possible payoff in the game, which at the same time is more than the cooperative payoff. Since, no matter what the other player does, it is always more advantageous for each individual player not to cooperate. Therefore, both players, if they are rational egoists, end up with the low non-cooperation payoff – at least as long as the game is not repeated. The reiterated Prisoner's Dilemma (RPD), in which the same players play through a sequence of Prisoner's Dilemmas, changes the situation, because defecting players can be punished with non-cooperation in the following rounds.

It can be shown that in the reiterated Prisoner's Dilemma there is no single best strategy. In order to find out if there exist certain strategies that are by and large more successful than other strategies and whether there are certain characteristics that successful strategies share, Robert Axelrod (1984) conducted a computer tournament with different strategies. Axelrod also fed the results of the tournament simulation into a population dynamical simulation, where more successful strategies would gradually out-compete less successful strategies in a quasi-evolutionary race. Famously, TIT FOR TAT emerged as the winner in Axelrod's tournament.[2]

The way Axelrod employed his model as a research tool was by running simulations and then generalizing from the results he obtained. These included recommendations such as that TIT FOR TAT usually is good choice for a strategy or that a strategy should not defect unmotivated itself, but should punish defections and should also be forgiving, etc. As subsequent research revealed, however, almost

---

[2] More detailed descriptions of the RPD-model and Axelrod's tournament can be found in Axelrod (1984), Binmore (1994, 1998) or Arnold (2008).

none of these conclusions was in fact generalizable (see Arnold (2013, 106ff., 126f.) with further references). For each of them there exist variations of the RPD-model where it does not hold and where following Axelrod's recommendations could be a bad mistake. The only exception are Axelrod's results about the collective stability of TIT FOR TAT, which he proved mathematically. The central flaw of Axelrod's research design is that it relies strongly on impressionistic conclusions and inductive generalizations from what are in fact contingent simulation results. This deficiency of Axelrod's model has convincingly been criticized by Binmore (1998, 313ff.). To give just one example: In Axelrod's tournament TIT FOR TAT won in two subsequent rounds. Axelrod concluded that TIT FOR TAT is a good strategy and that it is advisable to be forgiving. However, if one chooses the set of all two-state automata as strategy set – which is a reasonable choice because it contains all strategies up to a certain complexity level – then the unforgiving strategy GRIM emerges as the winner (Binmore 1994, 295ff.).

Axelrod's followers would usually be much more cautious about drawing general conclusions from simulations but they did not completely refrain from generalizing. In the ensuing research a historical pattern emerged where researchers would pick up existing models, investigate variants of these models, and eventually demonstrate that the previous results could not be generalized (Schüßler's and Arnold's simulations, which are discussed below, are examples for this pattern). Thus, Axelrod's research design became – despite its great deficiencies – a role model for simulation studies until today. As a justification for publishing yet another model, it would usually suffice to relate to the previous research. No reference to empirical research or just empirical applicability would be considered necessary. For example, Rangoni (2013) introduces his study of a variant of the RPD-model by mentioning that "Axelrod's work on the prisoner's dilemma is one of the most discussed models of social cooperation" and declaring "After more than thirty years from the publication of its early results, Axelrod's prisoner's dilemma tournament remains a cornerstone of evolutionary explanation of social cooperation", although – as will be discussed in the following – this "cornerstone of evolutionary explanation" has not been confirmed empirically in a single instance.

14.2.2 *The Empirical Failure of the RPD-Model*

Axelrod himself was confident that simulation studies like his yield knowledge that can be applied in the context of empirical application. In his book "The Evolution of Cooperation" (Axelrod 1984) he provided two case studies. One of these concerned biology. It was of highly speculative character as Axelrod honestly admitted and it has indeed never been confirmed since. Therefore, I am not going to discuss this particular case study here. Further biological research on the evolution of cooperation will briefly be outlined below.

The other one of Axelrod's case studies was a highly dramatic case study concerning the Live and Let Live System which emerged on some stretches of the deadlocked western front between enemy soldiers in World War One. However, as acknowledged by Axelrod his case study relies entirely on the prior historiographic work by Tony Ashworth (1980). Based on an extensive study of the historical sources, Ashworth had crafted a careful and highly differentiated explanation for the emergence, sustainment and eventual breakdown of the Live and Let Live on the western front. Axelrod's recasting of this story in game theoretical terms has nothing to add in terms of explanatory power, because the RPD-model is far too simple to account for the complicated network of causes for the Live and Let Live that Ashworth study had revealed (Arnold 2008, 180ff.). Even among game theorists it was disputed, whether there existed any straight-forward way to interpret the situation as a Prisoner's Dilemma at all (Schüßler 1997, 33ff.). Thus, if any particular scientific approach is to be credited with the successful explanation of the Live and Let Live in World War One, then it is not game theoretical modeling or computer simulations but the well-established methods of traditional historiography. Interestingly, though, this dramatic case-study did a lot to increase the popularity of Axelrod's simulation approach.

If Axelrod's attempts to apply his model to empirical case studies weren't particularly successful, then subsequent research could still demonstrate that the empirical application of these models is possible. The most noteworthy attempt to apply the RPD empirically was undertaken by Manfred Milinski, who sought to explain the seemingly cooperative behavior that shoal fishes show when inspecting a predator (1987). This paper is quoted time and again when it comes to giving an example for the empirical applicability of the RPD model. For example, Hoffman maintains in a research report about Axelrod's RPD framework that "This general framework is applicable to a host of realistic scenarios both in the social and natural worlds (e.g. Milinski 1987)." (Hoffmann 2000, 4.3). Milinski's 1987-paper, however, remains the sole example for the "host of realistic scenarios" to which this framework is supposedly applicable. The same paper by Milinski is quoted in Osborne's "Introduction to Game Theory" as an example for the empirical applicability of game theory (Osborne 2004, 445). Unfortunately, it was already by the late 1990s clear that Milinski's explanation of the predator inspection behavior did not work (Dugatkin 1997, 1998). The reason is that it is not possible to obtain the necessary empirical data to either confirm or disconfirm the RPD model in the case of the predator inspection behavior of sticklebacks. This is also more or less the conclusion at which Milinski and Parker arrive in a joint paper on the same topic that they published 10 years after the initial study by Milinski (Milinski and Parker 1997, 1245).

In a broad meta-study on the research on "Cooperation among Animals" Lee Allan Dugatkin (1997) does not find a single instance of animal cooperation where any of the many variants of the RPD model (Dugatkin lists more than two dozens of them in the beginning of his study) can successfully be applied. He summarizes the situation in a very thoughtful article as follows: "Despite the fact that game theory has a long standing tradition in the social sciences, and was incorporated in behavioral ecology 20 years ago, controlled tests of game theory models of cooperation are still relatively rare. It might be argued that this is not the fault of the empiricists, but rather due to the fact that much of the theory developed is unconnected to natural systems and thus may be mathematically intriguing but biologically meaningless" (Dugatkin and Reeve 1998, 57). The same frustration about empirically ungrounded model research is expressed by Peter Hammerstein: "Why is there such a discrepancy between theory and facts? A look at the best known examples of reciprocity shows that simple models of repeated games do not properly reflect the natural circumstances under which evolution takes place. Most repeated animal interactions do not even correspond to repeated games." Hammerstein (2003). It is safe to say that there exist no successful empirical application cases for the RPD in biology. But the fact that the modeling community still entertains the believe that there are such successful application cases, if not "a host of" them, clearly demonstrates how little, in fact, the community occupies itself with empirical matters.

14.3 Justificatory Narratives

If the simulations studies in this research tradition do not bear any explanatory value for empirical research, then the question naturally arises what they are good for. Some authors present explanations that are meant to justify the method. I will go through some of them before entering on the discussion of the general arguments in favor for the simulation method.

14.3.1 Axelrod's Narrative

Axelrod motivated the use of the Prisoner's Dilemma mostly by the fact that it already was an extremely popular game theoretical model that had already been used in experimental economic research. He compares the Prisoner's Dilemma to the E. Coli in biological research. Comparisons to the ever successful natural sciences are quite typical for the justificatory discourse of the modeling approaches in the social sciences. With the benefit of hindsight it can, however, be said that this comparison was slightly misleading. E. Coli is a great object of study in biology, because what one learns when studying E. Coli can often directly be transferred to other bacteria. Many bacteria are similar to E. Coli in important respects. The same is unfortunately not true for the RPD model, which is not at all a robust model (Arnold 2013, 127f.). Change the parameters of the simulation, the initial set of participating strategies or other aspects of the model only a bit and you can get qualitatively different results. Most likely, another strategy than TIT FOR TAT would turn out as winner, and maybe not even a friendly or cooperative strategy (Binmore 1994, 315).

One part of Axelrod's motivation is also a supposed advantage of the simulation approach to experimental approaches. Axelrod relates to the notorious problem of economic experimental research that the laboratory setting is usually highly artificial and that, therefore, any obtained results cannot easily be transferred to real life situations. He omits to mention, however, that computer simulations based on highly stylized models like the RPD share the same problem.

14.3.2 Schüßler's Narrative

Several years after Axelrod, Rudolf Schüßler (1997) published a book with game theoretical simulations. One part of this book directly relates to Axelrod. This part of Schüßler's book follows the pattern: Pick a well-known simulation, change the settings or other details of this simulations, produce "surprising" results and publish. If Axelrod had demonstrated with his simulation that the shadow of the future is crucial for the evolution of cooperation, Schüßler demonstrates with a modified simulations that this does not need to mean that the same partners must expect to meet again and again in order to sustain cooperation. In Schüßler's simulation cooperators succeed although cheaters can decide to break off the interaction at any time, thus avoiding punishment.[3]

Given Axelrod's previous simulations and conjectures this can appear surprising. But what is surprising? That a different simulation produces different results is prima facie anything but surprising. Given the almost complete modeling freedom – remember, there are no empirical constraints to be honored – and the volatility of the original model it would be surprising if no surprises could be produced. So why should we be interested in the results of another arbitrary simulation?

At this point Schüßlers narrative steps in. As Schüßler (1997, 91) writes "One of the central, classical assumptions of the normativistic sociology says that in an exchange society of rational egoists no stable cooperation can emerge (see Durkheim 1977; Parsons 1949). Alleged proofs for this thesis try to show that already simple analytical considerations suffice to draw this conclusion. The present simulation should be able to shake this firm conviction."[4] One may wonder whether this means that the simulation serves more than a purely didactic purpose. But be that as it may. It is in any case questionable whether the premises are correct. Do normativistic sociologists really rely on simple analytical considerations? Sociologists like Durkheim usually argue on the basis of thick narratives supported by empirical research. Highly abstract computer simulations like Rudolf Schüßler's simulations

---

[3] The details of this simulation are described in Schüßler (1997, 61ff.) and in a simpler form in Arnold (2008, 291ff.). For the curious: Schüßler achieves his effect, because the non-cooperators that break off the interaction are forced to pick a new partner from a pool that mostly contains non-cooperators from which it is impossible to rip a high payoff.

[4] This is my translation. The German original reads: "Eine der zentralen, klassischen Annahmen der normativistischen Soziologie besagt, daß in einer Austauschgesellschaft rationaler Egoisten keine stabilen Kooperationsverhältnisse entstehen können (vgl. Durkheim 1977, Parsons 1949). Angebliche Nachweise für diese These versuchen zu zeigen, daß bereits einfache, analystische Überlegungen zu diesem Schluß ausreichen. Die vorliegende Simulation sollte geeignet sein, diese Sicherheit zu erschüttern." (Schüßler, 1997, 91)

can at best prove logical possibilities. However, it is unlikely that this kind of discourse is vulnerable to proofs of logical possibilities. After all, a normativistic sociologist can easily claim that the seeming possibility of rational egoists to cooperate is an artifact of the simulations that strips away all concrete features of human nature, especially those of a psychological kind which make cooperation of egoists impossible in reality (Arnold 2013, 128ff.). (Generally, proofs of logical possibilities cannot disprove real impossibilities; e.g. a perpetuum mobile is logically possible but impossible in reality, because it contradicts the laws of nature. See Arnold (2013) for a detailed discussion of the category of logical possibility.)

Schüßler, who seems to be quite aware of the weaknesses of his argument, follows up with the remark that ultimately it is up to the scientist to decide whether this is sufficient or not (Schüßler 1997, 91). But as we have seen, proofs of logical possibility are simply not sufficient. And then again, it is an indefeasible claim that scientific knowledge is objective and that its validity is independent from the opinions and discretion of any particular person. If it were up to the discretion of the scientist to decide whether some theory or model is sufficient to decide a scientific question, we would not call that science any more.

It is noteworthy that Schüßler criticizes Axelrod quite strongly in the beginning of his book (Schüßler 1997, 33ff.), but then presents computer simulations of exactly the same brand as Axelrod's simulations. The same kind of performative self-contradiction is even more obvious in the following example.

14.3.3 The Story of "Slip Stream Altruism"

Although RPD simulations already fell out of fashion, I have myself published a book with RPD simulations as late as 2008. I felt uneasy about it at the time of writing the book and today I am even more convinced that the scientific method that I describe (but also criticize) in this book is fundamentally flawed. But the book was my PhD-thesis and I was not really given the free choice of topic – which is, of course, a widespread grievance of PhD-theses. So, I figured that the best I could make out of this situation was to follow the established pattern of research in this field, but also to examine it from an epistemological point of view and point out its deficiencies. The research pattern is that of producing a variant of an existing simulation model, finding "interesting" results and embedding them in a narrative that makes them appear "new", "surprising" or at least somehow noteworthy.

In the series of population dynamical simulations of the RPD that I conducted, there are quite a few simulations where naive cooperators, i.e. strategies that cooperate but other than TIT FOR TAT do not retaliate when the partner fails to reciprocate, can still survive with a low share of the population or – even more "surprising" – come out on top, i.e. with larger population share than even the retaliating cooperators (Arnold 2008, 109ff.). I used the term "slip stream altruism" as a catch phrase to describe this phenomenon, because the simulations prove the logical possibility that unconditional altruism (which some moralists consider to be the only form of altruism that deserves its name) can develop in the "slip stream" of tough, reciprocating strategies.

But is this phenomenon really surprising and did we really need a series of computer simulations to get the idea? As mentioned earlier, with unrestricted modeling freedom and a volatile base model like the RPD, one is liable to find all kinds of phenomena. There are not really any surprises. And just as in Schüßler's case there is a simple explanation for the phenomenon: Unconditional cooperators can come out on top, if the conditional cooperators that drive the non-cooperators to extinction are badly coordinated so that they inadvertently hurt each other (Arnold 2008, 113). So, the phenomenon that my simulation series yields acquires the appearance of being interesting, surprising or relevant mostly by the narrative and the rhetoric of "slip stream altruism" in which it is embedded.

I never took the story of slip stream altruism very seriously and, as I said earlier, I was already convinced that the simulation method as practiced by Axelrod and his followers leads to nothing at the time when I wrote the book down. (See, for example, my talk at the Models & Simulations in Paris 2006, several years before I wrote the book (Arnold 2006).) Given how strongly I criticize Axelrod-style simulations in the book, it may appear odd to the readers that I even bothered to conduct computer simulations of the same brand and describe them in the book. This was a tribute that I had to pay to the circumstances, however. Somewhat to my distress I later found that some readers liked the simulation series much better than my criticism of the method (Schurz 2011, 344, 356). Others, however, have understood that the main purpose of the book is a critical one (Zollman 2009).

14.3.4 The Social Learning Strategies Tournament

The last example of a justificatory narrative does not concern the RPD model, but a simulation enterprise that is similar in spirit to Axelrod's. The authors of this study explicitly refer to Axelrod for the justification of their approach (Rendell et al 2010b, 208–209). The model at the basis of the "Social Learning Strategies tournament" is a 100-armed bandit model (Rendell et al 2010a). Just like the RPD it is a highly stylized and very sparse model: The model assumes an environment with 100 cells representing foraging opportunities. The payoff from foraging is distributed exponentially: few high payoffs, many low or even zero payoffs. In each round of the game the players can choose between three possible moves: INNOVATE where they receive information about the payoff opportunity in a randomly picked cell; EXPLOIT where players forage one of their known cells to receive a payoff; OBSERVE where a player receives slightly imprecise information about the foraging opportunities that other players are exploiting. Arbitrarily many players can occupy one cell. The resources never expire, but the environment changes over time so that the players' information about good foraging opportunities gets outdated after a while. The payoffs drive a population dynamical model where players live and die and are replaced by new players depending on the success of the existing players.

The most important result of the tournament was that – under the conditions of this specific model – the best strategies relied almost entirely on social learning, i.e. playing OBSERVE. It almost did not make any sense at all to play INNOVATE.[5] Other than that the ratio between OBSERVE moves and EXPLOIT moves was crucial to success. Too few OBSERVE moves would lead to sticking with poor payoffs. Too many OBSERVE moves would mean that payoffs would not be gathered often enough which results in a lower average payoff. Finally, the right estimate of expected payoffs was important. The winning strategy and the second best strategy used the same probabilistic standard formula to estimate the expected payoff values (Rendell et al 2010b, 211).

The authors themselves make every effort to present their findings as a sort of scientific novelty. For that purpose they employ a framing narrative that links their model with an important research question, prior research and successful (or believed to be successful) past role models. The broader research question, mentioned in the beginning of the paper, to which the model is related is how cultural learning has contributed to the success of humans as a species: "Cultural processes facilitate the spread of adaptive knowledge, accumulated over generations, allowing individuals to acquire vital life skills. One of the foundations of culture is social learning,…" (Rendell et al 2010b, 208). Surely, this is a worthwhile scientific question.

As to the prior research they refer to theoretical studies. These, however, only "have explored a small number of plausible learning strategies" (Rendell et al 2010b). Therefore, the tournament was conducted which gathers a contingent but large selection of strategies. The tournament's results are then described as "surprising results, given that the error-prone nature of social learning is widely thought to be a weakness of this form of learning … These findings are particularly unexpected in the light of previous theoretical analyzes …, virtually all of which have posited some structural cost to asocial learning and errors in social learning." (Rendell et al 2010b, 212).

Thus, the results of the tournament constitute a novelty, even a surprising novelty. The surprising character of the results is strongly underlined by the authors of the study: "The most important outcome of the tournament is the remarkable success of strategies that rely heavily on copying when learning in spite of the absence of a structural cost to asocial learning, an observation evocative of human culture. This outcome was not anticipated by the tournament organizers, nor by the committee of experts established to oversee the tournament, nor, judging by the high variance in reliance on social learning …, by most of the tournament entrants." (Rendell et al 2010b, 212) Again, however, it is not surprising, but to be expected that one reaches results that differ form previous research if one uses a different model.

---

[5] This was partly due to an inadvertency in the design of the model, where OBSERVE moves could – due to random errors – serve much the same function as INNOVATE moves. The authors of the study did, however, verify that their results are not just due to this particular effect Rendell et al (2010a).

Axelrod's tournament plays an important role as historical paragon in the framing narrative: "The organization of similar tournaments by Robert Axelrod in the 1980s proved an extremely effective means for investigating the evolution of cooperation and is widely credited with invigorating that field."(Rendell et al 2010b, 208). But as mentioned earlier, the general conclusions that Axelrod drew from his tournament had already turned out not to be tenable and the research tradition he initiated did not really yield any empirically applicable simulation models. Nonetheless, the author's seem to consider it as an advantage that: "Axelrod's cooperation tournaments were based on a widely accepted theoretical framework for the study of cooperation: the Prisoner's Dilemma." (Rendell et al 2010b, 209). However, the wide acceptance of the Prisoner's Dilemma model says more about fashions in science than about the explanatory power of this model. Although not as widely accepted as the Prisoner's Dilemma, the authors are confident that "the basic generality of the multi-armed bandit problem we posed lends confidence that the insights derived from the tournament may be quite general." Rendell et al (2010b). But the generality of the problem does not guarantee that the conclusions are generalizable beyond the particular model that was used to describe the problem. Quite the contrary, the highly stylized and abstract character of the model raises doubts whether it will be applicable without ambiguity in many empirical instances. The generality of the model does not imply – nor should it, as I believe, lend any confidence in that direction to the cautious scientist – that it is of general relevance for the explanation of empirical instances of social and asocial learning. This simply remains to be seen. If anything at all then it is its robustness with respect to changes of the parameter values that lends some confidence in the applicability of the tournament's results. Robustness is of course only one of several necessary prerequisites for the empirical applicability of a model.

Summing it up, it is mostly in virtue of its framing narrative that the tournament's results appear as a novel, important or surprising theoretical achievement. If one follows the line of argument given here, however, then the model – being hardly empirically grounded and not at all empirically validated – represents just one among many other possible ways of modeling social learning. In this respect it is merely another grain of dust in the inexhaustible space of logical possibilities.

14.4 Discussion of Standard Arguments for Modeling

While the narratives discussed so far could be traced to their specific sources in the papers and books in which they appear, the following standard arguments for the supposed superiority of the simulation approach to studying the "evolution of cooperation" or for the use of formal models crop up in discussions and the less formal forms of scientific communication, but not so often in scientific papers. I have heard all of these arguments in discussions about the RPD simulation model more than once, but I cannot easily trace them back to printed sources. As I explained in the introduction, these arguments seem to me none the less to represent an attitude that effects the scientific work. Therefore, I believe that they deserve discussion.

14.4.1 Argument 1: Our Knowledge Is Limited, Anyway

Argument Our ability to gain knowledge is limited in the social sciences, anyway. Therefore, we have to be content with the kind of computer simulations we can make, even if they are not sufficient to generate empirical explanations.

Response No one says that we have to use computer simulations in the social sciences. If computer simulations do not work, other methods may still work. As explained earlier, the "Live and Let Live" in World War One cannot really be explained by RPD models, but historiographic methods still work perfectly well in this case.

Even if there exist no alternative methods, we should not accept the existing methods no matter how bad they are. The use of a particular scientific method is justified only, if the results it yields are better than mere speculation and by and large as good as or better than what can be achieved with alternative methods.

More generally, we should not mistake the failure of a paradigm – say, agent-based simulations or RPD-simulations of cooperation or rational choice theory or sociobiology – for the failure of a science. It is only from the keyhole perspective of the strict adherents to one particular paradigm that the limits of the paradigm appear as the limits of the science or of human cognition as such. In this respect the argument resembles the strategy of silent retreat to false modesty mentioned in the introduction. While it is laudable for a scientist to be modest about one's own claims of knowledge, scientific modesty becomes inappropriate when it gives up any claim of generating empirically falsifiable knowledge.

14.4.2 Argument 2: One Can Always Learn Something from Failure

Argument Even if Axelrod's approach ultimately turned out to be a failure, we can still learn important lessons from it. Failure is at least as important for the progress of science as success.

Response Unfortunately, it is not clear, whether the necessary lessons have already been learned. If Axelrod's computer tournament is still remembered as an "extremely effective means for investigating the evolution of cooperation" (Rendell et al 2010b, 208) by the scientific community then it seems that the lessons have not been learned. And even if the lessons have been learned (by some) then the many dozens of inapplicable simulations that have kept scientists busy in the aftermath of Axelrod's book have surely been a rather long detour.

14.4.3 Argument 3: Models Always Rely on Simplification

Argument Models, by their very definition, rely on simplifications of reality. If a model wouldn't simplify it would be useless as a model. After all, the best map of a landscape would be the landscape itself, but then it would be useless as a map. (A typical example is Zollman (2009) who relies on this argument in his criticism of mine. See also Green and Shapiro (1994, 191) who discuss a similar argument in the context of rational choice theory.)

Response On the other hand it is obvious that there must be some limit to how strongly a model may simplify reality. For otherwise any model could be a model for anything. So, where is the borderline between legitimate simplification and illegitimate oversimplification? A possible answer could be that a model is not oversimplified as long as it captures with sufficient precision all causally relevant factors of the modeled phenomenon with respect to a specific research question, i.e. all factors that are liable to determine the outcome of this question. In all other cases we should be very careful to trust an explanation based on that model alone.

At this point two replies are common: (1) That no one claims such an explanatory power for his or her own models. But then, what is the point of modeling, if models do not help us to explain anything? (2) That the research question did not require that all causally relevant factors have to be captured by one and the same model. However, if a model concentrates only on some causal factors, then these must at least be discernible empirically. Unfortunately, this is often not possible and certainly not with most RPD models. (See also Arnold (2014, 367f.).)

As far as RPD-simulations are concerned it appears clear to me that these are far too simplified to be acceptable representations of reality. One could object that they help us to understand the mechanism of reciprocal altruism as such. This is already one step back from claiming that RPD-models are an effective means for investigating the evolution of cooperation, because now it is merely claimed that they are illustrating a mechanism. However, for this purpose a single model would be sufficient. One does not need dozens of them. Plus, how and why reciprocal altruism works in principle has perfectly well been conceptualized by Robert Trivers (1971) many years earlier with a single simple equation.

14.4.4 Argument 4: No Alternatives to Modeling

Argument There is no real alternative modeling, anyway. If you try to do without models, merely relying on verbal explanations, you are just making use of implicit models that are never fully articulated. Surely, explicit modeling is better than relying on implicit models. Without models nothing could be explained. (See also Epstein (2008), who employs a variant of this argument.)

Response It is at least for the time being (the distant future of science may of course prove me wrong) practically impossible to express everything that can be expressed verbally in mathematical terms or with formal logic. This includes many of the causal connections that we are interested in when doing social sciences. Otherwise, how come that among the many books published about the causes, course and consequences of the First World War these days, there is no game theoretical or otherwise model-based study that could rival the conventional historical treatments? Otherwise, how come that lawyers, attorneys and judges – their job being to a large part one of logical reasoning, as one should think – do not use formal logic to express the legal connections they ponder over?

14.4.5 Argument 5: Modeling Promotes a Scientific Habit of Mind

Argument "To me, however, the most important contribution of the modeling enterprise – as distinct from any particular model, or modeling technique – is that it enforces a scientific habit of mind, which I would characterize as one of militant ignorance – an iron commitment to 'I don't know.' That is, all scientific knowledge is uncertain, contingent, subject to revision, and falsifiable in principle. (This, of course, does not mean readily falsified. It means that one can in principle specify observations that, if made, would falsify it). One does not base beliefs on authority, but ultimately on evidence. This, of course, is a very dangerous idea. It levels the playing field, and permits the lowliest peasant to challenge the most exalted ruler – obviously an intolerable risk." (Epstein 2008, 1.16)

Response Unfortunately, the modeling tradition discussed in this paper failed completely with respect to all the virtues that Epstein naively believes to be virtues promoted by modeling: It did not readily submit its results to empirical falsification. Where the few and far between attempts of empirical application failed, it did not learn from failure. The commitment to "I do not know" becomes a joke if modelers do not dare to come up with concrete empirical explanations or predictions any more. And as far as authority goes, the appeal to "scientific authority" in more or less subtle forms is a common rhetoric device in the modeler's discourse. (See also Moses and Knutsen (2012, 157), Green and Shapiro (1994, 195) and argument 7 below).

Generally, the scientific habit of mind does not at all depend on the use of models. Also, secondary virtues like clarity, explicitness and the like are by no means a prerogative of modelers. Computer simulation studies in particular can become dangerously unclear if the source code is not published or not well structured or not well commented.

14.4.6 Argument 6: Division of Labor in Science Exempts Theoreticians from Empirical Work

Argument There exists division of labor in science. Model builders are not responsible for the empirical application of their models, but they are mere suppliers. If the empirical scientists fail to test or otherwise make use of models, it is not the modelers that should be blamed.

Response But modelers need to take into account the conditions and restrictions that empirical research imposes, otherwise they run the danger of producing models that can never, not even under the most favorable circumstances, be applied empirically. In the case of the Axelrod-tradition it is clearly the modelers that must take the blame, because they failed to learn from the failures of early attempts at empirical application like Milinski's (1987). And they never worried about the restrictions under which empirical work struggles in the potential application fields of their models.

Now, one might say that this is also true for much of mathematics, and still mathematics has often proven to applicable, even in cases that no one had guessed before. But surely it is not a good research strategy to rely on later to come historical coincidences of science. Plus, there is an important difference between mathematics and models. Mathematics deals with general structures, while simulation-models like the RPD represent particular example cases (comparable to a concrete calculation in mathematics). From a technical point of view most models in the Axelrod tradition remain fairly trivial, while mathematics could – if worst comes to worst – still be justified by its high intellectual level which allows to ascribe an innate value to it.

14.4.7 Argument 7: Success Within the Scientific Community Proves Scientific Validity

Argument The scientific value of computer simulations in the social sciences cannot be disputed. There is a growing number of research projects, journals, institutes that is dedicated to social simulations. (Variants of this argument are: This book has been quoted so many times, it cannot be all wrong! Or, this article has been published in *Science*, the authors surely know what they are doing. See also Green and Shapiro (1994, 195), who discuss a similar argument.)

Response The scientific value of a method, theory, model or simulation is to be judged exclusively on the basis of its scientific merits, i.e. logical reasoning and empirical evidence, and not at all on the basis of its social success. As far as computer simulations are concerned, a survey by Heath et al (2009) on agent-based simulations revealed that the empirical validation of computer simulations is still badly lacking.

There is one grain of truth in this argument. For those questions, about which one does not know enough to judge the scientific arguments it is best to rely on the judgment of the socially approved specialists. But social success can never be used as an argument within a scientific dispute. After all, it is just the question whether the social success of a theory, model or paradigm was deserved from a scientific point of view.

14.4.8 Concluding Remarks

None of the arguments discussed above appear to be particularly pervasive in the first place. Never the less I believe they are worth being discussed, because – like the previously described narratives – they help to keep the spirits of the scientists up even in face of apparent failure. Just like social prejudices they need to be made explicit to be overcome.

14.5 History Repeats Itself: Comparison with Similar Criticisms in Neighboring Fields

Although this paper was mostly dedicated to the case of RPD-simulations of the evolution of cooperation, much of the criticism uttered here does not only concern this specific research tradition. In some points it overlaps with like-minded criticism of model oriented or "naturalistic" approaches in the social sciences. In this last part, I'd like to point out some of these overlaps.

In a fundamental, though still constructive criticism Green and Shapiro (1994) have described what they call the "Pathologies of Rational Choice Theory". The idea that people are by and large rational actors is in itself not necessarily connected to using mathematical models or simulations. But many of the pathologies that Green and Shapiro describe seem to be tied to a particular complex of ontological and methodological convictions lying at the base of the rational choice creed. Among these is a strong commitment to mathematical methods, which are prima facie considered to be more scientific than other methods. What is of interest in this context is what happens when these convictions are frustrated, which they must be, if on the basis of these convictions it is not possible to generate that amount of solid and empirically supported scientific results that had been promised and expected. Will the adherents of the school start to weaken or revise their fundamental convictions? Green and Shapiro (1994, 33ff.) found out that, rather then doing this, adherents of the school applied about any immunization strategy imaginable to protect their theoretical commitment. These strategies ranged form post-hoc theory development over projecting evidence from theory or searching exclusively for confirming evidence to arbitrary domain restrictions. The latter is of particular interest here, because it suggests a historical pattern that is analogous to the one observed in the history of the evolution of cooperation and which I have described as a retreat to false modesty.

According to Green and Shapiro (1994, 45) scientifically legitimate domain restriction is distinguished from arbitrary domain restriction by "specifying the relevant domain in advance by reference to limiting conditions", rather than "specifying as the relevant domain: 'wherever the theory seems to work'". This problem has – according to their analysis – been particularly acute in the so called "paradox of voter turnout", which consists in the fact that people vote at political elections even though the individual influence on the result is so marginal that any cost, even that of leaving the house for voting, should exceed the expected benefit. Now, rational choice theorists have never advanced any convincing explanation for this alleged paradox. Rather, they moved from the question of why people vote to much less ambitious explanations for turnout rate changes (Green and Shapiro 1994, 59). And even here they did not manage to advance more than quite unoriginal hypotheses concerning, for example, the relation between education and the inclination to vote.

In two respects this resembles my results about the scientific tradition of the evolution of cooperation. First of all with regards to the triviality of the results that the simulation-based approach produced in its later stage (like my "slip stream altruism"-story quoted above). Secondly, with respect to the stepping down from great scientific promises to such humble results. Had Axelrod believed that his simulation models have considerable explanatory power, many of his later followers (e.g. Schüßler) were so careful not to promise too much that one wonders what the simulation method is good for in the context of finding explanations for cooperative behavior, anyway. These coincidences between rational choice theory and RPD-simulations are not surprising, if one assumes that they represent typical immunization strategies of failing paradigms. One difference should be mentioned, though. In the case of rational choice it was largely an empirical failure of the theory, while in the case of the "evolution of cooperation" its was already the failure not to compare the models to empirical research.

Another connection can be pointed out between the criticism launched here and a more recent criticism of the naturalistic paradigm in the political sciences as part of the textbook on competing methodologies in social and political research by Moses and Knutsen (2012, 145–168). Moses and Knutson describe and (modestly) criticize the interconnected complex of ontological and methodological beliefs that makes up the naturalistic paradigm. This complex is composed of elements which are not unlike those that I have discussed as arguments and narratives in the two previous sections. One important element of these is the play with an assumed scientific authority (Moses and Knutsen 2012, 157ff.). Given the many imponderables that surround any theory in the social sciences, including those that profess to employ strictly scientific methods like formal models, Moses and Knutsen come to a similar result as I have: Namely, that this kind of professed scientism is largely a bluff.

References

Arnold, Eckhart. 2006. The dark side of the force. When computer simulations lead us astray and model think narrows our imagination. In *Pre conference draft for the models and simulations conference*, Paris, 12–14 June 2006. http://www.eckhartarnold.de/papers/2006_simulations/Simulations_preconference.html. Accessed 15 Sept 2014.

Arnold, Eckhart. 2008. *Explaining altruism. A simulation-based approach and its limits*. Heusenstamm: Ontos Verlag.

Arnold, Eckhart. 2013. Simulation models of the evolution of cooperation as proofs of logical possibilities. How useful are they? *Etica & Politica/Ethics & Politics* 15(2): 101–138.

Arnold, Eckhart. 2014. What's wrong with social simulations? *The Monist* 97(3): 361–379.

Ashworth, Tony. 1980. *Trench warfare 1914–1918. The live and let live system*. London: MacMillan Press Ltd.

Axelrod, Robert. 1984. *The evolution of cooperation*. New York: Basic Books.

Binmore, Ken. 1994. *Game theory and the social contract. I. Playing fair*. Cambridge, MA/London: MIT Press.

Binmore, Ken. 1998. *Game theory and the social contract. II. Just playing*. Cambridge, MA/London: MIT Press.

Dugatkin, Lee Alan. 1997. *Cooperation among animals*. New York: Oxford University Press.

Dugatkin, Lee Alan. 1998. Game theory and cooperation. In *Game theory and animal behavior*, ed. Dugatkin Lee Alan and Reeve Hudson Kern, 38–63. New York/Oxford: Oxford University Press.

Dugatkin, Lee Alan, and Kern Reeve Hudson. 1998. *Game theory and animal behavior*. New York/Oxford: Oxford University Press.

Durkheim, Émile. 1977. *Über die Teilung der sozialen Arbeit*. Frankfurt am Main: Suhrkamp.

Epstein, Joshua M. 2008. Why model? http://www.santafe.edu/research/publications/workingpapers/08-09-040.pdf. Accessed 15 Sept 2014. (Based on the author's 2008 Bastille Day keynote address to the Second World Congress on Social Simulation, George Mason University, and earlier addresses at the Institute of Medicine, the University of Michigan, and the Santa Fe Institute.)

Green, Donald P., and Ian Shapiro. 1994. *Pathologies of rational choice theory. A critique of applications in political science*. New Haven/London: Yale University Press.

Hammerstein, Peter. 2003. Why is reciprocity so rare in social animals? A protestant appeal. In *Genetic and cultural evolution*, ed. Hammerstein Peter, 83–94. Cambridge, MA/London: MIT Press.

Heath, Brian, Raymond Hill, and Frank Ciarallo. 2009. A survey of agent-based modeling practices (January 1998 to July 2008). *Journal of Artificial Societies and Social Simulation* 12(4):9. http://jasss.soc.surrey.ac.uk/12/4/9.html. Accessed 15 Sept 2014.

Hoffmann, Robert. 2000. Twenty years on: The evolution of cooperation revisited. *Journal of Artificial Societies and Social Simulation* 3(2). http://jasss.soc.surrey.ac.uk/3/2/forum/1.html. Accessed 15 Sept 2014.

Milinski, Manfred. 1987. Tit for tat in sticklebacks and the evolution of cooperation. *Nature* 325: 433–435.

Milinski, Manfred, and Geoffrey A. Parker. 1997. Cooperation under predation risk: A data-based ESS analysis. *Proceedings of the Royal Society* 264: 1239–1247.

Moses, Jonthon W., and Torbjørn L. Knutsen. 2012. *Ways of knowing. Competing methodologies in social and political research*. London: Palgrave MacMillan.

Osborne, Martin J. 2004. *An introduction to game theory*. New York: Oxford University Press.

Parsons, Talcott. 1949. *The structure of social action*. Glencoe: Free Press, first published: 1937 edition.

Popper, Karl R. 1971. *Logik der Forschung*. Tübingen: Mohr.

Rangoni, Ruggero. 2013. Heterogeneous strategy learning in the iterated prisoner's dilemma. *Etica & Politica/Ethics & Politics* 15(2): 42–57.

Rendell, Luke, R. Boyd, D. Cownden, M. Enquist, K. Eriksson, M.W. Feldman, L. Fogarty, S. Ghirlanda, T. Lillicrap, and Kevin N. Laland. 2010a. Why copy others? Insights from the social learning strategies tournament. *Science* 328: 208–213.

Rendell, Luke, R. Boyd, D. Cownden, M. Enquist, K. Eriksson, M.W. Feldman, L. Fogarty, S. Ghirlanda, T. Lillicrap, and Kevin N. Laland. 2010b. Supporting online material for: Why copy others? Insights from the social learning strategies tournament. *Science* 328: 2–53.

Schurz, Gerhard. 2011. *Evolution in Natur und Kultur*. Heidelberg: Spektrum Akademischer Verlag.

Schüßler, Rudolf. 1997. *Kooperation unter Egoisten: Vier Dilemmata*. München: R. Oldenbourg Verlag.

Trivers, Robert L. 1971. The evolution of reciprocal altruism. *The Quarterly Review of Biology* 46: 35–57.

Zacharias, Sebastian. 2013. *The Darwin revolution as a knowledge reorganisation. A historical-epistemological analysis and a reception analysis based on a novel model of scientific theories*. Ph.D. thesis, Max-Planck-Institute for the History of Science, Berlin.

Zollman, Kevin. 2009. Review of Eckhart Arnold, explaining altruism: A simulation-based approach and its limits. *Notre Dame Philosophical Reviews* 3. https://ndpr.nd.edu/news/23961-explaining-altruism-a-simulation-based-approach-and-its-limits. Accessed 15 Sept 2014.

Chapter 15
Artificial Intelligence and Pro-Social Behaviour

Joanna J. Bryson

15.1 Introduction

Collective agency is not a discrete characteristic of a system, but rather a spectrum condition. Individuals composing a collective must invest some resources in maintaining themselves as well as some in maintaining the collective's goals and structures. The question of how much to invest at which level of organisation is a complex one, for which there may be many viable solutions. For example, one might consider a married parent to be a member of three families—their parents', their partners' parents', and the new one they have created with their partner; a citizen of a village, state and country; an employee of at least one organisation, in which they may also be members of either orthogonal or nested teams; and a member of various other voluntary organisations. Some individuals will seek situations with more or fewer such memberships of collectives. Nevertheless, all of us constantly make choices—not always explicit—about how much attention and effort to devote to influencing the behaviour of each collective of which we are members.

Artificial intelligence is ordinarily seen as something quite separate from all the complexity of human social arrangements (Gunkel 2012). We picture AI as also having agency, like a human, then generally dismiss this vision as not possible, or at least not present. Such dismissal of AI is a mistake. *Intelligent* is not a synonym for *human*. Intelligence is just one attribute of humans, many other animals, and even plants (Trewavas 2005). In itself, intelligence does not determine personhood, nor is it sufficient for moral subjectivity. It is neither necessary nor sufficient for the autonomy that underlies moral agency. Mathematics is normally considered to require intelligence (Skemp 1961), yet calculators prove that arithmetic and geometry at

---

J.J. Bryson (✉)
Department of Computer Science, University of Bath,
Bath, UK
e-mail: jjb@alum.mit.edu

least can be conducted without a capacity for setting autonomous goals. Plants can autonomously pursue goals, and change their behaviour in response to their environment, but plants are not considered either moral patients deserving protection,[1] nor moral agents responsible for their actions. Therefore, artificial intelligence does not imply any sort of agency. Rather, like any other artefact, AI could be seen as an extension of human agency (Bryson and Kime 2011).

The purpose of this chapter is to examine how technology, particularly AI, is changing human collective behaviour and therefore both our collective and our individual agency. My intention is to be primarily descriptive, but there is of course a normative subtext, which I will attempt to make as explicit as possible. This primarily imposes on the final section of the chapter, Sect. 15.5, and results in some policy recommendations. This chapter's principal normative motivation is that society should better understand itself, so that it can better choose goals for the regulation and governance of AI, privacy and personal data. This is because by using our data, AI can generate predictions of our behaviour, which increases the utility of and propensity for investment at the collective level. These increases can result in changes not only to our societies, but as a consequence to the experience and meaning of being an individual.

In the following sections, I first further describe intelligence and the current state of AI. Next I describe current scientific understanding concerning why humanity is in its unique situation of knowing and therefore having responsibility, and how this relates to our tendency for collective and pro-social action. I will then describe a series of scientific results, some from social simulations, demonstrating the ease with which pro-sociality can evolve, and which elucidate the limits to which we and other species can and should invest in the collective. Finally, I close by using the models from the earlier sections to project consequences of the advances in AI on human culture and human collectives. These predictions will be based simply on extending my description of intelligence and AI to the models of social investment and examining their consequences.

15.2 Intelligence and the State of AI

To draw conclusions concerning the consequences of intelligence, we first need to define the term. For the purposes of this chapter, I will not attempt to capture its ordinary language meaning, but rather will introduce a simple, clear-cut computational definition of the term, which also relates to its characterisation in biology. Intelligence is capacity to

1. express an appropriate action,
2. in real time, and
3. in response to a perceived environment.

---

[1] Except where plants are seen as either a part of a broader ecosystem, or as a possession of a human.

Each of these components must be explained in turn. *Expressing an action* is necessary for intelligence to be judged—we will not consider any 'inner life' that is not demonstrated through some action, though action may include communication. *Appropriate* implies some goal, so any intelligent system has some metric by which its performance is judged. For biological systems, this is generally something related to survival.[2] For AI, we the makers define the goals. So for a calculator, it is sufficient to respond to button presses without noticing weather events. *In real time* is not a theoretical requirement of intelligence, but rather indicates that I am limiting my consideration to what also might be called cognitive systems (Vernon et al. 2007). It means that the agent exists in a dynamic environment, and can express action quickly enough that that action is generally still appropriate. "Generally" because of course very intelligent systems occasionally have traffic accidents—intelligence is not all-or-nothing, but rather varies in extent. I include the real-time requirement to focus on competences that find an appropriate action according to an agent's own sensing. This is to discriminate from processes like evolution or other abstract mathematical algorithms which may contribute to intelligence but do not produce a direct action outcome. Finally, *in response to a perceived environment* eliminates from consideration objects that act the same way at all times and just happen to sometimes be in an appropriate place and time when they do so. It also emphasises the importance of sensing to intelligence. Intelligence is judged by its actions as they relate to a context; the ability to perceive and discriminate contexts is therefore critical to intelligence.

Collective agency is not necessarily collective intelligence. Agency implies the capacity to be the author of environmental change. This change can be effected by a collective whether or not the authorship or motivation was achieved in a fully distributed way, as we might expect in collective intelligence (Williams Woolley et al. 2010). While intelligence originates change, that change can be effected by other agents that are not the original motivated entity. A captain may determine a team's strategy, a gardener may determine which wall an ivy will cover. On the other hand, observable collective intelligence is necessarily a form of collective agency. A swarm of insects may choose a new hive location (Marshall et al. 2009); a company may sue for changes in law enforcement (Rosenbaum 2014).

An Internet search is a highly intelligent process, requiring enormous capacity for perception—the perceptual ability to categorise billions of web pages based on a context set by search terms, and the action competence to serve one of these billions to your screen. But the agent responsible for the act, and that (principally[3]) benefits from that act is the human that requests the search. Here the expressed action of the individual user couldn't have been achieved without intelligence

---

[2] The arguments in this chapter hinge on inclusive fitness (Hamilton 1964; Gardner and West 2014) rather than individual survival, but I postpone that discussion here. It is appears in Sect. 15.3.

[3] Search companies record information about searches and the response of users to the web pages served, so those companies are also intelligent and motivated agents that benefit from the act of the search, but they do not originate it.

belonging to the corporation behind the search, but the motive force for the search is entirely individual.

A websearch is just one example of the AI-augmented individual capacities that have come to pervade twenty-first century life. Others include processes on phones that facilitate our communication, scheduling and even picture taking; AI in our word processors that increases our capacity to effectively communicate by checking grammar and spelling; filters on email that detect spam; filters on credit card expenditures that detect possible fraud; and filters on surveillance cameras that recognise faces, license-plate numbers, and even detect the emotions and intentions underlying human voices and gestures (Valstar and Pantic 2012; Griffin et al. 2013; Eyben et al. 2013; Kleinsmith and Bianchi-Berthouze 2013; Hofmann et al. 2014).

This intelligence enhances the agency of both individuals and corporations (see footnote 4) but has not produced a set of independent artificial actors competing with us for resources as imagined in science fiction. This lack of immediate, apparent, competitive threat, plus a heavy cultural investment in the privileges assumed to associate with human uniqueness, lead many to dismiss the possibility of AI, at the very time it is not only present but fundamentally changing our individual and social capacities.

AI has not yet caused significant change to our direct mechanical capacity for action. In terms of physically altering the world, AI requires a robot. The most prominent robots today are mechanisations of machines we can also use without AI, such as vacuum cleaners, cars and other tele-operated vehicles. We are now capable of acting much faster and at a much greater distance than we could before, but this is primarily due to improvements in telecommunication which are largely (though not entirely) independent of AI.

The way in which current AI fundamentally alters humanity is by altering our capacity for perception—our ability to sense what is in the world. Part of this is also due to communication. For example, we can now see what is happening very far away very quickly. But much more than this, we can remember and recall identical or even similar situations to one we presently observe. Other apes can do that too, but with language and subsequently writing, humans have had a special advantage which is that we can recall situations and actions we have not directly experienced. The reason that we can exploit similar rather than just identical previous contexts for recall is because we store this knowledge in abstract models. Abstraction saves storage space, but even more importantly allows for generalisation to new situations (Bishop 2006). In the simplest case we can find a 'near neighbour' context, treat the present one as the same, and expect similar outcomes (Lopez De Mantaras et al. 2005). Beyond this, we can use models to extrapolate to conditions we have not yet seen, so long as variation within the models tends to be continuous, and the new context does not differ extremely from our historical record. In such conditions we can generate novel variations on previous actions to meet the new conditions (Schaal and Atkeson 1998; Huang et al. 2013).

These processes are ordinarily referred to as machine learning. The reason I am describing them rather as perception is this: I want to emphasise that a great deal of intelligence is the problem of learning to recognise the categories of contexts in which a particular action is appropriate. Or another way to think of this is that with enough experience and a well-structured model for storing and recalling that experience, we can use the past to recognise the present and therefore predict and address the future.

There are two reasons that AI is generating staggering increases in humanity's available intelligence. First, the basic concepts of learning in general and machine learning in particular described above have been understood for decades (Hertz et al. 1991). In those decades our algorithms for building models have been steadily improving—the recently-trendy deep learning is just one of many fundamental improvements made over that time (Jacobs et al. 1991; McLachlan and Krishnan 2008; Hinton et al. 2006; Le Roux and Bengio 2008). Second, we have found ways to both acquire and store the data that makes up experience in digital format. Thus our models are better, bigger and over a vastly wider variety of human experience. For example, we no longer need to guess why and how people will vote or riot— enough of them happily broadcast their intentions and concerns on the Internet. The important thing to understand is that our models have become sufficiently good, that even where our data is biased, often we can compensate for that bias and still make accurate predictions (Beauchamp 2013; Wang et al. 2015; Rothschild et al. 2014).

15.3 Cooperation and Collective Agency

Prosthetic intelligence affects our lives in innumerable ways, most notably simply by allowing us to make more informed decisions, whether by providing more immediate access to restaurant reviews, health care advice or the day's weather forecast. But in this chapter, and in keeping with the rest of this book, I focus on an even more fundamental aspect of human behaviour—our propensity for cooperation and information sharing, and how exponential rates of improvement in our AI may affect these.

The human propensity for cooperation is often seen as unique (Sober and Wilson 1998; Henrich et al. 2001). There is no denying that humans are extraordinary in a number of ways: the extent and variety of our built culture, our language and written histories, and our recent domination of the planet's biomass (Haberl et al. 2007; Barnosky 2008). These indicators of uniqueness are not necessarily or even likely to be independent. For example, our propensity to share information might explain why we have accumulated the culture that allows us to dominate other species. Science considers the simplest viable explanation for any phenomena to be the most likely, so many researchers have been searching for a single-point explanation for human uniqueness.

Cooperation is however not at all unique to humans. Assuming only that we are talking about observed cooperative behaviour, not cooperative intent or forward planning, then cooperation is ubiquitous in nature. For the purpose of this chapter, I will define cooperation as the expression of altruistic behaviour among a collection of individuals. For altruistic I use a standard definition: an action which at least when executed is net costly to the actor, but provides net benefit to another agent. Although standard (Gintis et al. 2005), this definition is not universally accepted. Some biologists (and philosophers) are only happy to label an action 'altruistic' if over the entire lifetime of the individual its expected net value is costly, a situation which never occurs in nature (cf. Sylwester et al. 2014). However, the current understanding and explanation of cooperation in fields like economics and biological anthropology is that cooperation consists of costly actions that produce a public good. Even if the actor or their relatives are likely to get a disproportionate amount of that good, the fact that it facilitates communal benefits makes it cooperative (Burkart et al. 2014; Silva and Mace 2014; Taylor 2014). These types of explanations have been used to account for cooperation in nature—cooperation that often extends to one-way and even ultimate sacrifices by an individual agent for the collective good (Ackermann et al. 2008; Ferguson-Gow et al. 2014; Carter et al. 2014; Hobaiter et al. 2014).

By this definition, we can see that even the ultimately 'selfish' genes in fact exist entirely in cooperative contexts, collaborating with their competitors to compose multi-gene organisms (Dawkins 1976, 1982). The level of agency we are used to reasoning about as individual, that is macroscopic animals and plants, are the vehicles for hosts of competing replicators—genes, and arguably memes.[4] The vast majority of macroscopic life reproduces sexually, which is to say the individual agent is not replicated at all, but rather manages to replicate just (generally) half of its own genes in each of its offspring (Okasha 2012). However, these offspring are nearly always shared with another organism of the same species, and consequently necessarily share the vast majority of their replicators with both of their parents.

Cooperation between living individuals then is highly adaptive,[5] simply because copies of the same replicators that control the selection of the altruistic behaviour are very, very likely to reside in the individuals that receive the benefit (Hamilton 1964; Gardner and West 2014). This explanation of altruism is currently known as inclusive fitness, but has been mathematically related to the possibly more familiar concepts of kin selection and group selection (Gardner et al. 2011; Marshall 2011). To further complicate matters, social behaviour is in fact often controlled by replicators that are themselves socially communicated, whether in bacteria (Rankin et al. 2010), humans (Schroeder et al. 2014), or human institutions (Sytch and Tatarynowicz 2014).[6] What matters therefore is not overall relatedness, but a robust capacity of socialising behaviours to survive—presumably by replication—into the future.

---

[4] Memes are the hypothesised replicators for horizontal (non-genetic) transmission of behaviour. Like genes, they have yet to be precisely defined or measured (Mesoudi et al. 2004). It is also not yet clear the extent to which they change in frequency in accordance to Darwinian evolution (El Mouden et al. 2014). Nevertheless, memes are widely acknowledged as a useful abstraction for thinking about the transfer of traits expressed as behaviour between individuals by means other than biological reproduction.

[5] Adaptive in the biological sense of having been facilitating selection. The AI literature sometimes uses the term adaptive to mean plastic or mutable.

[6] Further, humans at least may choose to associate with those with similar gene structure even where they are not family members (Christakis and Fowler 2014).

To return to the conundrum of human uniqueness, my own hypothesis is that human uniqueness results not from a single cause but from a unique conjunction, at least in terms of extent, of two relatively common traits—a reliance on cognition, culture, and memory, found also in the other great apes and probably other long-lived species (Whiten et al. 1999; McComb et al. 2001; Krützen et al. 2005; Perry and Manson 2003); and a capacity for vocal imitation, something no other ape (or monkey) exhibits, but that has evolved several times apparently independently across a range of taxa[7] (Fitch 2000; Bispham 2006; Bryson 2008; Fitch and Zuberbühler 2013). Vocal imitation provides a communication medium sufficiently rich to support the redundancy necessary for an unsupervised learning process like evolution to operate across our vocalisations (Bryson 2009). Evolution over primate vocalisations, where selection is on both utility and memorability, could produce the system of human language (as per Smith and Kirby 2008; Wray 1998; Wray and Grace 2007). Our ape characteristics—long lives and memories, and predisposition to use culturally-acquired behaviour—allowed us to accumulate sufficient data to facilitate this process, and now allow the learning of complex languages.

Thus no one invented language. Language evolved as a public good, and with it an accumulating catalogue of complex, useful concepts—far more than one individual was otherwise likely to discover or invent for themselves (Dennett 2002). Language might be thought of as the first AI—it is an artefact that massively extends our individual levels of intelligence. As I introduced earlier and will argue more forcefully in the final section, taking our definition of AI to include the motivationless, locationless artefacts that are spoken and written language is a more useful and certainly less dangerous extreme than assuming something is not AI if it is not perfectly human-like. Regardless of whether you will accept language as AI, its intelligence-enhancing properties have consequences for the extent of our cooperation, as I discuss in the next section. Language and culture also may have spectacular consequences for human relatedness, as utilised in theoretical biology for computing the probability of altruistic acts due to inclusive fitness. Language and the culture that it facilitates increase the proportion of our replicators that are shared horizontally. This not only impacts the proportion of our relatedness, but also its plasticity, as humans can rapidly find and communicate ideas that discriminate as well as unite (Krosch and Amodio 2014).

The fact that our relatedness depends on socially-communicated replicators has significant ramifications for collective agency. Genes, individual animals, herds, families, villages, companies and religious denominations can all in some sense be said to be agents—they can all act in ways that effect change in the world. Many of these agencies are composed of others, and further at any level at which there can be seen to be action selection, there can also be seen to be evolutionary selection— at least some reenforcement for decisions taken, and some competition with other actors for limited resources (Wilson 1989; Keller 1999). Every such point of selection

---

[7] The capacity to recognise novel sounds and to learn novel contexts to express sounds should not be confused with the capacity for vocal imitation (Bryson 2009; Fitch and Zuberbühler 2013).

an agent faces presents them with an action-selection conundrum: how much resource (including, for entities that have it, attention) should that collection of opportunities and threats be allocated?

15.4 Factors Determining Investment in the Collective

15.4.1 Problem Specification and Methods

Before we can understand how AI may affect our identities and our societies, we need to form an understanding of how anything can affect these, and in what ways. In this section I address the question that both began this chapter and concluded the previous section: how do agents determine how much resource to devote to which level of the collectives in which they can have an effect? Because its answer hinges on perception and communication, this question will lead into the final discussion of AI's impact on our selves and our collectives.

Let us start by thinking about the problem in terms of a concrete case. An individual is living with a large number of others on a collectivised farm. This farm has been set in competition with other farms, so that whichever farm performs the best will be allocated more resources such as water, seed and fertiliser by the state. Unfortunately, as is often the case in collectivised farming, the system is not very efficient and not everyone is making enough money to have a family. Should our focal individual devote their time to raising their individual status within their own farm, so that their share of that farm's product is increased? Or should they devote their time to ensuring their farm will be more productive, so that the farm receives more income to distribute? Either strategy might reward the individual with the desired level of income. Also, the strategies are not entirely mutually exclusive: some time could be allocated to either, and if the individual is talented at managing then perhaps both could be achieved with the same actions.

In general in biology, wherever we have tradeoffs like these we find a diversity of solutions, with both different species and different individuals within species adopting different mixes of strategies (Darwin 1859). It is important to remember that while evolution is an optimising process, no species or individual is ever optimal. This is for two reasons: both because the world constantly changes, altering the criteria of 'optimum'; and because the number of possible strategies is inconceivably vast. The vast number of available strategies necessitates that any present solution is dependent not only on the optimising force of selection, but also on historical accidents that determined what available variation natural selection has been able to operate over.[8]

---

[8] The vast numbers of possible strategies is produced by a process called combinatorial explosion, which I explain in more detail in Sect. 15.4.2. The importance of having a varied set of available possible solutions in order for evolutionary selection to proceed is part of the 'Fundamental

Any such accident of variation may lead the locally-optimal strategy between two individuals to be different. For our farmer, the optimal decision for their strategising may depend on contexts local to the farm, such as opportunities for promotion based on the age of the management team, or might change by the year depending on the weather. In a good year, perhaps the best farms will be able to support a good standard of living for all employees, but in a drought it may be essential to be in the management tier. An individual farmer in a particular farm may have a better chance at promotion due to their charisma, or a better chance at a game-changing farming innovation due to their cleverness. The talents and position of close friends or family among fellow employees could also determine the better strategy.

As the example above illustrates, we are unlikely to determine a single optimal level of investment in a particular collective agency for any individual. However, we can describe a set of factors which influence the utility of investment at different levels, and describe models of how these relate to each other. These models can inform us about what strategies are most likely to be chosen, and how these probabilities might change when new technologies can be used to magnify or repress the impact of native characteristics. For example, if a new fertiliser is invented that allows all the farms to produce enough so every individual might be able to have a family, then this might eliminate the need to compete with other farms, and the farmer might best invest their time in ensuring equitable distribution within their own farm.

Factors contributing to individual versus group-level investment can be roughly decomposed into two categories:

1. Environmental: those factors exogenous to any of the agents' replicators, such as the weather, or that most individuals have very little influence over, such as international policy on banking or the environment.
2. Social: factors that influence how a collective can function, such as its capacity for communication, and the behavioural or genetic relatedness of its members (see discussion of inclusive fitness, above).

There is good evidence that variation in environmental context can determine the utility and structure of a collective. For example, spiteful, anti-social behaviour seems to increase in regions with a low GDP (Herrmann et al. 2008a) or scarce biomass (Prediger et al. 2013). Spite is the opposite of altruism—it is the willingness to pay a cost in to inflict a cost on others. This behaviour taken in isolation is necessarily maladaptive, as it hurts not only the individual but also another who almost certainly shares some measure of relatedness. It can only be accounted for if it covaries with some other attribute, for example if expressing spite increases social dominance and thus helps individuals in local competition (Rand et al. 2010; Powers

---

theorem of evolution' (Fisher 1930; Price 1972), and will be key in the final section of this chapter, Sect. 15.5.

et al. 2012).⁹ These results imply that more cooperative behaviour occurs when resources are more prevalent, but doesn't explain a mechanism. Perhaps the costs of a competitive strategy are less attractive when relative status is not essential to survival, or perhaps cooperation is a riskier strategy more often chosen when participants are better resourced.

The objective of this chapter however is to examine the impact of AI on collective agency. While AI certainly does and will continue to affect the workings of our financial markets, our capacity to damage or protect the environment, and so forth; predicting the consequences of this impact requires an understanding of economics and politics beyond the scope of this chapter. Here I focus on what I've just termed the social aspects of investment in the collective. I review what is known about the 'individual' (animal- or vehicle-level) decision to invest in public rather than private goods. Then in this chapter's final section, I examine how prosthetic intelligence might be expected to alter values in these equations to change our level of investment, and as a consequence, our identity.

Much of the evidence presented in this and the previous sections, including the papers just cited by Rand et al. (2010) and Powers et al. (2012), derives from formal models including social simulation. Given this chapter's context in this volume—where simulation has been presented by some (e.g. Arnold 2015) as somehow controversial—I will briefly revisit why and how simulations are now an accepted part of the scientific method.

The role of simulations in science has been at times confused, not only by occasional bad practice (as with any method), but also by claims by some of the method's innovators that simulations were a "third way" to do science (after induction and deduction, Axelrod 1997). However, more recently a consensus has been reached that simulation and modelling more generally are indeed a part of ordinary science (Dunbar 2002; Kokko 2007; Seth et al. 2012). The part that they are is theory building. Every model is a theory—a very-well specified theory. In the case of simulations, the models are theories expressed in so much detail that their consequences can be checked by execution on a computer. Science requires two things: theories that explain the world, and data about the world which can be used to compare and validate the theories. A simulation provides no data about the world, but it can provide a great deal of 'data' about a theory. First, the very process of constructing a simulation can show that a theory is incoherent—internally contradictory, or incomplete, making no account for some part of the system intended to be explained (Axelrod 1997; Whitehouse et al. 2012). Secondly, modelling in general can show us a fuller range of consequences for a theory. This allows us to make specific, formal hypotheses about processes too complex to entirely conceptualise inside a single human brain (Dunbar 2002; Kokko 2007). The wide-spread acceptance of simulations as a part of the scientific method can be seen by their inclusion in the highest levels of academic publication, both in the leading general science journals and in

---

⁹There is decent evidence that association with dominance is indeed the ultimate evolutionary explanation for spiteful behaviour, see for a review Sylwester et al. (2013).

the flagship journals for specific fields ranging from biology through political science.

Fortunately, a theory expressed formally as a simulation can also be expressed in the traditional, informal, ordinary-language way as well. This is the technique I use to describe the 'outcomes' (implications) of simulations throughout this chapter.

15.4.2 Models of Social Investment

In order for evolution to direct individuals to invest at a collective level two conditions need to hold. First, there needs to be some inclusive-fitness advantage for the replicators involved in this 'directing' (cf. Sect. 15.3 above.) Second, this advantage has to be discoverable, and discovered. As mentioned in the first part of this section, evolution optimises but never finds an optimum, partly because it cannot evaluate all possible candidates due to the infinite size of the candidate pool. The size of this pool derives from the fact that candidate 'solutions' are composed of combinations of available features. The number of possible combinations is exponentially related to the number of features: it is the number of features per candidate (f) raised to the number of possible values for these features (v), or f^v. This problem of combinatorics affects all forms of directed plasticity—that is, any system capable of change which has an evaluation criterion. In the Computer Sciences, this problem is known as combinatorial explosion, and characterises both AI planning and (machine) learning. But the same problem characterises both evolution and cognition, and by 'cognition' I also mean to include both learning and planning, where they are done by an individual over their or its lifetime.

To address the first condition first, inclusive-fitness (IF) benefit has proven a spectacularly complicated concept to reason about, although its fundamental veracity has been demonstrated time and again in both simulation and empirical data (Gardner and West, 2014, for a recent special issue). What makes IF difficult is not only the confound of memetic as well as genetic replicators, but also the problem of net benefit. We share genes with all life, nevertheless predation—and grazing—evolve (Folse and Roughgarden 2012; Ledgard 2001). We tend to favour those with whom we share more relatedness, yet our survival also depends on the stability of the ecosystem to which we are adapted. Still, since the focus of this chapter is on the impact of AI, I will neglect the Gaia-style analysis of ecosystemic agency (see instead Margulis and Hinkle 1997) and focus primarily on collectives consisting of a single species. Even here, IF leads to wildly counterintuitive effects, such as that promiscuity in socially-monogamous animals can lead natural selection to favour strategies that benefit the public good, such as mutual defence and conflict resolution (Eliassen and Jørgensen 2014).

Within species, families, or even swarms of clonal microbia, understanding IF requires consideration of the net benefit of collaboration. The costs of cooperation are not limited to the costs of the altruistic act, but also include the costs of cohabiting with close genetic relatives. These cohabitation costs include competition for resources ranging from food to shelter to mates, and increased exposure to biological threats such as disease and predation which will specialise to a particular species, immune system, and locale. In large animals the advantages of communal living have long puzzled biologists, with avoidance of predation via 'cover seeking' with a mob being a key hypothesis (Hamilton 1971). However this relationship is also not simple. Large populations also serve to attract predation and sustain disease (e.g. Bischof et al. 2014; Bate and Hilker 2013), though smaller group size does seem to increase predation risk (Shultz and Finlayson 2010). Recently in the megafauna literature there has been a new hypothesis: individuals in populations might benefit from information transmission, of which vigilance against predators is just a special case (Crockford et al. 2012; Chivers and Ferrari 2014; Hogan and Laskowski 2013; Derex et al. 2013). Transmission of behaviour may be at least as important as information about localised threats (Jaeggi et al. 2008; Dimitriu et al. 2014). Note that behaviour itself, when transmitted horizontally (that is, not by genes to offspring), must be transmitted as information via perception (Shannon 2001). But information is just one example of public goods held by non-human species. Others include territory (including food, shelter and even mating resources, Preuschoft and van Schaik 2000; Dunbar et al. 2009), physical shelters, even digestive enzymes (MacLean et al. 2010). Much of this cooperative production is performed by microbia, where in contrast to megafauna, genetic instructions for cooperative behaviour can be exchanged horizontally—even across species—and injected into the cellular organism to change a local population's behaviour (Rankin et al. 2010; Dimitriu et al. 2014).

Cooperation requires not only that the species affords some sort of cooperative behaviour (e.g. the genetic coding for collaboratively building a hive), but also the capacity to detect when it is a good time to invest in such an activity, and further who is or are the best partners with which to engage. This last is of particular interest, because we know that a variety of species appear to shift between cooperative phases of behaviour. Generalised reciprocity, first observed in Norwegian rats, is an increase in expression of altruistic behaviour that follows the observation of others engaged in cooperative acts (van Doorn and Taborsky 2012; Gray et al. 2015). This sort of behavioural flexibility might be thought useful for facilitating the spread of cooperation, since it allows potential cooperators to suppress cooperative behaviour in the presence of free riders that might exploit them. However such an interpretation may be biased. A better model might be more neutral, like our interpretation of the phase changes in collective behaviour exhibited by slime mould as an adaptation to localised environmental stress (Keller and Segel 1970; Leimgruber et al. 2014).

MacLean et al. (2010) have openly challenged the idea that cooperative behaviour (the creation of public goods) is always something to be maximised. They provide a case study of the production of digestive enzymes by the more altruistic of two isogenic yeast strains. The yeast must excrete these enzymes outside of their bodies (cell walls) as they can only directly absorb pre-digested food. The production of these enzymes is costly, requiring difficult-to-construct proteins, and the production of pre-digested food is beneficial not only to the excreting yeast but also to any other yeast in its vicinity.

In the case of single-cell organisms there is no choice as to whether to be free-riding or pro-social—this is determined genetically by their strain. But the two strategies are accessible to each other via a relatively common mutation. Natural selection performs the action selection for a yeast collective by determining what proportion of each strategy lives or dies. MacLean et al. (2010) demonstrate with both empirical experiments and models that selection operates such that the species as a whole benefits optimally. The altruistic strain in fact overproduces the public good (the digestive enzymes) at a level that would be wasteful if it were the only strategy pursued, while the free-riding strain underproduces. Where there are insufficient altruists free-riders starve, allowing altruists to invade. Where there are too few free-riders excess food accumulates, allowing free-riders to invade. Thus the greatest good—the most efficient exploitation of the available resources—is achieved by the species through a mixture of over-enthusiastic altruism and free riding. Why doesn't evolution just optimise the species as a whole to produce the optimal level of enzyme? Because the temporal cost (delay) associated with a single genome discovering a particular production level is greater than the temporal stability of that optimal value, which is of course determined by the dynamics of the ecosystem. In contrast, death and birth can be exceedingly rapid in microbia. A mixed population composed of multiple strategies, where the high and low producers will always over and under produce (respectively) and their proportions can be changed very rapidly is thus the best strategy for tracking the rate of environmental change—for rapidly responding to variation in opportunity.

Bryson et al. (2014) recently proposed that a similar dynamic may explain cultural variation in the extent of apparently anti-social, spiteful behaviour. This variation was originally observed by Herrmann et al. (2008a), but not explained. In the context of an anonymous economic game played in laboratories,[10] some proportion of nearly every population studied chose to punish (to pay a cost to penalise) altruists who were acting in a way that benefited the punishers. This sort of behaviour,

---

[10]These were public goods games (PGG). Participants were separated by partitions and were unable to directly interact with or identify other group members. They played games in groups of four, with each participant able to either keep all of the endowment received from the experimenter (20 experimental currency units; ECU) or contribute some portion of the endowment to the public good. At the end of a round, all contributions were combined and the sum multiplied by 1.6. The obtained amount was divided evenly amongst all of the group members, regardless of their contribution. The payoff of each participant was calculated by summing up the amount kept and the amount received from the public good. Ten rounds were played as described above, and also ten rounds with the addition of punishment: participants after seeing the contributions of other players to the pubic good and could decide how much they wished to spend on reducing the payoff to other players. Participants could spend up to 10 ECU punishing the other players. Each ECU spent on sanctioning resulted in 3 ECU being deducted from the payoff to the targeted individual. A participant's payoff was calculated by subtracting the amount of ECU spent on sanctioning and the deduction points received from other players from the payoff from the PGG. Received deductions were capped so as not to exceed PGG earnings. Participants did not receive information about who deducted points from their payoff, making punishment anonymous. At the end of the experiment, participants received real money in the local currency in exchange for the total ECU accumulated across all rounds. See further Sylwester et al. (2014); Herrmann et al. (2008b).

termed anti-social punishment (ASP), cannot be accounted for directly in evolutionary models, but must give some indirect benefit (as mentioned earlier, Sect. 15.4.1). Herrmann et al. (2008a) discovered that the propensity for ASP correlates with the gross domestic product (GDP) of the country where the experiments were conducted, and also with its rule of law as measured by the World Values Survey (Inglehart et al. 2004). Using the Herrmann et al data set, Sylwester et al. (2014) discovered that ASP results in a significant increase in variation in the level of investment in public goods, but not in any particular direction. In contrast, altruistic punishment (of free riders) produces a measurable increase in investment, while those receiving no punishment tend not to change their level of investment over repeated rounds of playing the game. This result is particularly striking because of the anonymous nature of the game—because individuals did not know who punished them, they could not tell whether they were being punished by those giving more or less than themselves.[11] Nevertheless, humans seem to be well-equipped to assess social context. We hypothesise that altruistic punishment is more likely to be coordinated, and coordinated punishment is taken as an indication of ingroup identity, signalling the construction of a collective, and this is what results in the increased investment. ASP in contrast signals a conflict over social status, which results in more varied behaviour, and therefore a greater potential rate of change for the society (Fisher 1930; Price 1972).

This series of hypothesised mechanisms for adjusting investment in different levels of agency is key to the purpose of this chapter—to consider how AI changes human collective agency. There are two points at which AI fundamentally changes our social capacities: detecting appropriate contexts for expressing cooperative behaviours an agent already knows, and the discovery or innovation of new cooperative behaviours with or without the contexts for their expression. Both of these points benefit by improved communication and superior perception.

Choosing appropriate partners is a particularly important part of detecting contexts for behaviour. Cooperative behaviour is most sustainable when the benefit received from the agent's cost will be high, and when there is similarly high benefit for low cost likely to be produced by the agent's collaborator(s). Thus where possible, cooperation often takes place in the context of a relationship where both the needs of the other and the likelihood of their reciprocation can be judged. Zahavi (1977) hypothesises that the time one agent spends with another is an honest signal of the value the first agent places on that relationship. Perry (2011) has used this bond-testing hypothesis to explain strange dysphoric games played amongst capuchins—monkeys well-known for both their intelligence and their aggressive coalition behaviour where coalitions are not necessarily formed with close relatives. Atkinson and Whitehouse (2011) suggest that time spent in mutual dysphoric situations underlies human religion, which serves the purpose of assuring human bonding across groups that require mutual support. Taylor (2014) has recently extended the bond-testing model, drawing attention to the fact that many human societies

---

[11] Those who gave the most or the least to the group could assess the nature of the punishment they received, but our results held even when these were excluded (Sylwester et al. 2014).

require temporally-expensive displays of investment in the lives of others with whom a family may have long-term economic relations, thus guaranteeing each other assistance in times of hardship. The time-costly displays (for example, constructing elaborate gifts) guarantee that a family is not making many shallow investments, but rather has only a few deeply-committed relationships.

In a more general and less specifically-human model than Taylor's, Roughgarden et al. (2006) propose that an explanation for physical intimacy (beyond what is necessary for procreation) may be that intimacy is a means of increased communication of physical status between potential coalition partners, allowing for the discovery of mutually-advantageous equilibria with respect to the extent of cooperative investment. The suggestion is that this intimacy goes beyond mere partner choice and timing to finding sufficient information about potential shared goals to afford new cooperative activities (Roughgarden 2012). Consider the implications of these results on the earlier discussion of human exceptionalism. Language has made humans the most extraordinary communicators in nature, and writing and AI have accelerated these effects. But our exceptional communication is not limited to deliberate or linguistic mechanisms—for an ape, even the amount of our communication by scent is exceptional (Stoddart 1990; Roberts and Havlicek 2011). This could well explain the exceptional extent of our cooperation.

To summarise, these models show that there will always be a tradeoff between investment in the individual and the collective. Individuals (at least some of them) must be sustained for the collective to exist, so investment can never go to the extreme of being fully collective. However there are a large number of situations in nature that are not zero sum—where altruism can evolve because the cost to the individual is lower than the benefit produced multiplied by the number of individuals helped, divided by their relatedness to the altruistic individual (Hamilton 1964). This idea of 'relatedness' is tricky though—it really depends only on how related the individuals are in whatever trait generates their social behaviour. Social behaviour may itself be transmitted socially, even in microbia (Rankin et al. 2010). Also, relatedness is judged based on the pool of others with which the individual competes. So if competition is imposed on a large scale such as when a government forces collective farms to compete between each other, two individuals in the same farm may seem more related than when a drought sets in and the members of the farm are set to competing with each other for survival (Lamba and Mace 2011; Powers et al. 2011).

We have also seen that investment strategies may vary within a population to the benefit of that population overall; provided that the various strategies are accessible to each other, again either by genetic or social transmission of the strategies. We have seen that selecting appropriate partners can increase the benefit-to-cost ratio, and thus support investing more heavily in cooperative, collective strategies. This selection is dependent on being able to perceive the needs and abilities of others. What would be the outcome for cooperative behaviour if we could exactly know the needs and interests—and predict the future behaviour—of our neighbours?

15.5 The Impact of AI on Human Cooperation and Culture

My main objective in this chapter is this: to convince you that AI is already present and constantly, radically improving; and that the threats and promises that AI brings with it are not the threats and promises media and culture have focussed on, of motivated AI or superintelligence that of themselves starts competing with humans for resources. Rather, AI is changing what collective agencies like governments, corporations and neighbourhoods can do. Perhaps even more insidiously, new affordances of knowledge and communication also change what even we as individuals are inclined to do, what we largely-unconsciously think is worth our while. 'Insidious' is not quite the right word here, because some of these effects will be positive, as when communities organise to be safer and more robust. But the fact that our behaviour can radically change without a shift in either explicit or implicit motivations—with no deliberate decision to refocus—seems insidious, and may well be having negative effects already.

As I indicated in Sect. 15.2, we are already in the process of finding out what happens when our ability to read and predict the behaviour of our fellows constantly improves, because this is the new situation in which we find ourselves, thanks to our prosthetic intelligence. Assuming the output of commercial AI remains available and accessible in price, then the models of the previous section tell us we should expect to find ourselves more and more operating at and influenced by the level of the collective. Remember that this is not a simple recipe for world-wide peace. There are many potential collectives, which compete for resources including our time. Also and it is possible to over-invest in many, perhaps most public goods. The models of Roughgarden and Taylor describe not systems of maximal cooperation, but rather systems of maximising individual benefit from cooperation. There are still physical and temporal limits to the number of people with whom we can best collaborate for many human goals (Dunbar 1992; Dunbar et al. 2009). We might nevertheless expect that our improved capacity to communicate and perceive can help us to achieve levels of cooperation not previously possible for threats and opportunities that truly operate at a species level, for example response to climate change or a new pandemic.

Our hopes should be balanced and informed though also by our fears. One concern is that being suddenly offered new capacities may cause us to misappropriate our individual investments of time and attention. This is because our capacity for cooperative behaviour is not entirely based on our deliberating intelligence or our individual capacity for plasticity and change. Learning, reasoning and evolution itself are facilitated by the hard-coding of useful strategies into our genetic repertoire (Depew 2003; Rolian 2014; Kitano 2004). For humans, experience is also compiled into our unconscious skills and expectations. These are mechanisms that evolution has found help us address the problems of combinatorics (see the first paragraph of Sect. 15.4.2). But these same solutions leave us vulnerable for certain pathologies. For example, a *supernormal stimulus* is a stimulus better able to trigger a behaviour than any that occurred in the contexts in which the paired association between stimulus and response was learned or evolved (Tinbergen and Perdeck 1950; Staddon 1975). Supernormal stimuli can result from the situation where, while the behaviour was being acquired, there was no context in which to evolve or learn a bound for the expression of that behaviour, so no limits were learned. The term was invented by Tinbergen to describe the behaviour of gull chicks, who would ordinarily peck the red dot on their parent's bill to be fed, but preferred the largest, reddest head they could find over their actual parents'. Natural selection limits the amount of red an adult gull would ever display, but not the types of artefacts an experimental scientist might create. Similarly, if a human drive for social stimulation (for example) is better met by computer games than real people, then humans in a gaming context might become increasingly isolated and have a reduced possibility to meet potential mates. The successful use of search engines—quick access to useful information—apparently causes a reduction in actual personal memory storage (Ward 2013). This effect may be mediated by the successful searcher's increased estimation of cognitive self worth. Though Ward describes this new assessment as aberrant, it may in fact be justified if Internet access is a reliable context.

The social consequences of most technology-induced supernormal stimuli will presumably be relatively transient. Humans are learning machines—our conscious attention, one-shot learning, and fantastic communicative abilities are very likely to spread better-adapted behaviour soon after any such benign exploitation is stumbled over. What may be more permanent is any shift between levels of agency in power, action, and even thought as a consequence of the new information landscape. The increased transparency of other people's lives gives those in control more control, whether those are parents, communities or school-yard bullies. But control in this context is a tricky concept, linked also with responsibility. We may find ourselves losing individual opportunities for decision making, as the agency of our collectives become stronger, and their norms therefore more tightly enforced.

The dystopian scenarios this loss of individual-level agency might predict are not limited to ones of governmental excess. Currently in British and American society, children (including teenagers) are under unprecedented levels of chaperoning and 'protection'. Parents who 'neglect' their children by failing to police them for even a few minutes can be and are being arrested (Brooks 2014). Lee et al. (2010 special issue) document and discuss the massive increase over the last two decades in the variety as well as duration of tasks that are currently considered to be parenting. Lee et al suggest that what has changed is risk sensitivity, with every child rather than only exceptional ones now being perceive as 'at-risk', by both parents and authorities. This may not be because of increased behavioural transparency afforded by technology and AI. Another possible explanation is simply the increased value of every human life due to economic growth (Pinker 2012). But what I propose here is that the change is not so much in belief about the possibility of danger, as the actuality of afforded control. Social policing is becoming easier and easier, so we need only to assume a fixed level of motivation for such policing to expect the amount of actual policing to increase.

Another form of AI-mediated social change that we can already observe is the propensity of commercial and government organisations to get their customers or users to replace their own employees. The training, vetting and supervision that previously had to be given to an employee can now be mostly encoded in a machine—and the rest transferred to the users—via the use of automated kiosks. While the machines we use to check out our groceries, retrieve our boarding cards and baggage tags, or purchase post office services may not seem particularly intelligent, their perceptual skills and capacities for interaction are more powerful and flexible than the older systems that needed to be operated by experts. Of course, they are still not trivial to use, but the general population is becoming sufficiently expert in their use to facilitate their replacement of human employees. And in acquiring this expertise, we are again becoming more homogenous in our skill sets, and in the way we spend that part of our time.

With AI public video surveillance our motions, gestures, and whereabouts can be tracked; with speech recognition our telephone and video conversations can be transcribed. The fact some of us but not others spew information on social media will rapidly be largely irrelevant. As better and better models are built relating any form of personal expression (including purchases, travel, and communication partners) to expected behaviour (including purchases, votes, demonstrations, and donations), less and less information about any one person will be needed to predict their likely behaviour (Jacobs et al. 1991; McLachlan and Krishnan 2008; Hinton et al. 2006; Le Roux and Bengio 2008).

Although I've been discussing the likely homogenising impact of increased AI and increased collective-level agency, collective agency is not necessarily egalitarian or even democratic. Again we can see this in nature and our models of the behaviours of animals very similar to us. In non-human primates, troops are described as either 'egalitarian', where any troop member can protest treatment by any other, and conflict is frequent but not violent; or as 'despotic', where interaction is limited by the dominance hierarchy, aggression is unilateral from dominant to subordinate, and fights while few are bloody (Thierry 2007). Which structure a species uses is partially determined by historic accident (phylogeny, Shultz et al. 2011), but also significantly by the species' ecology. If a species' preferred food source is defensible (e.g. fruit rather than insects) then a species will be more hierarchical, as it will under the pressure for safer spatial positions produced by the presence of predators (Sterck et al. 1997). The choice between social orders is not made by the individual monkeys, but by the dynamics of their ecological context.

Similarly, we cannot say exactly the power dynamics we expect to see as a consequence of increasing agency at collective, social levels. However a worrying prediction might be drawn from Rawls (1980), whose theory mandates that a 'veil of ignorance' is necessary to ensure ethical governance. Those in power should be under the impression that any law they make might apply to any citizen, including themselves. Can such ignorance be maintained in an age of prosthetic intelligence? If not, if those in power can better know the likely social position of themselves and their children or even the likely outcome of elections (Wang et al. 2015), how will this affect our institutions? As uncertainty is reduced, can we ensure that those in power will optimise for the global good, or will they be more motivated—and able—to maintain control?

The answers to these questions are not deterministic. The models presented in Sect. 15.4 make ranges of predictions based on interactions between variables, all of which can change. Our future will be influenced by the institutions and regulations we construct now, because these determine how easy it is to transition from one context into another, just as available variation partially determines evolution by determining what natural selection can select between (see footnote 9). Although many futures may be theoretically achievable, in practice the institutions we put in place now determine which futures are more likely, and how soon these might be attained.

Humans and human society have so far proved exceptionally resilient, presumably because of our individual, collective and prosthetic intelligence. But what we know about social behaviour indicates significant policy priorities. If we want to maintain flexibility, we should maintain variation in our populations. If we want to maintain variation and independence in individual citizens' behaviour, then we should protect their privacy and even anonymity. Previously, most people were anonymous due to obscurity. In its most basic form as absolute inaccessibility of information, obscurity may never occur again (Hartzog and Stutzman 2013). But previously, people defended their homes with their own swords, walls and dogs. Governments and other organisations and individuals are fully capable of invading our homes and taking our property, but this is a relatively rare occurrence because of the rule of law. Legal mandates of anonymity on stored data won't make it impossible to build the general models that can be used to predict the behaviour of individuals. But if we make this sort of behaviour illegal with sufficiently strong sanctions, then we can reduce the extent to which organisations violate that law, or at least limit their proclivity for publicly admitting (e.g. by acting on the information) that they have done so. If people have less reason to fear exposure of their actions, this should reduce the inhibitory impact on individuals' behaviour of our improved intelligence.

Already both American and European courts are showing signs of recognising that current legal norms have been built around assumptions of obscurity, and that these may need to be protected (Selinger and Hartzog 2014). Court decisions may not be a substitute though for both legislation and the technology to make these choices realistically available. Legislating will not be easy. In Europe there has been concern that the de facto mechanism of access to the public record has been removed as search engines have been forced not to associate newspaper articles with individuals' names when those individuals have asked to be disassociated from incidents which are entirely in the past (Powles 2014). As we do come to rely on our prosthetic intelligence and to consider those of our memories externalised to the Internet to be our own, such cases of who owns access to which information will become increasingly complex (Gürses 2010).

The evolution of language has allowed us all to know the concept of responsibility. Now we are moral agents—not only actors, but authors responsible for our creations. As philosophers and scientists we have also professional obligations with respect to considering and communicating the impacts of technology to our culture (Wittkower et al. 2013). AI can help us understand the rapid changes and ecological dominance our species is experiencing. Yet that same understanding could well mean that the rate of change will continue to accelerate. We need to be able to rapidly create, negotiate and communicate coherent models of our dynamic societies and their priorities, to help these societies establish a sustainable future. I have argued that the nature of our agency may fundamentally change as we gain new insights through our prosthetic intelligence, resulting in new equilibria between collective versus individual agency. I've also described scientific models showing how these equilibria are established, and the importance of individual variation to a robust, resilient, mutable society. I therefore recommend that we encourage both legislatures and individual citizens to take steps to maintain privacy and defend both group and individual eccentricity. Further, I recommend we all take both personal and academic interest in our governance, so that we can help ensure the desirability of the collectives we contribute to.

Acknowledgments Thanks to Lydia Harriss, Catrin Misselhorn, Miles Brundage and Evan Selinger for encouragement and discussions; David Gunkel and Will Lowe for debate; Misselhorn, Brundage, Selinger and Robin Dunbar for comments on an earlier draft; and Thomas König and the University of Mannheim SFB 884, The Political Economy of Reforms, for a quiet office to work in.

References

Ackermann, Martin, Bärbel Stecher, Nikki E. Freed, Pascal Songhet, Wolf-Dietrich Hardt, and Michael Doebeli. 2008. Self-destructive cooperation mediated by phenotypic noise. *Nature* 454: 987–990.
Arnold, Eckhart. 2015. How models fail? A critical look at the history of computer simulations of evolution in cooperation. In this volume.
Atkinson, Quentin D., and Harvey Whitehouse. 2011. The cultural morphospace of ritual form: Examining modes of religiosity cross-culturally. *Evolution and Human Behaviour* 32: 50–62.
Axelrod, Robert. 1997. *The complexity of cooperation: Agent-based models of competition and collaboration.* Princeton: Princeton University Press.
Barnosky, Anthony D. 2008. Megafauna biomass tradeoff as a driver of quaternary and future extinctions. *Proceedings of the National Academy of Sciences* 105: 11543–11548.
Bate, Andrew M., and Frank M. Hilker. 2013. Predator–prey oscillations can shift when diseases become endemic. *Journal of Theoretical Biology* 316: 1–8.
Beauchamp, Nick. 2013. Predicting and interpolating state-level polling using twitter textual data. In *New directions in analyzing text as data workshop*, ed. Ken Benoit, Daniel Diermeier, and Arthur Spirling, London School of Economics, September.
Bischof, Richard, H. Ali Dondas, Kabir Muhammad, S. Hameed, and Muhammad A. Nawaz. 2014. Being the underdog: An elusive small carnivore uses space with prey and time without enemies. *Journal of Zoology* 293(1): 40–48.
Bishop, Christopher M. 2006. *Pattern recognition and machine learning.* London: Springer.
Bispham, John. 2006. Rhythm in music: What is it? Who has it? And why? *Music Perception* 24: 125–134.

Brooks, Kim. 2014. *The day I left my son in the car.* Salon. http://www.salon.com/2014/06/03/the_day_i_left_my_son_in_the_car/. Accessed 3 June 2014.
Bryson, Joanna J. 2008. Embodiment versus memetics. *Mind & Society* 7: 77–94.
Bryson, Joanna J. 2009. Representations underlying social learning and cultural evolution. *Interaction Studies* 10: 77–100.
Bryson, Joanna J., and Philip P. Kime. 2011. Just an artifact: Why machines are perceived as moral agents. In *Proceedings of the 22nd International Joint Conference on Artificial Intelligence*, 1641–1646. Barcelona: Morgan Kaufmann.
Bryson, Joanna J., James Mitchell, and Simon T. Powers. 2014. Explaining cultural variation in public goods games. In *Applied evolutionary anthropology: Darwinian approaches to contemporary world issues*, ed. M.A. Gibson and D.W. Lawson, 201–222. Heidelberg: Springer.
Burkart, Judith M., O. Allon, F. Amici, Claudia Fichtel, Christa Finkenwirth, Heschl Adolf, J. Huber, K. Isler, Z.K. Kosonen, E. Martins, E.J. Meulman, R. Richiger, K. Rueth, B. Spillmann, S. Wiesendanger, and C.P. van Schaik. 2014. The evolutionary origin of human hyper-cooperation. *Nature Communications* 5: 4747.
Carter, Alecia J., S. English, and Tim H. Clutton-Brock. 2014. Cooperative personalities and social niche specialization in female meerkats. *Journal of Evolutionary Biology* 27: 815–825.
Chivers, Douglas P., and Maud C.O. Ferrari. 2014. Social learning of predators by tadpoles: Does food restriction alter the efficacy of tutors as information sources? *Animal Behaviour* 89: 93–97.
Christakis, Nicholas A., and James H. Fowler. 2014. Friendship and natural selection. *Proceedings of the National Academy of Sciences* 111(Suppl 3): 10796–10801.
Crockford, Catherine, Roman M. Wittig, Roger Mundry, and Klaus Zuberbühler. 2012. Wild chimpanzees inform ignorant group members of danger. *Current Biology* 22: 142–146.
Darwin, Charles. 1859. *On the origin of species by means of natural selection.* London: John Murray.
Dawkins, Richard. 1976. *The selfish gene.* Oxford: Oxford University Press.
Dawkins, Richard. 1982. *The extended phenotype: The gene as the unit of selection.* Oxford: W.H. Freeman & Company.
Dennett, Daniel C. 2002. The new replicators. In *The encyclopedia of evolution*, vol. 1, ed. Mark Pagel, E83–E92. Oxford: Oxford University Press.
Depew, David J. 2003. Baldwin and his many effects. In *Evolution and learning: The Baldwin effect reconsidered*, ed. Bruce H. Weber and David J. Depew. Cambridge, MA: Bradford Books/MIT Press.
Derex, Maxime, Marie-Pauline Beugin, Bernard Godelle, and Michel Raymond. 2013. Experimental evidence for the influence of group size on cultural complexity. *Nature* 503: 389–391.
Dimitriu, Tatiana, Chantal Lotton, Julien Bénard-Capelle, Dusan Misevic, Sam P. Brown, Ariel B. Lindner, and François Taddei. 2014. Genetic information transfer promotes cooperation in bacteria. *Proceedings of the National Academy of Sciences* 111: 11103–11108.
van Doorn, Gerrit Sander, and Michael Taborsky. 2012. The evolution of generalized reciprocity on social interaction networks. *Evolution* 66: 651–664.
Dunbar, Robin I.M. 1992. Time: A hidden constraint on the behavioural ecology of baboons. *Behavioral Ecology and Sociobiology* 31: 35–49.
Dunbar, Robin I.M. 2002. Modelling primate behavioral ecology. *International Journal of Primatology* 23: 785–819.
Dunbar, Robin I.M., Amanda H. Korstjens, Julia Lehmann, and British Academy Centenary Research Project. 2009. Time as an ecological constraint. *Biological Reviews* 84: 413–429.
El Mouden, Claire, Jean-Baptiste André, Oliver Morin, and Daniel Nettle. 2014. Cultural transmission and the evolution of human behaviour: A general approach based on the Price equation. *Journal of Evolutionary Biology* 27: 231–241.
Eliassen, Sigrunn, and Christian Jørgensen. 2014. Extra-pair mating and evolution of cooperative neighbourhoods. *PLoS One* 9: e99878.

Eyben, Florian, Felix Weninger, Nicolas Lehment, Björn Schuller, and Gerhard Rigoll. 2013. Affective video retrieval: Violence detection in Hollywood movies by large-scale segmental feature extraction. *PLoS One* 8: e78506.
Ferguson-Gow, Henry, Seirian Sumner, Andrew F.G. Bourke, and Kate E. Jones. 2014. Colony size predicts division of labour in attine ants. *Proceedings of the Royal Society B: Biological Sciences* 281: 20141411. doi: 10.1098/rspb.2014.1411.
Fisher, Ronald A. 1930. *The genetical theory of natural selection*. Oxford: Oxford University Press.
Fitch, W. Tecumseh. 2000. The evolution of speech: A comparative review. *Trends in Cognitive Sciences* 4: 258–267.
Fitch, W. Tecumseh., and Klaus Zuberbühler. 2013. Primate precursors to human language: Beyond discontinuity. In *The evolution of emotional communication: From sounds in nonhuman mammals to speech and music in man*, ed. Eckart Altenmüller, Sabine Schmidt, and Elke Zimmerman, 26–48. Oxford: Oxford University Press.
Folse, Henry J., and Joan Roughgarden. 2012. Direct benefits of genetic mosaicism and intraorganismal selection: Modeling coevolution between a long-lived tree and a short-lived herbivore. *Evolution* 66: 1091–1113.
Gardner, Andy, and Stuart A. West. 2014. Inclusive fitness: 50 years on. *Philosophical Transactions of the Royal Society B: Biological Sciences* 369: 20130356.
Gardner, Andy, Stuard A. West, and Geoff Wild. 2011. The genetical theory of kin selection. *Journal of Evolutionary Biology* 24: 1020–1043.
Gintis, Herbert, Samuel Bowles, Robert Boyd, and Ernst Fehr. 2005. Moral sentiments and material interests: Origins, evidence, and consequences, Chapter 1. In *Moral sentiments and material interests: The foundations of cooperation in economic life*, ed. Herbert Gintis, Samuel Bowles, Robert Boyd, and Ernst Fehr, 3–39. Cambridge, MA: MIT Press.
Gray, Peter B., Justin R. Garcia, Benjamin S. Crosier, and Helen E. Fisher. 2015. Dating and sexual behavior among single parents of young children in the United States. *Journal of Sex Research* 52: 121–128.
Griffin, Harry J., Min S.H. Aung, Bernardino Romera-Paredes, Ciaran McLoughlin, Gary McKeown, William Curran, and Nadia Bianchi-Berthouze. 2013. Laughter type recognition from whole body motion. In *Affective computing and intelligent interaction* (ACII), 2013 Humaine Association conference, Geneva, CH, 349–355.
Gunkel, David J. 2012. *The machine question: Critical perspectives on AI, robots, and ethics*. Cambridge, MA: MIT Press.
Gürses, Fahriye Seda. 2010. *Multilateral privacy requirements analysis in online social network services*. Ph.D. thesis, Katholieke Universiteit Leuven, Department of Computer Science.
Haberl, Helmut, K. Heinz Erb, Fridolin Krausmann, Veronika Gaube, Alberte Bondeau, Christoph Plutzar, Simone Gingrich, Wolfgang Lucht, and Marina Fischer-Kowalski. 2007. Quantifying and mapping the human appropriation of net primary production in earth's terrestrial ecosystems. *Proceedings of the National Academy of Sciences* 104: 12942–12947.
Hamilton, William D. 1964. The genetical evolution of social behaviour. *Journal of Theoretical Biology* 7: 1–52.
Hamilton, William D. 1971. Geometry for the selfish herd. *Journal of Theoretical Biology* 31: 295–311.
Hartzog, Woodrow, and Frederic Stutzman. 2013. Obscurity by design. *Washington Law Review* 88: 385–418.
Henrich, Joseph, Robert Boyd, Samuel Bowles, Colin Camerer, Ernst Fehr, Herbert Gintis, and Richard McElreath. 2001. Cooperation, reciprocity and punishment in fifteen small-scale societies. *American Economic Review* 91: 73–78.
Herrmann, Benedikt, Christian Thöni, and Simon Gächter. 2008a. Antisocial punishment across societies. *Science* 319: 1362–1367.
Herrmann, Benedikt, Christian Thöni, and Simon Gächter. 2008b. Supporting online material for antisocial punishment across societies. *Science* 319.
Hertz, John, Anders Krogh, and Richard G. Palmer. 1991. *Introduction to the theory of neural computation*. Redwood City: Addison-Wesley.

Hinton, Geoffrey, Simon Osindero, and Yee Teh. 2006. A fast learning algorithm for deep belief nets. *Neural Computation* 18: 1527–1554.

Hobaiter, Catherine, Anne Marijke Schel, Kevin Langergraber, and Klaus Zuberbühler. 2014. 'Adoption' by maternal siblings in wild chimpanzees. *PLoS One* 9: e103777.

Hofmann, Martin, Jürgen Geiger, Sebastian Bachmann, Björn Schuller, and Gerhard Rigoll. 2014. The TUM gait from audio, image and depth (GAID) database: Multimodal recognition of subjects and traits. *Journal of Visual Communication and Image Representation* 25: 195–206.

Hogan, Kelly E., and Kate L. Laskowski. 2013. Indirect information transfer: Three-spined sticklebacks use visual alarm cues from frightened conspecifics about an unseen predator. *Ethology* 119: 999–1005.

Huang, Bidan, Sahar El-Khoury, Miao Li, Joanna J. Bryson, and Aude Billard. 2013. Learning a real time grasping strategy. In *IEEE international conference on robotics and automation (ICRA)*, Karlsruhe, 593–600.

Inglehart, Ronald, Miguel Basánez, Jaime Díez-Medrano, Lock Halman, and Ruud Luijkx (eds.). 2004. *Human beliefs and values: A cross-cultural sourcebook based on the 1999–2002 values surveys*. México: Siglo XXI Editores.

Jacobs, Robert A., Michael I. Jordan, Steven J. Nowlan, and Geoffrey E. Hinton. 1991. Adaptive mixtures of local experts. *Neural Computation* 3: 79–87.

Jaeggi, Adrian V., Maria A. van Noordwijk, and Carel P. van Schaik. 2008. Begging for information: Mother–offspring food sharing among wild Bornean orangutans. *American Journal of Primatology* 70: 533–541.

Keller, Laurent. 1999. *Levels of selection in evolution*. Princeton: Princeton University Press.

Keller, Evelyn F., and Lee A. Segel. 1970. Initiation of slime mold aggregation viewed as an instability. *Journal of Theoretical Biology* 26: 399–415.

Kitano, Hiroaki. 2004. Biological robustness. *Nature Reviews Genetics* 5: 826–837.

Kleinsmith, Andrea, and Nadia Bianchi-Berthouze. 2013. Affective body expression perception and recognition: A survey. *IEEE Transactions on Affective Computing* 4: 15–33.

Kokko, Hanna. 2007. *Modelling for field biologists and other interesting people*. Cambridge: Cambridge University Press.

Krosch, Amy R., and David M. Amodio. 2014. Economic scarcity alters the perception of race. *Proceedings of the National Academy of Sciences* 111: 9079–9084.

Krützen, Michael, Janet Mann, Michael R. Heithaus, Richard C. Connor, Lars Bejder, and William B. Sherwin. 2005. Cultural transmission of tool use in bottlenose dolphins. *Proceedings of the National Academy of Sciences of the United States of America* 102: 8939–8943.

Lamba, Shakti, and Ruth Mace. 2011. Demography and ecology drive variation in cooperation across human populations. *Proceedings of the National Academy of Sciences* 108: 14426–14430.

Le Roux, Nicolas, and Yoshua Bengio. 2008. Representational power of restricted Boltzmann machines and deep belief networks. *Neural Computation* 20: 1631–1649.

Ledgard, Stewart F. 2001. Nitrogen cycling in low input legume-based agriculture, with emphasis on legume/grass pastures. *Plant and Soil* 228: 43–59.

Lee, Ellie, Jan Macvarish, and Jennie Bristow. 2010. Risk, health and parenting culture. *Health, Risk & Society* 12: 293–300.

Leimgruber, Kristin L., Adrian F. Ward, Jane Widness, Michael I. Norton, Kristina R. Olson, Kurt Gray, and Laurie R. Santos. 2014. Give what you get: Capuchin monkeys (Cebus apella) and 4-Year-old children pay forward positive and negative outcomes to conspecifics. *PLoS One* 9: e87035.

Lopez De Mantaras, Ramon, David McSherry, Derek Bridge, David Leake, Barry Smyth, Susan Craw, Boi Faltings, Mary Lou Maher, Michael T. Cox, Kenneth Forbus, Mark Keane, Agnar Aamodt, and Ian Watson. 2005. Retrieval, reuse, revision and retention in case-based reasoning. *The Knowledge Engineering Review* 20: 215–240.

McComb, Karen, Cynthia Moss, Sarah M. Durant, Lucy Baker, and Soila Sayialel. 2001. Matriarchs as repositories of social knowledge in African elephants. *Science* 292: 491–494.

McLachlan, Geoffrey, and Thriyambakam Krishnan. 2008. *The EM algorithm and extensions*, vol. 382, 2nd ed. Hoboken, NJ: Wiley.

MacLean, R. Craig, Ayari Fuentes-Hernandez, Duncan Greig, Laurence D. Hurst, and Ivana Gudelj. 2010. A mixture of "cheats" and "co-operators" can enable maximal group benefit. *PLoS Biology* 8: e1000486.

Margulis, Lynn, and Gregory Hinkle. 1997. The biota and gaia. *In slanted truths*, 207–220. New York: Springer.

Marshall, James A.R. 2011. Ultimate causes and the evolution of altruism. *Behavioral Ecology and Sociobiology* 65: 503–512. doi:10.1007/s00265-010-1110-1.

Marshall, James A.R., Rafal Bogacz, Anna Dornhaus, Robert Planqué, Tim Kovacs, and Nigel R. Franks. 2009. On optimal decision-making in brains and social insect colonies. *Journal of the Royal Society Interface* 6: 1065–1074.

Mesoudi, Alex, Andrew Whiten, and Kevin N. Laland. 2004. Is human cultural evolution Darwinian? Evidence reviewed from the perspective of the origin of species. *Evolution* 58: 1–11.

Okasha, Samir. 2012. Social justice, genomic justice and the veil of ignorance: Harsanyi meets Mendel. *Economics and Philosophy* 28: 43–71.

Perry, Susan. 2011. Social traditions and social learning in capuchin monkeys (Cebus). *Philosophical Transactions of the Royal Society, B: Biological Sciences* 366: 988–996.

Perry, Susan, and J.H. Manson. 2003. Traditions in monkeys. *Evolutionary Anthropology* 12: 71–81.

Pinker, Steven. 2012. *The better angels of our nature: The decline of violence in history and its causes*. London: Penguin.

Powers, Simon T., Alexandra S. Penn, and Richard A. Watson. 2011. The concurrent evolution of cooperation and the population structures that support it. *Evolution* 65: 1527–1543.

Powers, Simon T., Daniel J. Taylor, and Joanna J. Bryson. 2012. Punishment can promote defection in group-structured populations. *Journal of Theoretical Biology* 311: 107–116.

Powles, Julia. 2014. What we can salvage from 'right to be forgotten' ruling. *Wired,* 15 May.

Prediger, Sebastian, Björn Vollan, and Benedikt Herrmann. 2013. Resource scarcity, spite and cooperation. *German Institute of Global and Area Studies (GIGA) working papers* 227. Hamburg.

Preuschoft, Signe, and Carel P. van Schaik. 2000. Dominance and communication: Conflict management in various social settings, Chapter 6. In *Natural conflict resolution*, ed. Filippo Aurel and Frans B.M. de Waal, 77–105. Berkeley, CA: University of California Press.

Price, George R. 1972. Fisher's 'fundamental theorem' made clear. *Annals of Human Genetics* 36: 129–140.

Rand, David G., Joseph J. Armao, Mayuko Nakamaru, and Hisashi Ohtsuki. 2010. Anti-social punishment can prevent the co-evolution of punishment and cooperation. *Journal of Theoretical Biology* 265: 624–632.

Rankin, Daniel J., Eduardo P.C. Rocha, and Sam P. Brown. 2010. What traits are carried on mobile genetic elements, and why? *Heredity* 106: 1–10.

Rawls, John. 1980. Kantian constructivism in moral theory. *The Journal of Philosophy* 77: 515–572.

Roberts, S. Craig, and Havlicek Jan. 2011. Evolutionary psychology and perfume design. In *Applied evolutionary psychology*, ed. S. Craig Roberts, 330–348. Oxford: Oxford University Press.

Rolian, Campbell. 2014. Genes, development, and evolvability in primate evolution. *Evolutionary Anthropology: Issues, News, and Reviews* 23: 93–104.

Rosenbaum, Sara. 2014. When religion meets workers' rights: Hobby Lobby and Conestoga Wood Specialties. *Milbank Quarterly* 92: 202–206.

Rothschild, David, Sharad Goel, Andrew Gelman, and Douglas Rivers. 2015. The mythical swing voter. In *Collective intelligence*, MIT. Unpublished preprint presented at conference. Available on arXiv:1406.7581.

Roughgarden, Joan. 2012. Teamwork, pleasure and bargaining in animal social behaviour. *Journal of Evolutionary Biology* 25: 1454–1462.

Roughgarden, Joan, Meeko Oishi, and Erol Akçay. 2006. Reproductive social behavior: Cooperative games to replace sexual selection. *Science* 311: 965–969.

Schaal, Stefan, and Christopher G. Atkeson. 1998. Constructive incremental learning from only local information. *Neural Computation* 10: 2047–2084.
Schroeder, Kari Britt, Gillian V. Pepper, and Daniel Nettle. 2014. Local norms of cheating and the cultural evolution of crime and punishment: A study of two urban neighborhoods. *PeerJ* 2: e450.
Selinger, Evan, and Woodrow Hartzog. 2014. Obscurity and privacy. In *Routledge companion to philosophy of technology*, ed. Joseph Pitt and Ashley Shew. New York: Routledge.
Seth, Anil K., Tony J. Prescott, and Joanna J. Bryson (eds.). 2012. *Modelling natural action selection*. Cambridge: Cambridge University Press.
Shannon, Claude E. 2001. A mathematical theory of communication. *ACM SIGMOBILE Mobile Computing and Communications Review* 5: 3–55.
Shultz, Susanne, and Laura V. Finlayson. 2010. Large body and small brain and group sizes are associated with predator preferences for mammalian prey. *Behavioral Ecology* 21: 1073–1079.
Shultz, Susanne, Christopher Opie, and Quentin D. Atkinson. 2011. Stepwise evolution of stable sociality in primates. *Nature* 479: 219–222.
Silva, Antonio S., and Ruth Mace. 2014. Cooperation and conflict: Field experiments in Northern Ireland. *Proceedings of the Royal Society B: Biological Sciences* 281: 20141435.
Skemp, Richard R. 1961. Reflective intelligence and mathematics. *British Journal of Educational Psychology* 31: 45–55.
Smith, Kenny, and Simon Kirby. 2008. Cultural evolution: Implications for understanding the human language faculty and its evolution. *Philosophical Transactions of the Royal Society, B: Biological Sciences* 363: 3591–3603.
Sober, Elliott, and David Sloan Wilson. 1998. *Unto others: The evolution and psychology of unselfish behavior*. Cambridge, MA: Harvard University Press.
Staddon, John Eric R. 1975. A note on the evolutionary significance of "supernormal" stimuli. *The American Naturalist* 109: 541–545.
Sterck, E.H.M., D.P. Watts, and C.P. van Schaik. 1997. The evolution of female social relationships in nonhuman primates. *Behavioral Ecology and Sociobiology* 41: 291–309.
Stoddart, David Michael. 1990. *The scented ape: The biology and culture of human odour*. Cambridge: Cambridge University Press.
Sylwester, Karolina, Benedikt Herrmann, and Joanna J. Bryson. 2013. Homo homini lupus? Explaining antisocial punishment. *Journal of Neuroscience, Psychology, and Economics* 6: 167–188.
Sylwester, Karolina, James Mitchell, and Joanna J. Bryson. 2014. Punishment as aggression: Uses and consequences of costly punishment across populations. To be resubmitted.
Sytch, Maxim, and Adam Tatarynowicz. 2014. Friends and foes: The dynamics of dual social structures. *Academy of Management Journal* 57: 585–613.
Taylor, Daniel J. 2014. Evolution of the social contract. Ph.D. thesis, University of Bath, Department of Computer Science.
Thierry, Bernard. 2007. Unity in diversity: Lessons from macaque societies. *Evolutionary Anthropology* 16: 224–238.
Tinbergen, N., and A.C. Perdeck. 1950. On the stimulus situation releasing the begging response in the newly hatched herring gull chick (Larus argentatus argentatus Pont.). *Behaviour* 3: 1–39.
Trewavas, Anthony. 2005. Green plants as intelligent organisms. *Trends in Plant Science* 10: 413–419.
Valstar, Michel F., and Maja Pantic. 2012. Fully automatic recognition of the temporal phases of facial actions. *IEEE Transactions on Systems, Man, and Cybernetics, Part B: Cybernetics* 42: 28–43.
Vernon, David, Giorgio Metta, and Giulio Sandini. 2007. A survey of artificial cognitive systems: Implications for the autonomous development of mental capabilities in computational agents. *IEEE Transactions on Evolutionary Computation* 11: 151–180.
Wang, Wei, David Rothschild, Sharad Goel, and Andrew Gelman. 2015. Forecasting elections with non-representative polls. International Journal of Forecasting. In press.

Ward, Adrian F. 2013. Supernormal: How the internet is changing our memories and our minds. *Psychological Inquiry* 24: 341–348.

Whitehouse, Harvey, Ken Kahn, Michael E. Hochberg, and Joanna J. Bryson. 2012. The role for simulations in theory construction for the social sciences: Case studies concerning divergent modes of religiosity. *Religion, Brain & Behavior* 2: 182–224.

Whiten, Andrew, Jane Goodall, William C. McGew, Toyoaki Nishida, Vernon Reynolds, Yukimaru Sugiyama, Caroline E.G. Tutin, Richard W. Wrangham, and Christophe Boesch. 1999. Cultures in chimpanzees. *Nature* 399: 682–685.

Williams Woolley, Anita, Christopher F. Chabris, Alex Pentland, Nada Hashmi, and Thomas W. Malone. 2010. Evidence for a collective intelligence factor in the performance of human groups. *Science* 330: 686–688.

Wilson, David Sloan. 1989. Levels of selection: An alternative to individualism in biology and the human sciences. *Social Networks* 11: 257–272. Special issue on non-human primate networks.

Wittkower, D.E., Evan Selinger, and Lucinda Rush. 2013. Public philosophy of technology: Motivations, barriers, and reforms. *Techné: Research in Philosophy and Technology* 17: 179–200.

Wray, Alison. 1998. Protolanguage as a holistic system for social interaction. *Language and Communication* 18: 47–67.

Wray, Alison, and George W. Grace. 2007. The consequences of talking to strangers: Evolutionary corollaries of socio-cultural influences on linguistic form. *Lingua* 117: 543–578.

Zahavi, Amotz. 1977. The testing of a bond. *Animal Behaviour* 25: 246–247.

Index

A
Action
  capacity for, 4, 181, 221
  simple, 206, 210, 211
  simulation, 12, 13
Actor network theory (ANT), 48, 49
Adaptivity, 5, 53, 54, 56, 222, 225
Affordance, 12, 296
Agency
  affective aspects of, 8–9
  collective, 3–21, 46, 149–166, 185–202, 219–234, 281, 283, 285–290, 294, 298
  distributed, 48–57
  intentional, 10, 13, 189, 197, 198
  massively shared, 187, 197–201
  minimal, 226, 228
Agent-based model (ABM), 47, 252, 255–256
Agent(s)
  collective, 14, 18, 20, 223, 226, 229, 232, 234
  superordinate, 229
Alienation/alienated, 191, 195–200
Altruistic behaviour/altruism, 285, 286, 292
Androids, 81, 84–90, 92, 94, 97–101
Artefact, 45, 48, 50, 52, 54, 56, 282, 287, 297
Artificial agent/system, 4, 6, 8, 10, 17–20, 57, 63, 64, 66, 72, 74–77, 206, 217, 224
Artificial intelligence, 63, 65, 76, 81, 92, 96, 98, 99, 138, 213, 255, 281–300
Artificial life, 21, 85, 90, 101, 220
Assistive technology at workplaces (ATW), 109–128

Attractiveness, 47, 83
Automata, 90–94, 264
Autonomous, 4, 5, 10, 45, 52, 63, 77, 93, 95, 100, 119, 132, 139, 140, 143, 144, 220, 223–225, 227, 229, 255, 282
Autonomy, 4–8, 10–12, 121, 126–127, 132, 143, 219–221, 223, 281
  basic, 5–6, 8, 10–12
Autopoiesis, 219–223

B
Bacterium, 221, 222, 226, 228
Behavioral cooperation, 31–35, 37
Belief-desire-intention model (BDI), 16, 65–69, 75–77, 213, 217
Biology, 51, 219, 220, 264, 266, 272, 282, 287, 288, 291
Body, 16, 17, 27, 37, 66, 74, 87, 95, 139, 193, 194, 196, 220, 221, 226, 227, 229, 231–234
  living, 233, 234

C
Cell, 220, 227, 229, 269, 292, 293
Chimpanzee group hunting, 13, 171, 179, 182
Cognition, 54, 55, 67, 206, 207, 212, 214–215, 218–220, 230, 272, 287, 291
Cognitive
  architecture, 65–69, 72, 75–77
  cooperation, 29, 31–35, 37, 39, 40
Collaborative, 45–48, 53, 64, 78, 81, 199, 292

Collective
  action, 11, 18, 27, 51, 144, 156, 205–218
  agency, 3–21, 46, 149–166, 185–202, 219–234, 281, 283, 285–290, 294, 298
  intelligence, 283
  intentionality, 14–18, 21, 22, 27
  responsibility, 19, 143, 144
Common
  goal, 13, 29, 30, 37–39, 63, 64, 76, 171–175, 180–182, 186, 207
  knowledge, 15, 26–34, 37, 38, 153, 154, 195, 207
Communication, 9, 11, 45, 54–57, 63–65, 78, 81, 82, 100, 103, 137, 138, 185, 189, 199, 201, 215–217, 249–251, 254, 256, 262, 271, 283, 284, 287–289, 294–296, 298
Complex
  action, 157, 205, 210–212, 217
  behavior, 37, 210, 224, 233
Computer animations, 81, 82, 84
Consciousness, 210, 211, 230
Conversational partner/machine, 77, 78
Cooperation, 3–21, 25–42, 55, 63–78, 81, 100, 117, 124–126, 171, 173, 174, 180, 243, 252, 255, 256, 261–277, 285–288, 290–292, 294–300
  evolution of, 261–277
Cooperative
  behavior, 29, 39, 63, 76, 255, 256, 265, 277
  system, 63–78
Coordination, 11–13, 18, 29, 30, 33, 36, 46–48, 54, 55, 65, 75, 171, 181, 182, 199, 201, 206, 207, 209, 212, 214–216, 218
  emergent, 11–12
Cultural context, 100–102
Culture, 45, 83–89, 99, 101, 270, 282, 285, 287, 296–300
Cybernetics, 219, 220, 253
  second order, 219, 220
Cyber-physical systems (CPS), 45, 46, 54

D
Demons, 90–94
Disabilities, 116, 117, 127
Disease, 83, 84, 91, 187, 292
Dualism, 93, 169–172, 182, 230
Dyad, 30, 192, 193, 195, 230

E
Effectors, 212, 216
Embodied agent/system, 63, 69, 76
Embodiment, 102, 220, 223, 233
  biological, 223, 233
Emergence, 12, 21, 56, 84, 223–224, 239, 242–244, 252, 255, 265
  second order, 223
Emotion(s), 8–9, 17–18, 41, 68, 78, 113, 187, 215, 232
Empathy, 189
Empirical
  data, 39, 245, 246, 249–251, 265, 291
  falsification, 274
  research, 240, 250, 261, 264, 266, 267, 275, 277
Employment, 69, 110, 115–126, 128
Enabling mechanism, 149, 164
Enactive approach/enactivism, 219–220, 230, 232
Entrainment, 12
Ethics of technology, 49, 109–128
Evolution, 21, 45, 47, 49, 50, 221, 252, 255, 261–277, 283, 287, 288, 291, 293, 296, 299
  artificial, 221
Extended mind, socially, 231

F
Familiarity, 82, 83
Flock, 12, 20, 224, 225
Free will, 7, 90, 132, 210, 211
Function, 13, 15, 21, 67, 68, 70, 91, 123, 125, 172, 176, 178, 179, 187, 188, 221, 223, 226, 241, 289
  aesthetic, 81–103
Functionalism, 51, 63–68, 77, 78, 81, 83, 89, 91, 113, 128, 178, 195, 215, 220, 253, 254

G
Game theory, 241, 242, 255, 265–267, 274
Gamification, 114
Gilgamesh, 84, 85, 101
Goal
  directed agency, 13–14, 189, 191, 195
  directedness, 6, 8, 175, 176, 211, 215
  external, 6, 13, 14
  generation, 211
  internal, 6, 66
  shared, 27, 28, 31, 34, 35, 63, 64, 70, 73, 76, 170, 189–191, 195, 199, 201, 202, 213, 295
Golem, 90, 96
Group(s)
  action, 144, 191
  agency, 185, 186
  formation, 186, 190, 191, 200

Index

H
Human–computer
　interaction, 109
　interface, 231
Human-robot cooperation, 25, 36–41
Human uniqueness, 284, 285, 287

I
Imagination, 82, 100, 102
　imaginative, 21
Impaired persons, 115, 117, 119
Individuality, 221–226, 229, 233, 234
Individuation, 221, 226, 233
Information
　perception, 215
　processing, 54, 56, 211–213, 215, 216, 252, 255
Insect behavior, 13, 172, 225–227, 233, 283, 298
Intelligence, 5–6, 10–12, 21, 47, 51, 54, 55, 57, 63, 65, 76, 81, 85, 92, 96, 98, 99, 138, 213, 255, 281–300
Intention
　shared, 15, 26–35, 37–39, 41, 55, 169, 171, 186, 189, 190, 197, 198, 205, 207, 213, 231, 232
　we-, 14–17, 27–29, 37, 38, 186, 213
Intentionality, 6–9, 11, 13–18, 21, 22, 27, 50, 56, 136–138, 186, 205, 206, 208–211, 213, 217, 230–232
　higher-order, 7–11, 13–17
Interaction, 5, 8, 11–13, 18, 20, 29, 30, 37–40, 46–48, 50, 52–57, 63, 66–68, 70, 72–75, 77, 78, 102, 103, 109, 127, 137, 140, 143, 189, 197, 198, 201, 208, 209, 214, 220–234, 239–256, 266, 267, 298, 299
Interactional asymmetry, 222, 223, 225, 226, 229, 233, 234
Interactive behavior, 11, 12
Investment
　collective, 288–295
　group, 289
Irrationality, 159–161, 163, 165

J
Joint
　action, 12, 27–30, 35, 36, 38, 40, 45, 53, 54, 56, 153, 169, 171–177, 179, 181, 182, 193, 206, 207, 209, 231
　affordance, 12
　agency, 14, 46, 54
　attention, 29, 37, 64, 65, 73–77, 215
　commitment, 16, 17, 28, 33–36, 39–41, 187, 191–197, 199
　intention, 12, 14, 17, 56, 63, 64, 73, 76, 77
Justificatory narratives, 266–271

L
Law, legislation, 117, 299
Level of abstraction, 46, 51–52

M
Management science, 20, 239–256
Matching of plans, 165
Mathematics, 12, 241, 243, 247–251, 264, 266, 274–276, 281, 283, 286
Media, 185, 201, 296, 298
Metabolism, 220–222
Minimal rationality, 70
Model, 6, 12, 18, 20, 39, 41, 47, 49–52, 57, 63–65, 70, 75–77, 100, 102, 185, 187, 188, 213, 231, 241, 243–251, 254–256, 261, 263–276, 285, 290, 292, 294, 295
Moral agents, 132–134, 136, 138–143, 282, 299
Motor representations, 12, 162–165, 189, 209, 216
Movies/films, 82, 97–100
Multi-agent systems (MAS), 3, 10–20, 47, 54, 55, 223–229, 233
Multi-system agent, 223–229, 233
Mutual common knowledge, 26–34, 37, 38
Mutual cooperation, 13, 14
Mythology, 85, 90, 91

N
Naturalistic approach, 276
Network of responsibility, 135, 139, 142, 143
Normativity, 16, 17, 222, 223, 225–227, 229, 233, 234
Norms, 47, 118, 195, 222, 224, 227, 228, 242, 252, 255, 256, 297, 299
　social, 227, 228, 242, 255, 256

O
Operational closure, 220

P
Paradigm, 4, 20, 41, 65, 75, 76, 150, 154, 186–189, 219, 220, 222, 230, 231, 233, 239–242, 262, 272, 276, 277
　scientific, 20

Paradigmatic examples, 185–202, 215
Parallel *vs.* collective agency, 149–166
Partnership, 50, 55, 57, 63–78
Perception-action matching, 12
Perceptual crossing, 231
Personality, 85, 138
Personhood, 8–10, 136, 138, 281
Personification, 53–56, 97
Persons, 8–10, 19, 20, 30, 66, 69, 70, 77, 78, 88, 100, 120, 122–124, 132, 136, 137, 141, 142, 175, 199, 200, 231–234, 268, 298
Plan/planning
  interconnected, 16, 57, 155–160, 164, 165
  parallel, 16, 155–165
  shared, 191, 198–200
Plural agent, 18, 186, 190, 193–196
Pluralistic account, 185
Prisoner's dilemma, 255, 261, 263–266, 271
Prometheus, 85, 86, 95, 100
Propositional attitudes, 215
Pro-social behavior, 21, 281–300
Prosthetic intelligence, 285, 290, 296, 298–300
Puzzle-form, 189–201
Pygmalion, 86, 87, 95, 101

R
Rational choice theory, 263, 272, 273, 276, 277
Rational interaction, 13
Rationality, 7, 8, 70, 118, 159–161, 163, 165, 241, 242, 244
Reciprocal altruism, 273
Reductionism, 197
Reiterated Prisoner's Dilemma (RPD), 261, 263
Replication/replicator, 227, 286
Representation
  I-, 231
  We-, 231
Representationalism, 4, 6, 8, 12, 13, 17, 64, 66, 70, 72, 82, 84, 87, 99, 127, 156, 157, 161–165, 176, 189, 207–209, 212, 216, 227, 231, 243, 245, 247, 248, 250, 251, 273
Research method, 239–256
Responsibility, 18–20, 55, 110, 128, 131–144, 282, 297, 299

Retreat to false modesty, 272, 277
Robo-ethics, 141
Robot, 3, 25–41, 45, 63, 81, 131–144, 220, 284
Robotics, 3, 45, 54–56, 84, 131, 220, 221
Robustness of a model, 271
Rules, 12, 15, 35, 36, 39, 40, 56, 72, 97, 116–118, 132, 138–141, 174, 216, 224, 225, 234, 242, 244, 255, 274, 294, 299

S
Schizophrenia, 233
Selection, 4, 54, 68, 188, 244, 270, 286–288, 291, 293, 295, 297, 299
Self-awareness, 136–138, 142
Self-consciousness, 9, 210, 211
Self-organization, 46, 219
Sense of belonging, 185, 189–202
Sentiment of attachment, 192
Simple behavior, 255
Simulation
  history of, 261–277
  in philosophy, 21
  method, 240, 245, 247, 248, 251–253, 255, 256, 266, 269, 277
  model, 243–251, 254, 255, 261, 262, 268, 271, 275, 277
  social, 4, 21, 47, 261, 275, 282, 290
Social
  cognition, 206, 212, 214–215, 218
  computing system, 46–48, 51, 55–57
  inclusion, 110, 115–117, 119, 120, 125, 128
  interaction, 18, 29, 55, 70, 74, 102, 103, 127, 137, 209, 214, 231, 232, 239–256
  learning, 269–271
  ontology, 186, 193
  sciences, 244, 265, 266, 272, 274–277
  simulation, 4, 21, 47, 261, 275, 282, 290
Software, 4, 6, 45–48, 51, 52, 54, 55, 57, 81, 134, 199, 221, 223, 243, 245, 250
Stegodyphus spiders, 172, 176
Subjectivity
  collective, 230–234
  first person, 234
  second person, 232
Superorganisms, 225, 226, 233

Swarm
  behavior, 11–12, 224, 225
  intelligence, 47, 51, 54, 55, 57
System
  artificial, 3–21, 33, 39, 63–66, 73, 77, 78, 81, 205–218, 220, 233
  autonomous, 225
  autopoietic, 221, 223
  neural, 231, 232
System dynamics, 20, 245, 252–255

T
Theoretical, 27, 36, 50, 75, 119, 149, 158–160, 187, 188, 197, 201, 202, 220, 221, 240, 243–248, 251, 256, 262, 265–267, 270, 271, 274, 276, 283, 287
  logic, 243–246
  vs. practical perspective, 158, 160
Theory
  of groups, 186, 189
  of mind, 64, 215, 216
Tool, 11, 13, 30, 45, 47, 49, 55–57, 63–78, 100, 112, 128, 188, 206, 208, 209, 212, 217, 240, 250, 251, 256, 261, 263
Types of agents, 3–11, 17, 20, 136, 176, 180

U
Umwelt, 229, 230
Uncanny valley, 8–9, 81–103
Unified theory, 188, 189
Uniform behavior, 11

V
Value, 6, 8, 14, 20, 52, 53, 67, 75, 114, 115, 118, 121–123, 138–140, 191, 194, 222, 227, 241, 245–249, 253, 255, 266, 270, 271, 275, 286, 290, 291, 293, 294, 297
Virtual reality, 66, 70, 74, 231

W
Work ethics, 109–128
Working conditions, 92, 115, 126, 127

If you have any concerns about our products,
you can contact us on
ProductSafety@springernature.com

In case Publisher is established outside the EU,
the EU authorized representative is:
**Springer Nature Customer Service Center GmbH
Europaplatz 3, 69115 Heidelberg, Germany**

Printed by Libri Plureos GmbH
in Hamburg, Germany